PLANT BIODIVERSITY AND BIOACTIVES

A Pathway to Sustainable Health and Environment

The Editors

Dr. Pratibha Arya obtained B.Sc., M.Sc. and Ph. D. degree from Kumaun University, Nainital. Her research interest includes Aquatic Hyphomycetes their bioactivities and ecological aspects. She has contributed several articles in National and International Conferences and Seminars and also published papers in National and International Journals. She holds the life membership of Indian Phyto pathological Society, India. She has also been awarded Young Scientist in Botany by Uttarakhand State Science and Technology Congress (UCOST), Dehradun, 2008. Her achievements include the **discovery of new species to science *Tetracladium nainitalense*,** Sati & Arya which have also been compiled and published in PLANT DISCOVERIES, 2009 by Ministry of Environment and Forests. Also, **discovery of new species to India *Tetracladium breve*,** Arya & Sati. With over 14 years of teaching and research experience Dr. P. Arya is working as Assistant Professor in the Department of Botany, Government Degree College, Talwari, Chamoli, affiliated by Sri Dev Suman Uttarakhand University. She also holds important academic and administrative positions at the College.

Dr. Nisha Dhoundiyal holds a Doctorate Degree (Ph.D.) and Master's Degree (M.Sc.) in Zoology from Kumaun University, Nainital, Uttarakhand. She has published more than ten research papers in respected national and international journals. She has over five years of experience as an Assistant Professor in both Private and Government Degree Colleges. Currently, she holds a position as an Assistant Professor in the Department of Zoology at Government Degree College Talwari– Tharali (Chamoli) affiliated by Sri Dev Suman Uttarakhand University. She has worked on various other projects, including reviewing numerous book chapters and journal papers.

Mr. Rajneesh Kumar is an Assistant Professor at Department of Higher Education, Government of Uttarakhand. He is Incharge of Economic Department at Government Degree College Talwari, Chamoli affiliated by Sri Dev Suman Uttarakhand University. He completed his Master's Degree in Economics from Jawahar Lal Nehru University (JNU), New Delhi and Undergraduate from Hansraj College, Delhi University (New Delhi). He has qualified UGC NET-JRF in 2020 in Economics. His areas of specialization are Development Economics, Statistics and Rural Economics.

Plant Biodiversity and Bioactives

A PATHWAY TO SUSTAINABLE HEALTH AND ENVIRONMENT

Editor-in-Chief

Dr. Pratibha Arya

Editors

Dr. Nisha Dhoundiyal

Mr. Rajneesh Kumar

2025

Daya Publishing House®

A Division of

Astral International Pvt. Ltd.

New Delhi – 110 002

ISBN: 978-93-5919-309-0

Publisher's Note:

Every possible effort has been made to ensure that the information contained in this book is accurate at the time of going to press, and the publisher and author cannot accept responsibility for any errors or omissions, however caused. No responsibility for loss or damage occasioned to any person acting, or refraining from action, as a result of the material in this publication can be accepted by the editor, the publisher or the author. The Publisher is not associated with any product or vendor mentioned in the book. The contents of this work are intended to further general scientific research, understanding and discussion only. Readers should consult with a specialist where appropriate.

Every effort has been made to trace the owners of copyright material used in this book, if any. The author and the publisher will be grateful for any omission brought to their notice for acknowledgement in the future editions of the book.

Published by : **Daya Publishing House®**
A Division of
Astral International Pvt. Ltd.
– ISO 9001:2015 Certified Company –
4736/23, Ansari Road, Darya Ganj
New Delhi-110 002
Ph. 011-43549197, 8130496929
E-mail: info@astralint.com
Website: www.astralint.com

Printed at : **Replika Press Pvt. Ltd.**

ACKNOWLEDGEMENT

We express our sincere gratitude to Dr. Yogendra Chandra Singh, Principal, Government Degree College, Talwari-Tharali (Chamoli), for his invaluable guidance and constant encouragement. We also extend our heartfelt thanks to our colleagues and well-wishers for their unwavering support and assistance in various aspects of this book. Finally, we are deeply grateful to all the contributors for sharing their valuable insights through their manuscripts, without which this publication would not have been possible.

PREFACE

The dynamic and interconnected world of plant biodiversity has always been at the heart of human civilization, shaping not only the ecological landscape but also offering invaluable resources for health, culture, and sustainability. As we face increasing environmental challenges and health concerns in the 21st century, it is more important than ever to tap into the rich potential of the natural world to find innovative solutions. The book Plant Biodiversity and Bioactives: Pathways to Sustainable Health & Environment brings together a diverse collection of research and insights that explore the intricate relationship between plants, their bioactive compounds, and their role in fostering both human well-being and environmental preservation.

This book features a wide range of topics that delve into the diverse uses of plants, from their medicinal properties to their role in sustainable agriculture and ecological conservation. The contributions of leading researchers and experts from varied disciplines provide a holistic view of how plant biodiversity not only enriches our lives but also offers pathways to solving some of the most pressing issues of our time.

The first section of the book focuses on the promising intersection of plant-based knowledge and cutting-edge technologies like artificial intelligence. It begins with chapters that explore the potential of AI in advancing drug discovery from medicinal plants, followed by studies on the role of supermolecules in cardiovascular treatments and the cultural importance of ethnobotanical research in preserving indigenous knowledge. These chapters reflect the profound impact plants have in shaping modern medicine and highlight how traditional practices continue to inform innovative solutions.

The second section shifts focus to the bioactive compounds present in plants, their pharmacological benefits, and how they can lead to sustainable health solutions. Detailed analyses of plants like Mulethi (Glycyrrhiza glabra) showcase the potential of natural substances in health promotion, while research on endophytic fungi and their antagonistic effects on plant pathogens demonstrates how plants can also contribute to sustainable agricultural practices.

The final sections of the book move towards a broader ecological perspective, highlighting the significance of plant biodiversity in environmental conservation. Chapters on the diversity of aquatic fungi in the Himalayas, the conservation of endangered species like Lilium polyphyllum, and the influence of environmental degradation on sustainable development reflect the crucial role plants play in maintaining ecological balance.

In addition to these scientific explorations, the book also touches on the cultural dimensions of plant biodiversity, showcasing how traditional knowledge systems are integral to preserving both the environment and the health of local communities. From the Chipko Movement in Uttarakhand to the environmental consciousness embedded in Hindi literature, these narratives underscore the intertwined nature of human culture and nature.

This book is a testament to the ever-growing significance of plants in our quest for sustainable solutions. It is our hope that the readers whether they be scientists, policymakers, or curious individuals will find inspiration in these pages, igniting a deeper appreciation for the wealth of knowledge and potential that plants offer for a healthier and more sustainable world.

Dr. Pratibha Arya
Dr. Nisha Dhoundiyal
Mr. Rajneesh Kumar

CONTENTS

CHAPTER 1

ADVANCING DRUG DISCOVERY WITH AI: FROM MEDICINAL PLANTS TO NOVEL THERAPEUTIC AGENTS

Hritwik Baghel, Ram Kumar Sahu, Abdul Faruk

Department of Pharmaceutical Sciences,
HNB Garhwal University Srinagar-Garhwal-246174, Uttarakhand

Abstract

The title explores the transformative role of artificial intelligence (AI) in drug discovery, particularly in the identification of bioactive compounds from medicinal plants. Traditional drug discovery methods have often been labor-intensive, time-consuming, and environmentally unsustainable. AI, through techniques such as machine learning, deep learning, and molecular docking, presents a more efficient and scalable solution. By leveraging vast phytochemical databases, AI-based virtual screening can predict bioactivity, pharmacokinetic properties, and drug-likeness of compounds. This approach accelerates the discovery process while minimizing the need for exhaustive experimental testing. The chapter also discusses the application of AI in predicting Absorption, Distribution, Metabolism, Excretion, and Toxicity (ADMET) profiles, which is critical in reducing failure rates in later stages of drug development. With advancements in AI algorithms, such as reinforcement learning and hybrid models, the potential for discovering novel therapeutic agents from medicinal plants continues to expand, offering sustainable and precise solutions for future drug development.

Keywords: AI-based virtual screening, medicinal plants, bioactive compounds, drug discovery, machine learning, deep learning, molecular docking, ADMET, phytochemical databases, sustainable drug development.

Introduction

The Need for AI in Phytopharmaceutical Research

Medicinal plants have long been a cornerstone of traditional medicine, offering a vast array of bioactive compounds that have been used in order to treat numerous diseases. These plants contain complex mixtures of Secondary metabolites, like alkaloids, flavonoids, terpenoids, and glycosides, many of which possess therapeutic properties. Despite their potential, identifying bioactive molecules from medicinal plants is an arduous task. Traditional methods of screening bioactives, which rely heavily on experimental techniques, are time-consuming, resource-intensive, and often inefficient.

Historically, drug discovery from plants followed a trial-and-error approach. Scientists would isolate compounds from plant extracts and evaluate their biological activity through in vitro or in vivo assays. While effective in some instances, this method had its limitations. For instance, the extraction, isolation, and structural elucidation of bioactive molecules could take months or even years to complete. Additionally, many promising compounds might go unnoticed due to the limitations of experimental screening techniques, which are often biased toward certain types of chemical structures.

In this context, implementation of artificial intelligence (AI) into the area of phytopharmaceutical research offers a game-changing opportunity. AI, considering its ability to process vast amounts of data quickly and accurately, has the potential to revolutionize the way bioactive molecules are identified and developed into drugs. This chapter explores the role of AI-based virtual screening for the identification of bioactive compounds from medicinal plants, highlighting the need for such advanced technologies in modern drug discovery.

The Limitations of Traditional Screening Methods

Traditional drug discovery from medicinal plants faces multiple challenges. The major limitations is time and cost associated with experimental screening. For instance, researchers may need to screen thousands of plant extracts to identify a single bioactive compound with therapeutic potential. This process involves laborious steps, including extraction, isolation, and testing of each compound. Moreover, the hit-to-lead ratio in traditional screening is often low, meaning that a large number of compounds must be tested to identify a few promising candidates.

Furthermore, many bioactive molecules present in plants may exist in minute quantities, making their isolation and characterization difficult. In some cases, the bioactivity of a compound may not be detected due to its low concentration or the presence of other compounds that interfere with the assay results. Traditional methods also require the use of expensive and sometimes hazardous chemicals, contributing to the environmental footprint of the drug discovery process.

In contrast, AI-based virtual screening offers a faster, more cost-effective alternative to traditional methods. By leveraging computational models, AI can predict the bioactivity of compounds from their chemical structure, thus reducing the need for extensive laboratory experiments. This not only accelerates the methods of discovery, but also helps researchers to investigate a wider spectrum of plant species and chemicals compound.

AI in Virtual Screening: A Game-Changer for Phytopharmaceuticals

Artificial intelligence has advanced considerably in the pharmaceutical sector, particularly in drug discovery. One prominent application is virtual screening, a computational method that helps identify potential drug candidates from extensive chemical libraries. This technique employs algorithms to predict the likelihood of a molecule interacting with a biological target, such as a protein or enzyme, based on its chemical structure. By doing so, researchers can focus on the most promising compounds for further experimental testing.

In the context of medicinal plants, AI-driven virtual screening is used to forecast the bioactivity of phytochemicals, streamlining the drug discovery process. Medicinal plants offer a vast array of structurally diverse compounds, many of which remain underexplored. Through the application of AI, researchers can quickly assess the therapeutic potential of thousands of plant-derived molecules, identifying those most likely to be effective against specific diseases.

AI models, including deep learning and machine learning algorithms, can analyze large datasets of phytochemicals and their known bioactivities, revealing patterns that may not be evident through traditional methods. These models can also anticipate the bioactivity of new or untested compounds, guiding researchers toward the most promising candidates for further investigation. This approach not only conserves assets and time but also enhances the likelihood of discovering novel bioactive molecules with therapeutic potential.

The Role of AI in Accelerating Drug Discovery from Medicinal Plants

The use of AI in virtual screening offers significant potential to speed up the drug discovery process in multiple ways. Firstly, AI can process vast amounts of data from phytochemical databases, such as PubChem, ChEMBL, and others, to find compounds with structures similar to known bioactive molecules. This enables researchers to swiftly identify compounds with likely therapeutic properties, minimizing the need for extensive experimental trials.

Second, AI can be used to predict the pharmacokinetic and pharmacodynamic properties of bioactive compounds, such as their absorption, distribution, metabolism, excretion, and toxicity (ADMET). By predicting these properties early in the drug discovery process, AI can help researchers filter out compounds that are likely to fail in later stages of development, thus reducing the risk of costly failures.

Third, AI-driven virtual screening is used to used to explore the chemical diversity of plant species that have not been extensively studied. Many medicinal plants used in traditional medicine have not been subjected to rigorous scientific investigation, and their phytochemical composition remains largely unknown. AI can help researchers explore the therapeutic potential of these plants by predicting the bioactivity of their constituent compounds, thus expanding the pool of potential drug candidates.

AI-Powered Platforms and Tools in Virtual Screening

Several AI-powered platforms and tools have been developed to facilitate virtual screening in drug discovery. These platforms use a variety of machine learning algorithms, such as neural networks, support vector machines, and random forests, to predict the bioactivity of molecules. Some of the most commonly used AI-based virtual screening tools include AutoDock, DeepChem, and ChemProp.

AutoDock is a widely used molecular docking software that allows researchers to predict the binding affinity of small molecules to biological targets. By incorporating AI algorithms, AutoDock can predict the binding modes of bioactive compounds with high accuracy, thus helping researchers identify potential drug candidates from large chemical libraries.

DeepChem is an open-source platform that uses deep learning algorithms to predict the bioactivity of molecules. DeepChem allows researchers to train AI models on large datasets of chemical structures and their known bioactivities, enabling the rapid identification of potential drug candidates.

ChemProp is another AI-based tool that uses machine learning to predict the properties of small molecules. ChemProp can be used to predict a variety of molecular properties, including bioactivity, ADMET properties, and drug-likeness. This makes it a valuable tool for virtual screening in drug discovery.

Overview of AI Techniques in Virtual Screening

The exponential development of artificial intelligence (AI) has transformed various fields of science, and drug discovery is no exception. When it comes to screening bioactive molecules from medicinal plants, AI-based techniques offer an innovative, efficient, and powerful alternative to traditional methods, which are often labor-intensive and time-consuming. This chapter explores the primary AI techniques employed in virtual screening for bioactive compounds, with a focus on how they accelerate the discovery of potential therapeutic agents derived from medicinal plants.

Machine Learning in Virtual Screening

Machine learning (ML) is an element of AI that serves a vital role in virtual screening. It relies on algorithms capable of learning from data and making predictions or decisions without being explicitly programmed for specific tasks. In drug discovery, ML models can analyze large datasets of chemical compounds to predict the likelihood of a compound being bioactive.

Supervised Learning

Supervised learning is one of the most common types of machine learning used in virtual screening. In this approach, models are trained using labeled datasets where the input (chemical features) and output (bioactivity) are known. Supervised learning algorithms can then predict the bioactivity of novel compounds based on their chemical features.

- **Random Forest (RF)**: RF is an ensemble learning method often applied to virtual screening. It operates by constructing a multitude of decision trees during training and outputting the class (bioactive or non-bioactive) that is the mode of the classes predicted by individual trees. RF has been widely used due to its ability to handle large datasets, resist overfitting, and provide accurate predictions. A study by Vesel et al. (2021) demonstrated that RF models could predict bioactive compounds in medicinal plants with high precision, identifying potential anti-inflammatory agents.
- **Support Vector Machines (SVM)**: SVM is another popular supervised learning technique used in virtual screening. SVM works by finding a hyperplane that best separates bioactive from non-bioactive compounds in a high-dimensional space. Its flexibility allows it to handle both linear and non-linear relationships between features, making it effective for complex datasets. Research by Chen et al. (2020) has shown that SVM models can effectively classify plant-derived compounds with antibacterial properties.

Deep Learning

Deep learning (DL), a subset of machine learning, is particularly powerful due to its ability to model intricate patterns in data. Deep learning employs neural networks with multiple layers (hence "deep") that can automatically learn to extract features from raw data, making it highly effective for virtual screening.

- **Convolutional Neural Networks (CNNs)**: Originally developed for image recognition, CNNs have been adapted for drug discovery by encoding molecular structures as images or graphs. CNNs can learn to recognize bioactive patterns within molecular structures, helping to identify promising plant-derived compounds. A recent study by Xu et al. (2023) utilized CNNs to predict the binding affinity of phytochemicals from traditional Chinese medicine, yielding highly accurate results in identifying potential antiviral agents.

- **Recurrent Neural Networks (RNNs)**: RNNs, particularly Long Short-Term Memory (LSTM) networks, are used in virtual screening for sequential data like molecular dynamics simulations. These networks can capture temporal dependencies, making them suitable for predicting how bioactive molecules interact with their targets over time. Hwang et al. (2021) successfully employed LSTM networks to screen flavonoids from medicinal plants, predicting their effectiveness against specific enzymes.

Unsupervised Learning

Unlike supervised learning, unsupervised learning techniques work with unlabeled data. These techniques are useful in exploring large chemical databases to discover patterns or groupings that may not be immediately obvious. In virtual screening, unsupervised learning can be used for clustering compounds based on structural similarities or other features.

- **Clustering Algorithms**: Methods like K-means clustering and hierarchical clustering are often used to group phytochemicals with similar structures. This enables researchers to identify clusters of molecules that are likely to exhibit similar biological activities. Zhang et al. (2022) utilized hierarchical clustering to group natural compounds from medicinal plants, discovering several new compounds with anti-cancer potential based on their structural similarities to known bioactive compounds.
- **Principal Component Analysis (PCA)**: PCA is a dimensionality reduction technique used to simplify large datasets by reducing the number of variables while retaining essential information. It is particularly useful in virtual screening to reduce the complexity of high-dimensional chemical descriptors. PCA has been successfully applied in several studies, including one by Liu et al. (2020), which used PCA to identify lead compounds from traditional herbs, effectively narrowing down a large library of compounds for further screening.

Reinforcement Learning in Drug Discovery

Reinforcement learning (RL) is a relatively new method in virtual screening but has demonstrated significant potential. In RL, an agent improves its decision-making by interacting with its environment and receiving positive or negative feedback based on the actions it performs. In the context of drug discovery, RL can help optimize molecular structures to achieve the desired bioactivity.

- **Molecular Optimization with RL**: An example of RL in virtual screening is optimizing bioactive molecules to improve their pharmacokinetic properties. By defining a reward function that favors molecules with higher predicted bioactivity, RL algorithms can iteratively modify molecular structures to improve their drug-likeness. A study by Popova et al. (2018) demonstrated that RL models could generate novel molecules with enhanced bioactivity by exploring the chemical space of plant-derived compounds.

Hybrid Approaches

In practice, many virtual screening models combine multiple AI techniques to improve accuracy and robustness. For example, deep learning models may be combined with clustering algorithms to first group compounds and then apply neural networks for activity prediction. Similarly, reinforcement learning can be integrated with deep learning to optimize molecular structures during the screening process.

One notable example is a study by Yang et al. (2023), where a hybrid model combining CNNs and SVMs was used to screen for potential anti-diabetic compounds from medicinal plants. The hybrid approach outperformed single models, identifying several novel compounds for further in vitro testing.

Phytochemical Databases and AI Integration

The identification and utilization of bioactive compounds from medicinal plants hold great potential for drug discovery. However, the sheer complexity of natural products and the vast biodiversity of plant species make the discovery process both time-consuming and resource-intensive. The implementation of artificial intelligence (AI) into phytochemical screening has provided a robust solution to this challenge, particularly through the use of phytochemical databases. These databases serve as repositories of information on the chemical structures, bioactivities, and therapeutic potentials of compounds derived from plants. When coupled with AI algorithms, they enable virtual screening and prediction of bioactivity in ways that traditional approaches cannot match.

Role of Phytochemical Databases in Drug Discovery

Phytochemical databases are essential for cataloging the wealth of chemical diversity present in plants. They house detailed information about bioactive compounds, including their molecular structure, properties, and known biological activities. These databases provide an invaluable resource for identifying potential drug candidates based on structural similarity, pharmacological profiles, and interaction with specific biological targets.

Some widely used phytochemical databases include **PubChem**, **ChEMBL**, **KNApSAcK**, and **NPASS**. These databases store both experimental and predicted data, offering researchers a comprehensive platform to explore compounds' pharmacodynamics and pharmacokinetics. **PubChem**, for instance, is a public resource for chemical molecules and their biological activities, containing millions of bioactive small molecules, which can be screened for drug discovery purposes. Similarly, **ChEMBL** offers bioactivity data for drug-like molecules, allowing for the virtual screening of natural products in the search for therapeutic agents.

By integrating data from these databases with AI-based algorithms, researchers can predict the bioactivity of phytochemicals and simulate their interactions with biological targets, significantly accelerating the drug discovery process.

AI-Powered Virtual Screening in Phytochemical Research

Virtual screening is a computational technique used to identify potential drug candidates from large libraries of compounds. When applied to phytochemicals, this technique uses AI and machine learning algorithms which helps in prediction of likelihood compound's therapeutic potential, reducing the need for costly and time-consuming experimental testing.

AI algorithms, such as **deep learning** and **support vector machines (SVM)**, have shown tremendous potential in virtual screening processes. These algorithms are trained on existing data from phytochemical databases and can recognize patterns that indicate biological activity. For example, deep learning networks can be used to predict how a phytochemical compound will interact with a protein target based on the structural data contained in databases like **BindingDB** or **ZINC**. This approach allows researchers to virtually test thousands of compounds in a fraction of the time it would take using traditional laboratory techniques.

By utilizing AI in this way, researchers can rapidly identify lead compounds for further investigation. For example, AI-based virtual screening has been applied to medicinal plants for identifying anti-cancer agents, anti-inflammatory compounds, and antimicrobial molecules. These virtual screens offer a cost-effective and efficient means to tap into the vast chemical diversity that plant species offer.

Integration of AI with Phytochemical Databases

The integration of AI with phytochemical databases hinges on several factors, including data availability, quality, and the appropriate selection of AI algorithms. A critical step in this process is the **curation and preprocessing of data**. Phytochemical databases must be carefully curated to ensure that the data used in AI models are accurate, consistent, and comprehensive. Inaccurate or incomplete data can lead to faulty predictions, reducing the efficacy of virtual screening.

Another challenge lies in the **representation of molecular structures**. In AI models, molecular structures are typically represented as numerical features (descriptors), such as molecular weight, hydrophobicity, or electronegativity. These features are derived from the chemical structures stored in phytochemical databases and are used as inputs for machine learning models. For instance, **SMILES (Simplified Molecular Input Line Entry System)** strings, a text-based representation of molecular structures, are commonly used in virtual screening. Machine learning models process these representations to identify patterns that correlate with bioactivity.

Moreover, the use of AI-driven **molecular docking simulations** is becoming increasingly common in the integration of phytochemical databases. Molecular docking is a technique that predicts the orientation of a molecule when bound to a protein target. When combined with AI algorithms, docking simulations can rapidly screen phytochemicals for their binding affinities to specific drug targets, offering insights into the mechanisms of action of bioactive compounds.

AI and Phytochemical Data Analytics

Another powerful application of AI is in **data mining** and **pattern recognition** within large datasets. Phytochemical databases often contain extensive and complex data on a wide range of plant species and compounds. AI can analyze these data to uncover relationships between chemical structures and biological activities that may not be immediately obvious through manual analysis. This ability to discern patterns and predict bioactivity based on structural features is a significant advantage of AI in virtual screening.

For example, **random forest** algorithms and **neural networks** are particularly effective in recognizing bioactivity patterns within phytochemical datasets. These methods can process large amounts of data from phytochemical databases to generate predictive models that rank the likelihood of specific compounds exhibiting therapeutic effects.

Future Directions

As AI continues to evolve, its integration with phytochemical databases will become even more sophisticated. The development of more advanced machine learning models, such as **generative adversarial networks (GANs)**, will allow for the generation of entirely new phytochemical structures based on known bioactive compounds. Additionally, the integration of **multi-omics data** which is "genomics", "proteomics" and "metabolomics" into phytochemical databases will enhance the predictive power of AI models, enabling a more holistic approach to drug discovery from medicinal plants.

Applications of AI-Based Virtual Screening in Medicinal Plants

The application of artificial intelligence (AI)-based virtual screening in the identification of bioactive molecules from medicinal plants has revolutionized the field of drug discovery. This approach offers an efficient, cost-effective, and accelerated pathway to identifying potential therapeutic compounds from the rich reservoir of phytochemicals present in medicinal plants. The convergence of AI and medicinal plant research has opened new horizons for the discovery of drugs that can address unmet medical needs, while also providing sustainable solutions by reducing the need for large-scale plant extractions and laboratory experimentation. In this chapter, we will explore how AI-driven virtual screening has been applied in the discovery of bioactive compounds from medicinal plants, highlighting successful case studies and discussing the potential of this approach in future drug development.

Traditional Challenges in Bioactive Compound Discovery

Historically, the process of discovering bioactive molecules from medicinal plants involved labor-intensive methods such as extraction, fractionation, and biological testing. These techniques required significant time and resources, and often led to the isolation of well-known or previously studied compounds, leaving many potential bioactives undiscovered. Additionally, medicinal plants contain a vast array of compounds, and identifying the ones with therapeutic potential has been a daunting task. Traditional screening methods are not only time-consuming but also environmentally unsustainable due to the extensive plant material needed for experimentation. Furthermore, this approach often suffers from the low hit rates of bioactive compounds, resulting in a limited number of leads that can be further developed into drugs.

To overcome these challenges, AI-based virtual screening has emerged as a powerful tool that can predict the biological activity of plant-derived compounds through computational models, drastically reducing the need for wet-lab experimentation. AI algorithms, particularly machine learning (ML) techniques, have the capacity to sift through vast phytochemical databases, analyze molecular structures, and predict interactions with biological targets, thus enhancing the efficiency of drug discovery from medicinal plants.

AI in Virtual Screening for Medicinal Plants

AI-based virtual screening leverages advanced computational algorithms to analyze phytochemicals' molecular properties and predict their interaction with biological targets. The process typically involves three key steps: (1) constructing a library of phytochemical compounds, (2) applying AI models to predict their binding affinity or interaction with a specific target, and (3) validating these predictions through biological assays.

A key technique employed in virtual screening is machine learning, which includes several sub-methods like deep learning, support vector machines, and random forest algorithms. These models are trained on known bioactive compounds and their interactions with targets, allowing them to accurately predict the activity of unknown compounds. Additionally, molecular docking simulations, driven by AI, can be used to predict and visualize how plant-derived molecules bind to target proteins. This computational method not only speeds up the screening process but also provides crucial insights into the molecular mechanisms behind the action of bioactive compounds.

Case Studies of AI-Based Virtual Screening in Medicinal Plants

Several successful applications of AI-driven virtual screening in the discovery of bioactive molecules from medicinal plants have been reported. For instance, a study by Medina-Franco et al. (2020) employed AI-based virtual screening to identify potential inhibitors of SARS-CoV-2 main protease from a library of plant-derived compounds. The researchers used molecular docking and deep learning techniques to screen a vast collection of phytochemicals, ultimately identifying several promising candidates with antiviral activity. This study highlights the power of AI in rapidly identifying bioactives from plants that could be developed into therapeutic agents against emerging infectious diseases.

Another notable application is the discovery of anti-inflammatory compounds from *Curcuma longa* (turmeric). Using AI-based virtual screening, researchers identified several curcuminoids with strong binding affinity to inflammatory mediators, including cyclooxygenase-2 (COX-2). The AI model predicted these compounds' ability to inhibit the enzyme, a key target in the treatment of inflammation-related disorders such as arthritis and cardiovascular diseases (Singh et al., 2022). The identified compounds were later validated through in vitro assays, confirming the accuracy of AI predictions.

Similarly, researchers have applied AI-based virtual screening to discover anticancer agents from *Catharanthus roseus*, a plant known for its rich alkaloid content. By leveraging AI techniques, such as random forest and deep learning models, researchers identified novel alkaloids with high affinity for the tubulin protein, a key target in cancer therapy (Patil et al., 2021). These findings demonstrate AI's potential in expanding the pool of plant-derived compounds with anticancer activity, accelerating the discovery of novel drugs that can be further developed for clinical use.

AI in Predicting Drug-Like Properties and ADMET Profiles

In addition to identifying potential bioactive molecules, AI-based virtual screening also plays a crucial role in predicting drug-like properties and the absorption, distribution, metabolism, excretion, and toxicity (ADMET) profiles of phytochemicals. The drug development process is often hindered by poor pharmacokinetics and adverse toxicological effects, which can cause promising compounds to fail during clinical trials. AI algorithms are now capable of predicting these parameters early in the discovery process, allowing researchers to prioritize compounds with favorable ADMET profiles for further development.

A recent study applied AI-based virtual screening to evaluate the ADMET properties of flavonoids from *Camellia sinensis* (green tea). The AI model was trained on a dataset of known flavonoids and their pharmacokinetic profiles, enabling it to predict the bioavailability and toxicity of novel compounds (Chen et al., 2023). This approach significantly reduced the time and cost associated with traditional ADMET testing and allowed the researchers to focus on a smaller subset of compounds with the highest potential for drug development.

Future Prospects and Opportunities

The integration of AI in virtual screening for bioactive molecules from medicinal plants offers exciting opportunities for the future of drug discovery. With the continued development of more sophisticated AI models and the expansion of phytochemical databases, researchers will be able to explore the vast diversity of plant compounds more efficiently. The combination of AI with other cutting-edge technologies, such as quantum computing and multi-omics data integration, could further enhance virtual screening's predictive power, paving the way for the discovery of novel therapeutics with greater precision.

Moreover, AI-based virtual screening supports sustainable drug development by minimizing the environmental impact of traditional plant extraction methods. By identifying the most intriguing bioactive molecules in the early stages of discovery, researchers can reduce the need for large-scale plant harvesting and promote the conservation of biodiversity.

AI-Powered Prediction of Drug-Like Properties

In drug discovery, one of the most critical steps is evaluating whether a molecule has favorable drug-like properties. These properties, collectively known as ADMET (Absorption, Distribution, Metabolism, Excretion, and Toxicity), help determine if a compound will be effective as a therapeutic agent. Traditional methods for testing ADMET properties are time-consuming and expensive. However, the application of artificial intelligence (AI), particularly machine learning (ML), has transformed this process, offering faster, more cost-effective, and more accurate predictions.

Absorption and Distribution: Key Considerations in Drug Effectiveness

Absorption refers to how a drug enters the bloodstream after administration, while distribution refers to how it is transported throughout the body. Both processes are essential in determining a compound's bioavailability and therapeutic efficacy. For bioactive compounds derived from medicinal plants, AI-based models can predict how effectively they will be absorbed and distributed by analyzing molecular descriptors like solubility, polarity, and molecular size.

Machine learning algorithms, such as neural networks and decision trees, are commonly used for these predictions. These models are trained using extensive datasets of known compounds and their ADMET properties. As a result, they can identify whether a new bioactive compound is likely to be absorbed through the gastrointestinal tract or cross key biological barriers, such as the blood-brain barrier. This ability is particularly valuable during early drug discovery stages, allowing researchers to quickly evaluate large collections of plant-based molecules.

Metabolism and Excretion: Ensuring Drug Stability and Elimination

After absorption and distribution, drugs undergo metabolism, mainly in the liver, where they are converted into metabolites. These metabolites must be safe and non-toxic for the drug to be suitable for therapeutic use. Once metabolized, the drug is excreted, typically through the kidneys or liver. AI models can predict how plant-derived bioactive compounds will be metabolized and excreted, helping to ensure that they do not build up in harmful concentrations.

AI systems can analyze molecular structures and predict their interactions with liver enzymes, particularly cytochrome P450, which is crucial in drug metabolism. These predictions can identify any potential toxic by-products created during metabolism. Furthermore, AI models can estimate the excretion rate and pathways of these compounds, helping to predict their duration of action and safety profile.

Toxicity Prediction: Safeguarding Against Adverse Effects

Toxicity is a major reason why drug candidates fail in clinical trials. Therefore, it is essential to predict potential toxicity early in the drug development process. AI-based models have proven effective in this area, relying on large datasets that contain toxic and non-toxic properties of numerous compounds to make predictions.

A commonly used AI approach for toxicity prediction is Quantitative Structure-Activity Relationship (QSAR) modeling. QSAR models link a compound's chemical structure to its biological function., including potential toxicity. Machine learning algorithms like support vector machines and random forests are often used to build QSAR models that predict toxicity based on molecular features. This method is highly valuable in phytopharmaceutical research, where numerous plant-derived compounds must be evaluated for adverse effects.

AI Tools for ADMET Prediction: Streamlining Drug Discovery

Several AI-powered tools and platforms are available for predicting ADMET properties, which helps improve the efficiency of drug discovery. **SwissADME** is a commonly used tool that predicts a range of pharmacokinetic properties such as absorption, solubility, and drug-likeness. **pkCSM** is another widely used platform that predicts pharmacokinetics and toxicity using graph-based signatures.

These tools are especially beneficial in phytopharmaceutical research, where large numbers of bioactive compounds from medicinal plants need to be screened. By automating the prediction of ADMET properties, AI platforms enable researchers to focus on compounds with the highest likelihood of success, saving both time and resources.

Challenges and Opportunities

The integration of AI-based virtual screening for bioactive molecules from medicinal plants presents both significant opportunities and notable challenges. As artificial intelligence continues to evolve and revolutionize drug discovery, it is crucial to acknowledge the hurdles that need to be addressed for the technology to reach its full potential. These challenges, however, also open doors for innovative solutions and advancements in the field.

Data Availability and Quality

One of the foremost challenges in AI-based virtual screening is the **availability and quality of phytochemical data**. Artificial intelligence models require vast and well-curated datasets to make accurate predictions about the bioactivity of plant-derived compounds. However, many phytochemical databases are incomplete, lacking comprehensive data on numerous medicinal plants. Only a small fraction of the estimated hundreds of thousands of medicinal plants have been thoroughly studied, and even fewer have detailed information about their bioactive constituents.

Moreover, data inconsistency across various sources poses an additional challenge. Phytochemical databases may store data in different formats, making it difficult for AI models to process the information efficiently. This lack of standardized data formats hinders the capability of machine learning models to learn patterns accurately and reliably.

To overcome these challenges, there is a need for improved data curation and the development of unified data standards. Creating open-access, well-annotated phytochemical databases would enhance the efficacy of AI models in drug discovery. Collaboration among researchers, data scientists, and database curators can help standardize the data collected from medicinal plants, ultimately improving prediction accuracy.

Algorithmic Complexity and Prediction Accuracy

Another significant challenge is the **algorithmic complexity** required to simulate the interactions between bioactive compounds and biological targets. While AI models such as deep learning and neural networks have shown promise, they often struggle with the inherent complexity of biological systems. Medicinal plants contain a vast array of compounds, and these molecules often interact with multiple biological targets simultaneously. This multi-target effect is common in traditional plant-based medicines but remains difficult to model accurately using current AI tools.

In addition to this, the black-box nature of many AI algorithms reduces the transparency of their predictions. For example, deep learning models, while powerful, are often difficult to interpret. Researchers may struggle to understand how specific features in the data lead to certain predictions, making it harder to trust AI's results in critical areas like drug discovery.

To address these concerns, efforts are being made to develop more interpretable AI models that provide insights into the mechanisms behind their predictions. Improved interpretability will enhance researchers' confidence in AI-based predictions and facilitate their use in practical applications, such as drug discovery from medicinal plants.

Integration with Emerging Technologies

An exciting opportunity lies in integrating AI with other **emerging technologies**. The field of **multi-omics data integration**, which combines genomics, proteomics, and metabolomics, holds great promise for enhancing AI-based virtual screening. Multi-omics data provides a more comprehensive view of the biological processes influenced by bioactive compounds. When combined with AI, this approach allows for a deeper understanding of how plant-derived molecules interact with biological systems.

Furthermore, the emerging field of **quantum computing** has the potential to overcome current computational limitations in virtual screening. Quantum computers are capable of solving complex problems much faster than classical computers. As quantum computing technology matures, it may significantly enhance the ability of AI models to screen and predict bioactive compounds from medicinal plants more accurately and efficiently.

While the integration of these technologies presents an exciting frontier for drug discovery, their implementation is still in its early stages. Researchers must continue to explore how quantum computing and multi-omics data can complement AI-based virtual screening to fully unlock the potential of medicinal plants in drug development.

Ethical and Regulatory Considerations

The growing use of AI in pharmaceutical research raises important ethical and regulatory concerns. As AI tools become more prevalent in drug discovery, it is crucial to ensure that the use of these technologies adheres to established safety and ethical standards. One key concern is the regulatory approval of AI-discovered drugs. Traditional drug approval processes are not well-suited to the unique challenges posed by AI-driven drug development. Regulatory agencies will need to adapt to new workflows that incorporate AI at every stage of the drug discovery pipeline.

Moreover, there are concerns about data privacy **and** intellectual property. As AI systems process vast amounts of data from various sources, including proprietary databases, issues related to data ownership and privacy can arise. Developing clear regulations around the ethical use of data in AI-driven drug discovery is essential to avoid potential legal and ethical pitfalls.

Future Opportunities

Despite the challenges, the future of AI-based virtual screening for bioactive molecules from medicinal plants is filled with opportunities. Continued advancements in AI algorithms, data curation, and emerging technologies will significantly enhance the efficiency and accuracy of virtual screening processes. By addressing current limitations, AI has the potential to transform the discovery of plant-based drugs, unlocking new therapeutic compounds that have remained hidden due to the limitations of traditional research methods.

Furthermore, as AI models become more transparent and interpretable, their adoption in the pharmaceutical industry will likely increase. Researchers will be able to better understand the predictions made by AI algorithms, leading to more informed decisions in the drug discovery process.

AI in Sustainable Drug Development

The growing demand for environmentally friendly and sustainable practices in pharmaceutical research has led to the emergence of AI-based approaches in drug development. Traditional drug discovery methods are often resource-intensive, costly, and time-consuming, with a high rate of failure in the later stages of development. Artificial intelligence offers a more sustainable alternative, significantly reducing the need for extensive laboratory testing and experimentation. By harnessing the power of AI, pharmaceutical companies can develop drugs more efficiently, with less environmental impact, while also accelerating the discovery of new therapeutic compounds from medicinal plants.

Reducing Resource Consumption in Drug Discovery

Most significant contributions of Artificial Intelligence in drug development is its ability to minimize resource consumption. Conventional drug discovery involves screening large libraries of compounds, often using chemical synthesis and biological assays that require considerable amounts of materials, energy, and labor. In contrast, AI-based virtual screening allows researchers to predict the bioactivity of compounds using computational models, thereby eliminating the need for large-scale physical experiments.

AI models can process and analyze vast datasets from phytochemical databases, enabling the identification of potential drug candidates *in silico*. This approach reduces the reliance on chemical reagents and minimizes the waste generated from laboratory experiments. By integrating AI into the early stages of drug discovery, companies can develop more sustainable workflows that decrease their environmental footprint.

Accelerating Drug Discovery from Medicinal Plants

AI plays an Important role in accelerating the discovery of plant-based drugs, which is an essential aspect of sustainable pharmaceutical development. Medicinal plants offer a vast repository of bioactive compounds with therapeutic potential, but identifying these compounds using traditional methods can be time-consuming. AI-based virtual screening expedites this process by predicting the interactions between phytochemicals and biological targets, allowing researchers to focus on the most promising candidates for further development.

Moreover, AI-driven molecular docking simulations help optimize the structure-activity relationships (SAR) of bioactive compounds, enabling researchers to modify and enhance their properties for better efficacy and safety. By streamlining the discovery of plant-based drugs, AI not only contributes to the sustainable use of natural resources but also supports the conservation of biodiversity by reducing the need for excessive plant harvesting.

Promoting Green Chemistry and Sustainable Practices

AI's role in sustainable drug development also involves the advancement of green chemistry. The principles of green chemistry focus on designing chemical processes and products that minimize or eliminate the use and production of harmful substances. AI can optimize synthetic pathways in drug production, ensuring that bioactive compounds are manufactured in an environmentally sustainable manner.

For instance, AI algorithms can predict the most efficient chemical reactions and conditions to synthesize a particular compound, reducing the need for harmful reagents and minimizing waste production. Additionally, AI can help design biodegradable drugs by predicting their environmental impact, ensuring that new pharmaceuticals do not contribute to long-term pollution.

Future Directions in AI-Driven Sustainable Drug Development

The future of sustainable drug development lies in the continued advancement of AI technologies. As AI algorithms become more sophisticated, they will be able to predict not only the efficacy and safety of new drugs but also their environmental impact. Integrating AI with multi-omics data and green chemistry principles will further enhance the sustainability of pharmaceutical research.

Moreover, AI can aid in the development of circular pharmaceutical economies, where waste products from drug manufacturing can be recycled and repurposed. By designing drugs with sustainability in mind from the outset, AI will help shape a future where pharmaceutical development is both economically viable and environmentally responsible.

Future Directions and Conclusion

The application of AI-based virtual screening for bioactive molecules from medicinal plants is set to evolve significantly in the coming years. Future directions in this field will likely focus on improving the accuracy and efficiency of AI models through advancements in algorithms, data integration, and computational power.

Advancements in AI Algorithms

As machine learning algorithms become more sophisticated, there will be a growing emphasis on developing interpretable AI models that can explain their predictions more clearly. Improved transparency in AI decision-making will build trust among researchers and regulatory agencies, facilitating the adoption of AI-driven approaches in drug discovery. Additionally, the use of generative models like generative adversarial networks (GANs) could enable the creation of entirely new bioactive compounds based on existing phytochemicals, thus expanding the scope of drug discovery.

Multi-Omics and Quantum Computing Integration

Another key area of development will be the integration of multi-omics data which is "genomics", "proteomics" and "metabolomics" into AI-powered virtual screening processes. By combining diverse biological datasets, researchers can achieve a more comprehensive understanding of how bioactive compounds influence biological pathways.

Quantum computing is also expected to play a transformative role in the future of virtual screening. With its capacity to solve complex problems at unprecedented speeds, quantum computing could significantly enhance the ability of AI to screen vast phytochemical libraries and predict bioactivity with higher accuracy.

Conclusion

In conclusion, while the challenges of AI-based virtual screening are significant, the opportunities for innovation and discovery are vast. With advancements in AI algorithms, data integration, and emerging technologies like quantum computing, the future of drug discovery from medicinal plants is poised for tremendous growth. By addressing current limitations, AI-based virtual screening can become a cornerstone of sustainable drug development, unlocking new therapeutic potentials from nature.

References:

1. Xu, J., Zhang, Y., and Wang, T., "CNN-Powered Virtual Screening of Phytochemicals in Traditional Chinese Medicine," Nature Communications, vol. 14, no. 1, 2023.
2. Yang, Q., Shen, Z., and Xiao, J., "Hybrid AI Models for Virtual Screening of Anti-Diabetic Compounds from Medicinal Plants," Journal of Computational Chemistry, vol. 45, no. 7, 2023.
3. Yuan, Q., Shen, Z., and Xiao, J., "Quantum computing in drug discovery: Opportunities for AI-enhanced virtual screening," Nature Reviews Drug Discovery, vol. 22, no. 4, pp. 274-290, 2023.
4. Chen, J., Wang, Y., Zhou, J., and Zhang, H., "AI-based prediction of ADMET properties for flavonoids from Camellia sinensis: A sustainable approach to drug discovery," Journal of Phytopharmacology, vol. 10, no. 1, pp. 45-58, 2023.
5. Zhang, Y., Singh, R., and Gupta, V., "AI-based virtual screening identifies anti-inflammatory compounds from Curcuma longa," Pharmacognosy Research, vol. 14, no. 3, pp. 132-140, 2022.
6. Basu, A., Choudhury, A., and Mukherjee, A., "Artificial intelligence in drug discovery: Opportunities, challenges, and the road ahead," Journal of Chemical Information and Modeling, vol. 62, no. 3, pp. 1234-1246, 2022.
7. Kumar, P., Singh, S., and Dhiman, M., "Regulatory perspectives on artificial intelligence in drug discovery: Current trends and future directions," Pharmaceutical Technology, vol. 34, no. 6, pp. 67-74, 2022.
8. Vesel , P., Krej ová, D., and Ková , J., "Random Forest Modeling for Predicting Bioactivity in Medicinal Plants," Journal of Medicinal Chemistry, vol. 64, no. 5, 2021.
9. Chen, H., Wang, Y., and Li, X., "Phytochemical data availability in AI-powered virtual screening: Challenges and prospects," Bioinformatics Advances, vol. 7, no. 2, pp. 345-352, 2021.

10. Hwang, S., Shin, S., and Choi, J. Y., "LSTM Networks for Predicting Bioactivity of Flavonoids," Journal of Chemical Information and Modeling, vol. 61, no. 6, 2021.
11. Cao, Y., Romero, J., Olson, J. P., and Aspuru-Guzik, A., "Quantum chemistry in the age of quantum computing," Chemical Reviews, vol. 121, no. 16, pp. 10856-10915, 2021.
12. Zhang, L., Wang, T., and Zhang, Y., "Multi-omics data integration in AI-based virtual screening: Towards a more comprehensive understanding of bioactive molecules," Computational Biology and Chemistry, vol. 84, p. 107195, 2020.
13. Medina-Franco, J. L., Martínez-Mayorga, K., and Méndez-Lucio, O., "Virtual screening and molecular modeling of plant-derived compounds against SARS-CoV-2 main protease," Phytochemistry Letters, vol. 36, pp. 23-29, 2020.
14. Ekins, S., Puhl, A. C., and Zorn, K. M., "Exploiting machine learning for end-to-end drug discovery and development," Nature Biotechnology, vol. 37, no. 3, pp. 302-307, 2019.
15. Vamathevan, J., Clark, D., Czodrowski, P., and others, "Applications of machine learning in drug discovery and development," Nature Reviews Drug Discovery, vol. 18, no. 6, pp. 463-477, 2019.
16. Chen, H., Engkvist, O., Wang, Y., Olivecrona, M., and Blaschke, T., "The rise of deep learning in drug discovery," Drug Discovery Today, vol. 23, no. 6, pp. 1241-1250, 2018.
17. Valli, M., Pivatto, M., and Queiroz, E. F., "Natural products in drug discovery: advances and opportunities," Future Medicinal Chemistry, vol. 10, no. 13, pp. 1581-600, 2018.
18. Popova, M., Mikhal'chik, E. V., and Sokolov, S. Y., "Reinforcement Learning for Molecular Optimization in Drug Discovery," Nature Machine Intelligence, vol. 1, no. 6, 2018.
19. Dias, D. A., Urban, S., and Roessner, U., "A historical overview of natural products in drug discovery," Metabolites, vol. 2, no. 2, pp. 303-336, 2012.
20. Gaulton, A., Hersey, A., and Piquemal, J.-P., "ChEMBL: a large-scale bioactivity database for drug discovery," Nucleic Acids Research, vol. 40, no. D1, pp. D1100–D1107, 2012.
21. Kirchmair, J., Williamson, M. J., Tyzack, J. D., Tan, L., Bond, P. J., Bender, A., and Glen, R. C., "Computational prediction of metabolism: Sites, products, SAR, P450 enzyme dynamics, and mechanisms," Journal of Chemical Information and Modeling, vol. 52, no. 3, pp. 617-648, 2012.
22. Daina, A., Michielin, O., and Zoete, V., "SwissADME: a free web tool to evaluate pharmacokinetics, drug-likeness and medicinal chemistry friendliness of small molecules," Scientific Reports, vol. 7, no. 1, p. 42717, 2017.

23. Pires, D. E., Blundell, T. L., and Ascher, D. B., "pkCSM: Predicting Small-Molecule Pharmacokinetic and Toxicity Properties Using Graph-Based Signatures," Journal of Medicinal Chemistry, vol. 58, no. 9, pp. 4066–4072, 2015.

CHAPTER 2

RECENT ADVANCEMENT OF SUPERMOLECULES IN THE EVOLUTION OF CARDIOVASCULAR DRUGS

Balbir Singh, Hasandeep Singh, Brahmjot Singh, Disha, Navkaran Kumar, Amandeep Kaur, Neena Bedi, and Sarabjit Kaur*

Department of Pharmaceutical Sciences,
Guru Nanak Dev University, Amritsar

Abstract

The cardiovascular diseases are the leading cause for morbidity and mortality globally. Therefore, there is a great need for the development of novel and effective therapies. Supramolecular chemistry involves the coalition of two or more chemical species through intermolecular forces such as hydrogen bonding, van der Waals forces and electrostatic interactions, etc. which leads to the organized entities of complex nature. Various studies have reported the role of supramolecular drug delivery in various cardiovascular diseases such as myocardial infarction, atherosclerosis and ischaemic heart disease. A cell free therapy of peptide amphiphile nanostructuresand nanoscale filaments was designed to mimic the activity of potent angiogenic signalling protein vascular endothelial growth factor (VEGF) and this VEGF-mimetic filament induce phosphorylation of VEGF receptors and also promotes proangiogenic behavior in endothelial cells. Furthermore, the liver toxicity of liver X receptor agonist GW3965 (LXR) used in treatment of atherosclerosis was reduced by forming nanostructures of peptide amphiphile nanofibers which contained apolipoprotein A1–derived targeting peptide 4F to serve as nanocarriers for LXR delivery. The treatment reduces the formation of plaque suggesting the potential of supramolecularnanostructures in atherosclerosis. In addition, the side effects of amiodarone resulting from its accumulation in off-target tissues were reduced by nanoparticledrug carrier composed of L-lysine cross-linked succinyl-β-cyclodextrin and also enhanced theselective delivery of the drug to cardiac tissue by 250%. Therefore, supramolecular combinations not only produce dynamically stable nanostructures but also precisely control the therapeutic agents which offer new insights for the systematic and targeted treatment of cardiovascular diseases.

Keywords: Supramolecules, Cardiovascular diseases, Nanoformulations

Introduction

Cardiovascular diseases (CVDs) are the prominent cause of deaths globally with about 17.7 million deaths in 2015 which is expected to rise to 23.6 million in 2030 (WHO, 2017). The terminology "cardiovascular diseases" includes pulmonary embolism, deep vein thrombosis, myocardial infarction (MI) and coronary heart disease and all of these leads to ischaemia and tissue death. The treatment option available for the management of cardiovascular diseases is only oral medicines or invasive surgery. Moreover, there is an enormous gap in this area of medicine for novel therapeutics and improved patient compliance (Chandarana et al., 2018). The increasing risk and prevalence of CVDs all over the world requires novel therapeutic approaches for their management (Mansukhani et al., 2019).

Supramolecules are unique nanostructures having dynamic covalent and non-covalent intermolecular interactions such as hydrophobic interactions, van der Waal forces, electrostatic interactions, π-π stacking and hydrogen bonding (Caulder and Raymond, 1999; Wang et al., 2018a). The supramolecular assembly was initially invented from the groundbreaking work of Lehn on host-guest self-assembly in 1987 (Lehn, 1988). It initiates from the biological systems and widely used in various biomedical fields (Cui and Xu, 2017). These types of assemblies are very commonly found in nature such as actin in eukaryotic cytoplasm and the supramolecular structure of cell membrane phospholipids (Kim et al., 2019). In biological systems, the supramolecular assemblies are involved in various functions including molecular transport and release, environmental demarcation along with cell–extracellular interactions and communications (Ariga et al., 2019).

Therefore, from the inspiration of natural supramolecules, researchers have developed several supramolecular self-assembled biomaterials such as hollow capsules, hydrogels, coating films and nanoparticles for the diagnosis and treatment of diseases. The supramolecular materials include monodisperse molecules such as polysaccharides, proteins and various biomass materials and complex aggregates (particles, micelles) (Song et al., 2015). The supramolecules are able to accurately control the diagnostic or therapeutically active components and stimulus response characteristics to physiological indicators providing the dynamically stable nanostructures in the field of biology and medicine (Wang et al., 2018b).

Myocardial Infarction

Cardiovascular disorders are one of the major causes of death globally, from which myocardial infarction (MI) is most common, affecting more than 7 million individuals each year (Guo et al., 2021). MI can be defined as irreversible myocardial muscle fibre damage caused by prolonged ischaemia due to sustained imbalance of perfusion, supply and demand. The traditional view suggests that cell death occurs through two mechanistically and morphologically distinct modes: i) apoptosis (a programmed process associated with cell shrinkage, fragmentation into apoptotic bodies, and removal by professional phagocytes) and ii) necrosis (a passive, unregulated process characterized by cell swelling and loss of membrane integrity) (Konstantinidis et al., 2012). Apoptosis results in clean removal of dead cells and does not induce an inflammatory reaction, whereas necrosis leads to release of danger signals that trigger intense inflammation. Descriptive studies suggest that both apoptosis and necrosis are implicated in death of cardiomyocytes following MI (Nicotera and Melino, 2004). Ultrastructural investigations documenting cell swelling and membrane disruption in large numbers of infarct cardiomyocytes, and the intense inflammatory reaction triggered by MI suggest an abundance of necrotic cardiomyocytes. On the other hand, large numbers of apoptotic cardiomyocytes have been identified in infarcted hearts using both ultrastructural and histochemical strategies (Frangogiannis, 2011).

Supramolecular Hydrogel in improving cardiac function after myocardial infarction

Recent studies indicate that direct intramyocardial injection of biomaterials including injectable hydrogels have shown benefit in preventing cardiac maladaptive repair and reducing cardiac dysfunction (Yoshizumi et al., 2016; Rodell et al., 2016; Vu et al., 2015; Rodell et al., 2015;Annabi et al., 2014;; Ban et al., 2014). Intramyocardial injectable biomaterials can not only provide structural support for weakened myocardial wall and enhance the compliance of ventricular wall, but also provide platforms for controlled deliveries of drugs or biofactors. Recently, self-assembled supramolecular hydrogels formulated from peptide derivatives have shown significant potential in cancer cell inhibition (Feng et al., 2017; Wang et al., 2016; Versluis et al., 2016), detection of important analytes, drug delivery, vaccine production and regenerative medicine (Lock et al., 2017; Lock et al., 2016; Han et al., 2017; Ren et al., 2015; Xu et al., 2016). Furthermore, supramolecular hydrogels of peptides were injected into tissues or organs to treat cardiovascular diseases such as MI and limb ischemia in various studies and showed great therapeutic benefits.

Previously Chen et al., utilized the peptide derivatives of Cur-FFE-ssERGD and Nap-FFGGG-NO together at a ratio of 4:1 to co-assemble into a mixed component hydrogel. The mixed component hydrogel has the ability to release curcumin through the hydrolysis of ester bond between curcumin and peptide. Thenthe release profile of curcumin from gel was monitored which revealed the stable release of curcumin during the 24 h experimental period, and the release speed of curcumin was about 25.33 μg mLh-1 at physiological temperature conditions (37°C). There was about 8.78% of curcumin being released from the gel during 24 h experimental period. The classic Griess reaction nitric oxide (NO) assay was also used to determine the release curve of NO from the gel and the mixed component gel could controllably and constantly release NO over 48 h in presence of β-galactosidase. During this time period, about 26.14% of the total amount of NO got released from the gel when large concentration of β-gal (0.2 U mL-1) wasadded. Meanwhile, there was only a tiny amount of NO (about 3.39%) released from the gel at the low concentration of β-gal (0.02 U mL-1), and no NO release from the gel in the absence of β-gal. These observations indicated that both curcumin and NO could be constantly released from the mixed component hydrogel. This study showed that the combinational treatment of curcumin and NO could remarkably reduce collagen deposition, improve cardiac function, ameliorate adverse myocardium remodelling, suppress apoptosis and hypertrophy as well as attenuate the expression of matrix metalloproteinases (MMPs),transforming growth factor-β1 (TGF-β1) and upregulate the expression of Silent information regulator 1 (SIRT1) than curcumin alone (Rizk et al., 2014), because of the synergistic action of both molecules and the angiogenesis promotion ability of NO. L-arginine, an NO donor, showed defence against diabetic cardiac fibrosis and decreased the expression of collagen-1 along with TGF-β1 through upregulating SIRT1 expression. Co-delivery of curcumin and NO in a controllable manner by hydrogels may be considered as a promising option for treatment of cardiovascular disease (Chen et al., 2017).

Folic Acid-Derived Hydrogel for Acute Myocardial Infarction

Induced pluripotent stem (iPS) cells, which are able to differentiate into multiple cell types, have the potential to generate cardiovascular lineage cells for myocardial repair after ischemic damage. Therefore, supramolecular hydrogels of self-assembling peptides emerge as promising biomaterials for regenerative medicine and tissue engineering (Boekhoven and Stupp, 2014). The supramolecules formed by pure peptides and peptide amphiphiles have been used directly or in combination with stem cells to treat cardiovascular diseases such as MI and limb ischemia (Webber et al., 2010). Supramolecular hydrogels formed by short peptides are versatile for large scale preparation and practical applications. In order to construct supramolecular hydrogelators of short peptides, many aromatic capping groups are linked with short peptides through covalent bonds, in order to drive self-assembly by introducing π-π interactions intomolecules. Until now, naphthyl-, pyrenyl-, fluorenyl-, phenothiazinyl- and several therapeutic agents containing aromatic rings have been demonstrated as powerful aromatic capping groups to construct hydrogelators of short peptides (Bowerman and Nilsson, 2010). In a study, the essential vitamin, folic acid (FA) was used to modify short peptides to form supramolecular hydrogels. The designed FA-modified peptides and their degradation products are highly biocompatible because they are constituted by natural amino acids and vitamins. The short peptides containing di-phenylalanine (-FF-) sequence are excellent hydrogelators, and folic acid monomers are capable of forming stable tetramers through hydrogen bonds,therefore, theyare helpful for self-assembly (Sakai et al., 2006; Wang et al., 2011). Previously, amethod was developed for glutathione molecule to prepare supramolecular hydrogels by self-assembly of folic acid-modified peptide and was suitable for cell encapsulation and transplantation. The iPS cells labelled with CM-DIL were transplanted into the hearts of mice having MI with or without FA hydrogel encapsulation.The results of the study suggested that FA hydrogel significantly improved the retention and survival ofiPS in MI hearts post injection, leading to augment of therapeutic efficacy of iPS including better cardiac function and much less adverse heart remodelling, by subsequent differentiation towards cardiac cells and stimulation of neovascularization. The study reported a novel supramolecular hydrogel based on FA-peptides capable of improving the therapeutic capacity of iPS, which possess big potential in the treatment of MI (Li et al, 2018).

Supramolecular Self-Assembled Nanofibers in Myocardial Infarction Therapy

The stimulation of hepatocyte growth factor (HGF) signalling to activate the Met receptor produces antiapoptosis and anti-inflammatory effects and promotes angiogenesis, which can be applied forthe treatment of pulmonary fibrosis, myocardial infarction, and so on (Madonna et al.,2010; Gallo et al., 2015; zhang et al., 2018; Shimamura et al., 2020). HGF maintains the latent form of a single-chain protein known as the hepatocyte growth factor precursor [pro-HGF] in normal states and is specifically activated through proteolytic cleavage into a double-chain form at the site of tissue injury to perform its biological activity. Acute myocardial infarction (AMI) can cause acute ischemia and a hypoxia environment could cause adventitia fibroblasts to transform into myofibroblasts, promoting the secretion of an inactive pro-HGF (Binder et al., 2015; Fu et al., 2020; Short et al., 2004). The supramolecular self-assembly of synthetic peptides has a great prospect in the fields of nanomedicine, tissue engineering, and regenerative medicine (Rong et al., 2018; Liu et al., 2020; Feng et al., 2020; Medina et al., 2017; Zhou et al., 2018; Xing etal., 2018; Zhang et al., 2018; Cai et al., 2017; Wang et al., 2018; Cui and Xu, 2017; Shigemitsu et al., 2018; Mershin et al., 2010; Newcomb et al., 2016). Synthetic peptides derived from the segments of native N-terminals HGF β have been used for identification of activation pockets for the allosteric activation of the pro-HGF signalling in a reported mechanistic work. Recently, utilizing a special phage display selection method, Lazarus and co-workers developed a series of zymogen peptide activators (ZAPtides) of pro-HGF that precisely binds to the activation pocket and avoid the irreversible proteolytic cleavage mechanism to activate pro-HGF-dependent Met signalling pathway for cell survival and migration (Landgraf et al., 2014) as shown in Fig 1. Inspired by these studies, thefunctional peptide nanofibers were designed that integrate a self-assembly motif and a peptide activator of the pro-HGF, which can form a crosslinked fibrous network that increases the probability of contact between the peptide activator and β chain of HGF to improve the efficiency of activation. The results showed that self-assembly nanofibers can successfully activate the HGF and engage the Met receptor to initiate signalling, allowing pro-angiogenesis and antiapoptosis post-AMI, effectively reducing the area of myocardial infarction and saving cardiac function. The oxygen/glucose deprivation (OGD) due to AMI resulted in vascular adventitia fibroblasts whichdifferentiate into myofibroblasts and secrete proHGF. In the meantime, there is upregulation of specific Met receptor of hepatocyte growth factor (HGF) in endothelial cells during AMI. Nonetheless, the poor diagnosis of AMI revealed that the pro-HGF is not effectively activated and improving the activation efficiency of pro-HGF may play a progressive role in treatment of AMI.

From a series of zymogen activator peptides of pro-HGF (ZAPtides), (IVGGYPWWMDV, compound 2 (Comp.2) as a bioactive motif with high affinity and favorable activation efficiency was chosen to bind the allosteric region for activating the pro-HGF. Additionally, we selected Nap-FF as the self-assembly segment due to its well-known, excellent selfassembly capacity (Zhou et al., 2018; Xing etal., 2018; Zhang et al., 2018; Cai et al., 2017; Wang et al., 2018). Through the standard solid-phase peptide synthesis (SPPS), the designed self-assembling functional peptide (Nap-FFEG-IVGGYPWWMDV, Comp.1) was synthesized, in which glutamic acid (E) was used for modulating the balance of hydrophilicity-hydrophobicity and glycine (G) acted as a linker to connect the two moieties. A control peptide with a scrambled functional sequence (Nap-FFEG-YPGGVWVWMDI, Comp.3, was synthesized for comparison with the activity of Comp.1. In this study, we combined the self-assembly motif with the bioactive zymogen activator peptide (IVGGYPWWMDV) to obtain Comp.1, which can strongly activate the pro-HGF. Through the incorporation of self-assembly motifs, Comp.1 can self-assemble into supramolecular nanofibers with a β-sheet secondary structure, which may make the bioactive peptide segment distribute in a dense and orderly manner on the surface of nanofibers and possibly solve the unordered aggregation and random distribution of bioactive peptides in an aqueous solution. Endothelial cells play a vital role in angiogenesis (Nazarnezhad et al., 2020). As a key factor in the function of blood vessels, the viability of endothelial cells is decisive to the reconstruction of blood flow for rescuing cardiomyocytes, reducing the infarct size, and improving cardiac function. The in vitro results of the TUNEL staining and the expression of apoptosis-related cleavedcaspase-3 showed that Comp.1 could more effectively protect endothelial cells and improve cell survival under the OGD condition. On the other hand, it is crucial for the revascularization after infarction to form the neovascularization. Through the tube formation experiment, scratch test, and determination of the expression level of angiogenesis-related VEGFA, we confirmed that Comp.1 had the most significant effect on the promotion of human umbilical vein endothelial cell (HUVEC) angiogenesis. To verify that Comp.1 indeed activated the pro-HGF and initiated its downstream signaling pathways, we detected the distribution of Comp.1 at the cellular level and measured the protein expression of the Met receptor, ERK1/2, and AKT. The results showed that, at 1 h of incubation, the fluorescent puncta of Comp.1, rather than Comp.2 and Comp.3, were observed to accumulate on the cell membrane, which suggested that Comp.1 could better bind to the pro-HGF and then move to the Met receptor on the HUVEC. Moreover, the phosphorylations of the Met, AKT, and ERK1/2 were significantly upregulated with the treatment of Comp.1. Taken together, this can prove that Comp.1 possesses a higher affinity than Comp.2 to bind to and activate the pro-HGF. Subsequently, we created an AMI mouse model to further verify the function of Comp.1 in vivo. The results of the ex vivo imaging showed that the retention time of Comp.1 in the heart was much longer than that of Comp.2, indicating that the self-assembled nanofibers could amplify the action time of small molecule peptides. In addition, the ex vivo imaging also showed that Comp.1 was interestingly accumulated in the area of the myocardial infarction. For this result, we speculate that the vascular leakage induced by the injured vessels in the infarct area may have been conducive to the penetration and retention of nanofibers. In terms of the treatment results, the Masson staining showed that Comp.1 could significantly alleviate the infarct area and inhibit the degree of fibrosis. The results of the echocardiography, including LVEF and FS, showed that treatment with Comp.1 could effectively maintain ventricular systolic function after AMI. Moreover, the immunofluorescence staining indicated that the numbers of neovascularization in the infarct marginal area were significantly increased after the treatment of Comp.1. These results collectively indicate that Comp.1 has the superior potential to reconstruct microcirculation, protect cardiomyocytes, and decrease myocardial fibrosis after myocardial infarction to avoid adverse cardiac events.

Therefore, in a study conducted by Guo et al, supramolecular nanofibersself-assembled with compound 1 (Comp.1, Nap-FFEG-IVGGYPWWMDV) were designed thathave a strong ability to activate pro-HGF and initiate HGF-Met signalling. Previous studies have showed the enhanced ability ofComp.1 in activating pro-HGF to perform antiapoptosis and pro-angiogenesis. The results of *in vivo* studies have also confirmed the increased retention time and accumulation of Comp.1 in the infarct area of heart shown in Fig 2. In addition, there was a significant increase of angiogenesis by Comp.1 in the marginal area of AMI, resulting in reduced myocardial fibrosis, and protecting cardiac function. Hence, the structure of bioactive peptides through supramolecular self-assembly was optimizedwhichamplify their therapeutic effect by improving efficiency and providing new strategy for AMI therapy (Guo et al, 2021).

Treatment of Hypertension

Hypertension is one of the strongest risk factorsfor almost all different cardiovascular diseases developed during life, including coronary heart disease, left ventricular hypertrophy, heart valve diseases, cardiac arrhythmias including atrial fibrillation, cerebral stroke and renal failure aortic syndromes. The relationship between blood pressure and cardiovascular diseases along with renal morbidity and mortality have been addressed in a large number of observational studies (Lewington et al., 2002). In previous studies, it is indicated also by the particularly high cardiovascular risk in patients with an elevated systolic and a normal or low diastolic BP (isolated systolic hypertension) (Kannel et al., 1981). The concept is based on the fact that only a small fraction of the hypertensive population has an isolated elevation of blood pressure, with majority exhibiting additional cardiovascular risk factors. So, the prevention of cardiovascular disease and treatment recommendations for mildly elevated blood pressure should be related to quantification of total cardiovascular risk (Kjeldsen, 2018). Blood pressure (BP)-related absolute risk of cardiovascular disease (CVD) revealed an exponential association between increased BP and risk for coronary heart disease (CHD) and stroke. It identified a relatively small increment in absolute risk of CVD with increasing BP levels at lower BP values (Fuchs, 2018). Prevention of age-related increase in BP would substantially reduce the vascular consequences usually attributed to aging. Individuals who develop hypertension late in life are not at increased risk for CVD earlier in life (Olsen et al., 2016). It is important to focus greater attention on initiatives for prevention of the typical age-related increase in BP in addition to control the high BP in those with established hypertension. Even partial improvement in the age-related increase in BP would eliminate a large proportion of the existing burden of BP-related CVD (Fuchs and Whelton, 2020).

Supramolecular interactions between losartan and hydroxypropyl-β-CD

Less soluble supramolecular complex between the losartan potassium (Los) and hydroxypropil-β-cyclodextrin (HPβCD) were characterized by phase-solubility, NMR techniques (1H and 2D-ROESY) and isothermal titration calorimetry (ITC) in order to attain physical-chemical knowledge of the system. In addition, the hypertensive effect of composition Los/HPCD was evaluated aiming to obtain a more efficient oral pharmaceutical composition.

Losartan potassium (Los) is a high soluble orally active, non-peptide drug used in the treatment of hypertension and related disorders (Kaplan, 1999; Mcintyre et al., 1997; Oparil et al., 2001). It has been demonstrated to have a greater effect than previous peptide receptor antagonists and angiotensin converting enzyme (ACE) inhibitors because of its enhanced specificity, selectivity, and tolerability. Los is metabolized in the body, forming a pharmacologically active carboxylic acid metabolite EXP3174 that presents low bio-availability of 33% (19–62%) with an approximate one-hour post-administration plasma peak concentration and a mean half-life of 2 h (Mcintyre et al., 1997). Thus, it is hypothesized that by sequestering a highly watersoluble drug, such as Los, through supramolecular non-covalent host-guest complexes formation using modified cyclodextrin, such as hydroxypropil-β-cyclodextrin (HPβCD), lower water Los solubility and more efficient angiotensin II AT1 receptor antagonistic oral pharmaceutical composition can be obtained.

Angiotensin Receptor Antagonism and Neprilysin Inhibition:

The angiotensin receptor blocker valsartan and the neprilysin inhibitor prodrug sacubitril was mixed in a 1:1 ratio sodium supramolecular complex. The synthesis ofValsartan/Sacubitrilwas carried out by mixing2 drugs following addition ofaqueous sodium hydroxide solution to promote crystallization (Ksander et al., 1995; Feng et al., 2012). The crystal structure of valsartan/ sacubitrilhas 6 sacubitriland 6 valsartan moieties in their anionic forms, 18 penta- and hexa-coordinated sodium cations, 15 water molecules, and istermed as sodium supramolecular complex. The complex breaks into valsartan and sacubitril after administrationand are further converted to active inhibitor of neprilysin LBQ657with the help of esterases. Ruilope et al.evaluated the effect of valsartan/ sacubitrilover an8-week treatment schedule in mild to moderate hypertension (mean sitting DBP of 90 to 109 mmHg after washout of antihypertensive medications or 95 to 109 mmHg in untreated patients). Eighty-seven percent of the patients in whole study were white, and 57% were male. Treatment with sacubitrilalone and all doses of valsartan/sacubitrilenhanced the circulating concentrations of cGMP. No significant effect on plasma renin activity was observed by sacubitril alone. In addition, there was an increase in plasma renin activity by valsartan/sacubitril and bioequivalent doses of valsartan which suggests that the observed effect was solely due toloss-of-feedback inhibition of renin production during angiotensin receptor blockade. The plasma concentrations of aldosterone did not varyamong active treatment groups or from placebo. Consequently, Sacubitril is converted to LBQ657by esterasesandinhibitsthe enzyme neprilysin, which is responsible for the degradation of natriuretic peptides and many other vasoactive peptides.

Therefore, the combined angiotensin receptor antagonist and neprilysin inhibitor reports the pathophysiological mechanisms of heart failure*i.e.*activation of renin-angiotensin-aldosterone system and decreased sensitivity to natriuretic peptides. In the prospective comparison of ARNI with ACEI to determine Impact on Global Mortality and Morbidity in Heart Failure (PARADIGM-HF) trial, valsartan/sacubitril significantly decrease the mortality and hospitalization for heart failure, as well as blood pressure, compared to enalapriltreated patients with heart failure, reduced the ejection fraction, and elevated circulating level of brain natriuretic peptide or N-terminal pro-brain natriuretic peptide. Further clinical trials are evaluating the role of valsartan/sacubitril in the treatment of heart failure with preserved ejection fraction and hypertension (Hubers and Brown, 2016).

Arrhythmia

Cardiac arrhythmia refers to an irregular heart beat pattern, and there are various types including atrial as well ventricular arrhythmias. Atrial fibrillation is very common form of arrhythmia and is seen in patients in their old age, in conditions such as heart failure, and can also cause a variety of problems including stroke (Bravo-Valenzuela et al., 2018).

Amiodarone is a powerful anti-arrhythmic drug used for the management of several kind of cardiac arrhythmias. However, amiodarone possesses many undesired effects caused by unintentional tissue accumulation (Mujovi et al., 2020). Cardiac macrophages are the cells that are commonly involved in maintaining homeostatic activities such asimmune response and flexibility of cardiac functions (Williams et al., 2018). It was demonstrated that amiodarone could be very effective when delivered to the heart by cardiac macrophages, an important step in reducing total volume and targeted tissue collection.

Therefore, supramolecule nanoparticle drug carrier composed of L-lysine cross-linked with succinyl-β-cyclodextrinwas assembled which showed that amiodarone has ability to bind through supramolecular host-guest interactions and high macrophage interactions. Pharmacokinetic analysis and a single cell level showed the accumulation of supramolecule nanoparticles in heart due to rapid absorption by cardiac macrophages. Moreover, from the biodistribution studies, it was revealed that there was an increased uptake of supramoleculesof amiodarone in macrophage rich tissues leading to 250% rise in the delivery of drug to heart muscles Thus, the supramolecule of CDNP-Amio (cyclodextrinnanoparticleloaded with amiodarone) is amenable totargeted delivery for antiarrhythmic agent's payload to macrophages of cardiac cells. The supramolecule was rapidly adopted bymacrophages*in vitro* during the treatment of 4 hours, a conclusion reflected in the plasma half-life of 53.7 ± 2.6 minutes *in-vivo* (Ahmed et al., 2019).

Atherosclerosis

Atherosclerosis is a thickening of arterial wall that results in the development of plaque and promote stenosis of coronary arteries. Inflammation has been involved in any stage of atherosclerosis, varying from begging of plaque to rupture (Khan et al., 2015). For a long time, it was reported that the main factors affecting the pathogenesis of atherosclerosis are high blood pressure and high blood cholesterol levels (Prilepskii et al, 2020). Evolution in biotechnology has provided various biomedical platforms for their use in targeted drug delivery systems for surgical and oncologic applications (Mansukhaniet al., 20).

Peptide Amphiphile Supramolecular Nanostructures as a Targeted Therapy for Atherosclerosis

Apolipoprotein A1 (ApoA1) is a major protein constituent of high-density lipoprotein (HDL) cholesterol and is efficient for elevated cholesterol efflux from atherosclerotic plaques in animals (Mendis et al., 2011). The atherosclerosis plaque targeted by covalent bonding to peptide amphiphile (PA) is an 18-amino acid ApoA1, known as 4F (Navab et al., 2005).PAs containing an apolipoprotein A1-mimetic peptide 4F, serves as an effective nanocarrier to target LXR agonist to atherosclerotic plaque *in vivo* (Guler and Garifullin, 2020). The systemic administration of ApoA1-PA nanofiber containing LXR agonist will target atherosclerosis lesions, deliver their therapeutic agents, and induce atherosclerosis plaque regression. The significant reduction in macrophage content within atherosclerotic lesion as compared to saline and non-targeted nanoparticle treatment, the effect of LXR agonist on reducing plaque was not quantified, and the mice did not continue a high- fat diet during the treatment, prevention accurate comparison between the efficacies of each nanocarrier system.

PA nanofibers include LXR agonist through hydrophobic collapse rather than covalent alternative covalent linkages. Self-assembly reversible nanofiber can undergo, some LXR agonist possibly lost during intravenous transport. Also reaching plaque, the LXR agonist may not readily release from the hydrophobic interior of the nanofibers to reach the intracellular nuclear target within macrophage foam cells (Mansukhani et al., 2019).

Monocyte Targeting Supramolecular for molecular diagnostic tool for Atherosclerosis

Conventional envisioning of atherosclerosis has concentrated on the occurrence of arterial lumen or the structure of plaque, and firstly gives the anatomic and physiological detail on causes behind stage of the disease (Buxton et al., 2011). The advancement of clinical biology of atherosclerosis recognizes the key inflammatory processes that leads to lesion initiation and progression. The emerging field of molecular imaging has control this information to target specific biochemical marker for accurate early diagnosis (Jaffer et al., 2006). Nanosized micelles are compatible for molecular level imagining tool because they are smaller enough to interact with receptor target and provide high surface area that can be covered with specific ligands to recognize tracking of their distribution. Studies showed that MCP-1 PAMs bind to the CCR2 receptor and accumulate in the aortic deposit in aortic arch in proportion to disease development in the ApoE knockout mouse as shown in Fig.3, offering a tool to both diagnose and differentiate between the stages of the atherosclerosis (Nakashima et al., 1994). Also, it is considered that CCR2 is a well-known receptor on monocytes, activated by endothelial cells and have also been reported to express some CCR2 within plaques and may also be contributing in targeting signal of MCP-1 PAMs *in vivo*. The nano-assemblies have the opportunity to advance atherosclerosis therapies by enhancing drug solubility through exploitation of the micelle's hydrophobic core, reducing the required dosage and potential for cytotoxicity via active targeting and focal delivery. This in turn increased the systemic circulation time of the carrier, and provides multifunctional approach by combining therapeutic and diagnostic agents within one system. Furthermore, these adapted modalities imaging for micelles could then be merged with anti-inflammatory drug for potential therapeutic application. In addition to the target molecular or cellular markers, micelles can be further optimized to assess structural components of vulnerable plaque by imaging fibrous cap thickness and the area of the necrotic core via incorporation of collagen or metalloproteinase-binding peptides (Newby, 2008). As monocytosis is characteristic to multiple inflammation disease, MCP-1 PAMs also have the potential to be employed in multiple applications (Yadav et al., 2010). PAMs may serve as a tool for discovery science as well, validating cellular abundance the complex etiology of atherosclerosis (Chung et al., 2015).

Bifunctional supramolecular nanofiber in atherosclerosis

The risk of atherosclerosis was elevated by lower circulation of IGF-1 level (Juul et al., 2002). The deficiency of IGF-1 receptor can accelerate progression of atherosclerosis (Higashi et al., 2016; Sivasubramaniyam et al., 2017).

The increased levels of circulating IGF-1 might serve as a promising therapeutic approach for the treatment of atherosclerosis. Various research groups have designed supramolecular nanofibers possessing anti-inflammatory properties and IGF-1 bioactivity. Besides, it was also demonstrated that IGF-1 mimetic peptide could decrease the systemic inflammation and enhanced the cholesterol efflux of macrophages through induction of ABCA1 and ABCG1which further leads to the inhibition of foam cell formation and also development of atherosclerosis lesions. The formation of foam cell from macrophages and VSMC is a sign of atherosclerosis (Moore et al., 2018).Furthermore, there was a marked increase in the expression of both ABCA1 and ABCG1 by IGF-1 or IGF-1 mimetic peptides which accounts for decrease in atherosclerosis progression in ApoE-deficient mice (Sukhanovet al., 2007).

The alteration in lipid metabolism is also a significant contributor to the development of atherosclerosis. In addition, the lipid profile and body weight of mice were not changed by administration of IGF-1 mimetic peptides which indicates that the anti-atherogenic effect of IGF-1 mimetic peptides is independent of lipid metabolism but other factors such as anti-inflammatory, increased cholesterol efflux, and inhibiting formation of macrophage-like SMCs, acts as major contributors to its anti-atherogenic effect. The designed bifunctional supramolecular nanofibers combined the bioactivity of IGF-1 and anti-inflammatory property of Npx, and effectively inhibit the development of atherosclerosis (Shang et al., 2020).

Conclusion

Although, a large number of unconventional and non-conventional treatment options are available for treatment of various cardiovascular diseases such as myocardial infarction, atherosclerosis and ischaemic heart disease, but all of these are associated with one or more problems in their systematic and targeted delivery. Supramolecular combinations are an alternative strategy for treatment of CVDs and are also comparatively safer. Supramolecular combinations help to produce dynamically stable nanostructures which further precisely control the therapeutic agents and offer new insights for the systematic and targeted treatment of cardiovascular diseases. Thus, supramolecular based drug delivery systems can prove to be really helpful in cardiovascular problems and is the new ray of hope for treatment of CVDs.

References

Ahmed M. S, Rodell C. B, Hulsmans M, Kohler R. H, Aguirre A. D, Nahrendorf M, &Weissleder R. A supramolecular nanocarrier for delivery of amiodarone anti-arrhythmic therapy to the heart. Biocnjug chem.2019;30(3):733-740.

Annabi N, Tamayol A, Uquillas J. A, Akbari M, Bertassoni L. E, Cha C, &Khademhosseini A.25th anniversary article: Rational design and applications of hydrogels in regenerative medicine. Adva Mater.2014; 26(1):85-124.

Ariga K, Ahn E, Park M, & Kim B. S. Layer-by-layer assembly: recent progress from layered assemblies to layered nanoarchitectonics. Chem Asian J. 2019;14(15):2553-2566.

Ban K, Park H. J, Kim S, Andukuri A, Cho K. W, Hwang J. W & Yoon Y. S. Cell therapy with embryonic stem cell-derived cardiomyocytes encapsulated in injectable nanomatrix gel enhances cell engraftment and promotes cardiac repair. ACS nano. 2014;8(10):10815-10825.

Binder A, Ali A, Chawla R, Aziz H. A, Abbate A &Jovin I. S . Myocardial protection from ischemia-reperfusion injury post coronary revascularization. Expert Rev Cardiovasc Ther. 2015;13(9):1045-1057.

Boekhoven J &Stupp S. I. 25th anniversary article: supramolecular materials for regenerative medicine. Adva Mater. 2014;26(11):1642-1659.

Bowerman C. J & Nilsson B. L. A reductive trigger for peptide self-assembly and hydrogelation. J Am Chem Soc. 2010;132(28):9526-9527.

Bravo-Valenzuela N. J, Rocha L. A, Nardozza L. M. M & Júnior E. A. Fetal cardiac arrhythmias: Current evidence. Ann Pediatr Cardiol.2018;11(2):148.

Buxton D. B, Antman M, Danthi N, Dilsizian V, Fayad Z. A, Garcia M. J&Weissleder R. (2011). Report of the National Heart, Lung, and Blood Institute working group on the translation of cardiovascular molecular imaging. Circulation. 2011;123(19):2157-2163.

Cai Y, Shen H, Zhan J, Lin M, Dai L, Ren C & Yang Z. Supramolecular "Trojan Horse" for nuclear delivery of dual anticancer drugs. J Am Chem Soc.2017;139(8):2876-2879.

Chandarana M, Curtis A & Hoskins C. The use of nanotechnology in cardiovascular disease. Appl Nanosci. 2018 8(7):1607-1619.

Chen G, Li J, Song M, Wu Z, Zhang W, Wang Z&Ou C. A mixed component supramolecular hydrogel to improve mice cardiac function and alleviate ventricular remodeling after acute myocardial infarction. Adv Funct Mater. 2017;27(34): 1701798.

Chung E. J, Mlinar L. B, Nord K, Sugimoto M. J, Wonder E, Alenghat F. J & Tirrell M. Monocyte targeting supramolecular micellar assemblies: a molecular diagnostic tool for atherosclerosis. Adv Healthc Mater.2015;4(3):367-376.

Cui H & Xu B. Supramolecular medicine. Chem Soc Rev .2017;46(21):6430-6432.

Cui H & Xu B. Supramolecular medicine. Chem Soc Rev. 2017;46(21):6430-6432.

Feng L, Karpinski P. H, Sutton P, Liu Y, Hook D. F, Hu B&Ziltener C. LCZ696: a dual-acting sodium supramolecular complex. Tetrahedron Lett. 2012;53(3):275-276.

Feng M, Li X, Hou X, Chen J, Zhang H, Song S& Shi C. Specific angiogenic peptide binding with injectable cardiac ECM collagen gel promotes the recovery of myocardial infarction in rat. J Biomed Mater Res A.2020;108(9):1881-1889.

Feng Z, Wang H, Zhou R, Li J & Xu B. Enzyme-instructed assembly and disassembly processes for targeting downregulation in cancer cells. J Am Chem Soc. 2017; 139(11):3950-3953.

Frangogiannis N. G. Pathophysiology of myocardial infarction. Compr Physiol. 2011;5(4):1841-1875.

Fu H, Fu J, Ma S, Wang H, Lv S,& Hao Y. An ultrasound activated oxygen generation nanosystem specifically alleviates myocardial hypoxemia and promotes cell survival following acute myocardial infarction. J Mater Chem B . 2020;8(28):6059-6068.

Fuchs F. D &Whelton P. K. High blood pressure and cardiovascular disease. Hypertension. 2020;75(2):285-292.

Fuchs F. D, Fuchs & Pires. Essentials of hypertension. Cham, Switzerland: Springer.2018.

Gallo S, Sala V, Gatti S & Crepaldi T. Cellular and molecular mechanisms of HGF/Met in the cardiovascular system. Clin. Sci.2015;129(12):1173-1193.

Guler M. O & Garifullin R. Design and Synthesis of Peptides for Developing Biomaterials.2020

Guo W, Feng W, Huang J, Zhang J, Fan X, Ma S & Chen M . Supramolecular Self-Assembled Nanofibers Efficiently Activate the Precursor of Hepatocyte Growth Factor for Angiogenesis in Myocardial Infarction Therapy. ACS Appl Mater Interfaces. 2021;13(19):22131-22141.

Han J, Duan P, Li X & Liu M. Amplification of circularly polarized luminescence through triplet–triplet annihilation-based photon upconversion. J Am Chem Soc. 2017;139(29):9783-9786.

Higashi Y, Sukhanov S, Shai S. Y, Danchuk S, Tang R, Snarski P &Delafontaine P . Insulin-like growth factor-1 receptor deficiency in macrophages accelerates atherosclerosis and induces an unstable plaque phenotype in apolipoprotein E–deficient mice. Circulation, 2016;133(23):2263-2278.

Hubers S. A & Brown N. J. Combined angiotensin receptor antagonism and neprilysin inhibition. Circulation, 2016; 133(11):1115-1124.

Jaffer F.A, Libby P, &Weissleder R. Molecular and cellular imaging of atherosclerosis: emerging applications. J. Am. Coll. Cardiol, 2006;47(7):1328-1338.

Juul A, Scheike T, Davidsen M, Gyllenborg J, &Jørgensen T. Low serum insulin-like growth factor I is associated with increased risk of ischemic heart disease: a population-based case-control study. Circulation, 2002; 106(8):939-944.

Kannel W.B, Wolf P.A, McGee D.L, Dawber T.R, McNamara P, & Castelli W.P. Systolic blood pressure, arterial rigidity, and risk of stroke: the Framingham study. Jama,1981;245(12):1225-1229.

Kaplan N.M. Angiotensin II receptor antagonists in the treatment of hypertension. Am fam physician,1999:60(4):1185.

Khan R, Spagnoli V, Tardif J.C, &L'Allier P.L. Novel anti-inflammatory therapies for the treatment of atherosclerosis. Atherosclerosis, 2015:240(2):497-509.

Kim J, Narayana A, Patel S, & Sahay, G. Advances in intracellular delivery through supramolecular self-assembly of oligonucleotides and peptides. Theranostics, 2019;9(11):3191.

Kjeldsen S.E. Hypertension and cardiovascular risk: general aspects. Pharmacol. Res. 2018;129:95-99.

Kleinman C.S, & Nehgme R.A. Cardiac arrhythmias in the human fetus. Pediatr. Cardiol, 2004;25(3):234-251.

Konstantinidis K, Whelan R.S, &Kitsis R.N. Mechanisms of cell death in heart disease. Arterioscler Thromb Vasc Biol, 2012;32(7):1552-1562.

Ksander G.M, Ghai R.D, dejesus R, Diefenbacher C, Yuan A, Berry C, & Trapani A. Dicarboxylic acid dipeptide neutral endopeptidase inhibitors. J med. Chem. 1995:38(10);1689-1700.

Landgraf K. E, Steffek M, Quan C, Tom J, Yu C, Santell, L, & Lazarus R.A. An allosteric switch for pro-HGF/Met signaling using zymogen activator peptides. Net. Chem. Biol. 2014;10(7):567-573.

Lehn J.M. Supramolecular chemistry—scope and perspectives molecules, supermolecules, and molecular devices (Nobel Lecture). Angew Chem Int Edition in Engl. 1988; 27(1):89-112.

Lewington S. Prospective studies collaboration. Age-specific relevance of usual blood pressure to vascular mortality: a meta-analysis of individual data for one million adults in 61 prospective studies. Lancet. 2002; 360:1903-1913.

Li H, Gao J, Shang Y, Hua Y, Ye M, Yang Z, & Chen M. Folic acid derived hydrogel enhances the survival and promotes therapeutic efficacy of iPS cells for acute myocardial infarction. ACS Appl. Mater. Interfaces. 2018;10(29):24459-24468.

Liu S, Zhao M, Zhou Y, Li L, Wang C, Yuan Y, & Liu J. A self-assembling peptide hydrogel-based drug co-delivery platform to improve tissue repair after ischemia-reperfusion injury. Acta. Biomater. 2020;103:102-114.

Lock L.L, Li Y, Mao X, Chen H, Staedtke V, Bai R, & Cui H. One-component supramolecular filament hydrogels as theranostic label-free magnetic resonance imaging agents. ACS nano. 2017;11(1):797-805.

Lock L.L, Reyes C.D, Zhang P, & Cui H. Tuning cellular uptake of molecular probes by rational design of their assembly into supramolecular nanoprobes. J. Am. Chem. Soc. J AM CHEM SOC. 2016;138(10):3533-3540.

Madonna R, Rokosh G, De Caterina R, &Bolli R. Hepatocyte growth factor/Met gene transfer in cardiac stem cells—potential for cardiac repair. Basic Res. Cardiol. 2010;105(4):443-452.

Mansukhani N.A, Peters E.B, So M.M, Albaghdadi M.S, Wang Z, Karver M.R, & Kibbe M. R. Peptide amphiphile supramolecular nanostructures as a targeted therapy for atherosclerosis. Macromol. Biosci.2019;19(6):1900066.

Mansukhani N.A, Peters E.B, So M.M, Albaghdadi M.S, Wang Z, Karver M.R, & Kibbe, M. R. Peptide amphiphile supramolecular nanostructures as a targeted therapy for atherosclerosis. Macromolbiosci. 2019;19(6):1900066.

Mansukhani N.A, Peters E.B, So M.M, Albaghdadi M.S, Wang Z, Karver M. R, & Kibbe, M. R. Peptide amphiphile supramolecular nanostructures as a targeted therapy for atherosclerosis. Macromolbiosci, 219;19(6):1900066.

McIntyre M, Caffe S.E, Michalak R.A, & Reid J.L. Losartan, an orally active angiotensin (AT1) receptor antagonist: a review of its efficacy and safety in essential hypertension. Pharmacol. Ther. 1997;74(2):181-194.

Medina S.H, Michie M.S, Miller S.E, Schnermann M.J, & Schneider, J.P. Fluorous Phase Directed Peptide Assembly Affords Nano Peptisomes Capable of Ultrasound Triggered Cellular Delivery. Angew. Chem.Int. Ed. 2017;129(38):11562-11566.

Mendis S, Puska P, Norrving B, & World Health Organization.. Global atlas on cardiovascular disease prevention and control. World Health Organization.2011

Mershin A, Cook B, Kaiser L, & Zhang S. A classic assembly of nanobiomaterials. Net. Biotechnol. 2005;23(11):1379-1380.

Moore K. J, Koplev S, Fisher E.A, Tabas I, Björkegren J.L, Doran A.C, &Kovacic, J. C. Macrophage trafficking, inflammatory resolution, and genomics in atherosclerosis: JACC macrophage in CVD series (Part 2). J. Am. Col. Cardiol, 2018;72(18):2181-2197.

Mujovic N, Dobrev D, Marinkovi M, Russo V, &Potpara T.S. The role of amiodarone in contemporary management of complex cardiac arrhythmias. Pharmacol res. 2020;151:104521.

Nakashima Y, Plump A.S, Raines E.W, Breslow J.L, & Ross R. ApoE-deficient mice develop lesions of all phases of atherosclerosis throughout the arterial tree. Arteriosclerosis and thrombosis: J. vasc. boil. 1994;14(1):133-140.

NavabM, Anantharamaiah G.M, Reddy S.T, Hama S, Hough G, Grijalva V.R, &

Fogelman A.M. Apolipoprotein AI mimetic peptides. Arterioscler. thrombvasc. Boil. 2005;(7):1325-1331.

Nazarnezhad S, Baino F, Kim H.W, Webster T.J, &Kargozar S. Electrospun nanofibers for improved angiogenesis: Promises for tissue engineering applications. Nanomaterials. 2020;10(8):1609.

Newby A.C. Metalloproteinase expression in monocytes and macrophages and its relationship to atherosclerotic plaque instability. Arterioscler. Throm. and vasc boil. 2008;28(12):2108-2114.

Newcomb C. J, Sur S, Lee S.S, Yu J.M, Zhou Y, Snead M.L, &Stupp S. I. Supramolecular nanofibers enhance growth factor signaling by increasing lipid raft mobility. Nano letters. 2016;16(5):3042-3050.

Nicotera P, &Melino G. Regulation of the apoptosis–necrosis switch. Oncogene, 2004;23(16):2757-2765.

Olsen M.H, Angell S.Y, Asma S, Boutouyrie P, Burger D, Chirinos J.A, & Wang J.G. A call to action and a lifecourse strategy to address the global burden of raised blood pressure on current and future generations: the Lancet Commission on hypertension. The Lancet. 2016; 388(10060):2665-2712.

Oparil S, Aurup P, Snavely D, & Goldberg A. Efficacy and safety of losartan/ hydrochlorothiazide in patients with severe hypertension. Am. J. of cardiol. 2001;87(6):721-726.

Prilepskii A.Y, Serov N. S, Kladko D.V, &Vinogradov V.V. Nanoparticle-Based Approaches towards the Treatment of Atheroscler. Pharmaceutics. 2020;12(11):1056.

Ren C, Wang H, Mao D, Zhang X, FengzhaoQ, Shi Y, & Yang Z. When molecular probes meet self-assembly: an enhanced quenching effect. Angew Chem. 2015;127(16):4905-4909.

Riz S.M, El-Maraghy S. A, & Nassar N.N. A novel role for SIRT-1 in L-arginine protection against STZ induced myocardial fibrosis in rats. PloS one. 2014;9(12):e114560.

Rodell C.B, Lee M.E, Wang H, Takebayashi S, Takayama T, Kawamura T, & Gorman R C. Injectable shear-thinning hydrogels for minimally invasive delivery to infarcted myocardium to limit left ventricular remodeling. Circ. Cardiovas Interv. 2016; 9(10):e004058.

Rodell C.B, MacArthur Jr J.W, Dorsey S.M., Wade R. J, Wang L.L, Woo Y.J, & Burdick J. A. Shear thinning supramolecular hydrogels with secondary autonomous covalent crosslinking to modulate viscoelastic properties in vivo. Adv. Funct. Mater. 2015;25(4):636-644.

Rong S.L, Wang X.L, Wang Y.C, Wu H, Zhou X.D, Wang Z. K, & Gao D.L. Anti-inflammatory activities of hepatocyte growth factor in post-ischemic heart failure. Acta Pharmacol Sin. 2018;39(10):1613-1621.

Sakai N, Kamikawa Y, Nishii M, Matsuoka T, Kato T &MatileS..Dendrtic folate rosettes as ion channels in lipid bilayers.. J Am Chem Soc. 2006; 128(7):2218-2219.

Shang Y, Ma C, Zhang J, Wang Z, Ren C, Luo X & Fan G . Bifunctional supramolecular nanofiber inhibits atherosclerosis by enhancing plaque stability and anti-inflammation in apoE-/-mice. Theranostics, 2020;10(22):10231.

Shigemitsu H, Fujisaku T, Tanaka W, Kubota R, Minami S, Urayama K & Hamachi I. An adaptive supramolecular hydrogel comprising self-sorting double nanofibre networks. Nat Nanotechno.2018;13(2):165-172.

Shimamura M, Nakagami H, Sanada F, Morishita R. Progress of Gene Therapy in Cardiovascular Disease. Hypertension. 2020;76:1038-1044.

Short M, Nemenoff R. A, Zawada W. M, Stenmark K. R &Das M. Hypoxia induces differentiation of pulmonary artery adventitial fibroblasts into myofibroblasts. Am J Physiol Cell Physiol.2004; 286(2):C416-C425.

Sivasubramaniyam T, Schroer S. A, Li A, Luk C. T, Shi S. Y, Besla R & Woo M. Hepatic JAK2 protects against atherosclerosis through circulating IGF-1. JCI insight.2017;2(14).

Song G. B, Xu J, Zheng H, Feng Y, Zhang W. W, Li K & Zhang, H. Novel soluble dietary fiber–tannin self-assembled film: A promising protein protective material. J Agric Food Chem. 2015;63(24):5813-5820.

Sukhanov S, Higashi Y, Shai S. Y, Vaughn C, Mohler J, Li Y & Delafontaine P. IGF-1 reduces inflammatory responses, suppresses oxidative stress, and decreases atherosclerosis progression in ApoE-deficient mice. Arteriosclerthrombvasc boil. 2007;27(12):2684-2690.

Versluis F, Van Elsland D. M, Mytnyk S, Perrier D. L, Trausel F, Poolman J. M &Eelkema R. Negatively charged lipid membranes catalyze supramolecular hydrogel formation. J Am Chem Soc. 2016;138(28):8670-8673.

Vu T. D, Pal S. N, Ti L. K, Martinez E. C, Rufaihah A. J, Ling L. H & Kofidis T. (2015). An autologous platelet-rich plasma hydrogel compound restores left ventricular structure, function and ameliorates adverse remodeling in a minimally invasive large animal myocardial restoration model: a translational approach: Vu and Pal "Myocardial Repair: PRP, Hydrogel and Supplements". Biomaterials. 2015; 45: 27-35.

Wang C, Li Y, Ma Y, Gao Y, Dong D, Fang J & Zhang N. Thermoresponsive polymeric nanoparticles based on poly (2 oxazoline) s and tannic acid. Polym Sci A Polym Chem. 2018;56(14):1520-1527.

Wang C, Sang H, Wang Y, Zhu F, Hu X, Wang X& Cheng, Y. Foe to friend: supramolecular nanomedicines consisting of natural polyphenols and bortezomib. Nano letters. 2018;18(11):7045-7051.

Wang H, Feng Z, Qin Y, Wang J & Xu B. Nucleopeptide assemblies selectively sequester ATP in cancer cells to increase the efficacy of doxorubicin. Angewandte Chemie.2018; 130(18):5025-5029.

Wang H, Feng Z, Wang Y, Zhou R, Yang Z & Xu B. Integrating enzymatic self-assembly and mitochondria targeting for selectively killing cancer cells without acquired drug resistance. J Am Chem Soc. 2016;138(49):16046-16055.

Wang H, Yang C, Wang L, Kong D, Zhang Y & Yang Z. Self-assembled nanospheres as a novel delivery system for taxol: a molecular hydrogel with nanosphere morphology. ChemComm. 2011;47(15):4439-4441.

Webber M. J, Han X, Prasanna Murthy S. N, Rajangam K, Stupp S. I &Lomasney J. W. Capturing the stem cell paracrine effect using heparin presenting nanofibres to treat cardiovascular diseases. Tissue Eng Regen Med .2010; 4(8):600-610.

Williams J. W, Giannarelli C, Rahman A, Randolph G. J & Kovacic J. C. Macrophage biology, classification, and phenotype in cardiovascular disease: JACC macrophage in CVD series (part 1). J Am Coll Cardiol . 2018; 72(18):2166-2180.

Xing R, Yuan C, Li S, Song J, Li J & Yan X. Charge Induced Secondary Structure Transformation of Amyloid Derived Dipeptide Assemblies from β Sheet to Helix. Angew Chem Int Ed.2018; 57(6):1537-1542.

Xu T, Liang C, Ji S, Ding D, Kong D, Wang L & Yang Z. Surface-induced hydrogelation for fluorescence and naked-eye detections of enzyme activity in blood. Anal Chem.2016;88(14):7318-7323.

Yadav A, Saini V & Arora S. MCP-1: chemoattractant with a role beyond immunity: a review. Clinicachimica acta. 2010;411(21-22):1570-1579.

Yoshizumi T, Zhu Y, Jiang H, D'Amore A, Sakaguchi H, Tchao J& Wagner W. R.Timing effect of intramyocardial hydrogel injection for positively impacting left ventricular remodeling after myocardial infarction. Biomaterials. 2016;83:182-193.

Zhang C, Liu L. H, Qiu W. X, Zhang Y. H, Song W, Zhang L& Zhang X. Z. A transformable chimeric peptide for cell encapsulation to overcome multidrug resistance. Small, 2018;14(11):1703321.

Zhang Y, Xia M, Jin K, Wang S, Wei H, Fan C &Xiong W. Function of the c-Met receptor tyrosine kinase in carcinogenesis and associated therapeutic opportunities. Mol Cancer .2018; 17(1):1-14.

Zhou J, Du X, Chen X, Wang J, Zhou N, Wu D & Xu B. Enzymatic self-assembly confers exceptionally strong synergism with NF- B targeting for selective necroptosis of cancer cells. J Am Chem Soc.2018;140(6):2301-2308.

CHAPTER 3

ETHNOBOTANICALSTUDIES AS A TOOL FOR CULTURAL PRESERVATION IN THE UTTARAKHANDHIMALAYAS

S.K. Kuriyal[1], Pratibha Verma[1], Pooja Kuliyal[1] and Pratibha Baluni[2*]

1 Department of Botany, SDSUV Rishikesh Campus, Rishikesh
2 Department of Botany, G.D.C Dehradun Shahar, Dehradun

Abstract

Ethnobotanical studies are vital for preserving the traditional knowledge of indigenous communities, mainly regarding the use of local flora. This chapter emphasizes the implication of such studies in Uttarakhand Himalayas, where diverse plant species serve medicinal, nutritional and cultural purposes. Through case studies of key species like Brahma Kamal (*Saussurea obvallata),* Kutki (*Picorhiza kurooa*), Jatamansi (*Nardostachys jatamansi*), *Rhododendron spp., Hippophae rhamnoides* and *Cordyceps spp.,* the research highlights the integral role of traditional knowledge in biodiversity conservation and sustainable development. Recent findings highlight the need for urgent conservation efforts in the face of habitat loss and climate change. The study also examines threats to both plant species and traditional knowledge, urging more effective conservation strategies.

Keywords: Ethnobotany, Traditional Knowledge, Uttarakhand, Biodiversity Conservation, Medicinal Plants, Climate Change.

Introduction

Ethnobotany involves the study of the traditional knowledge and customs of indigenous communities related to plants and their uses including medicinal, agricultural and cultural applications. With increasing globalization, the need to preserve and document this knowledge has grown. Traditional knowledge is a cornerstone of cultural identity for many indigenous communities and a valuable resource for biodiversity conservation and sustainable development. The loss of such knowledge is often linked to environmental degradation, biodiversity loss and erosion of cultural heritage. Recent research has highlighted the increasing role of ethnobotanical knowledge is sustainable development and climate adaptation strategies, particularly for vulnerable communities (Maundu et al., 2021). These studies preserve medicinal, nutritional and cultural uses of plants, ensuring this information is available for future generations (Berkes, 2012; Molina et al., 2020). Ethnobotanical research reveals the symbiotic relationships between indigenous communities and their environment, supporting conservation strategies that are both effective and culturally relevant (Brown & White, 2017; Patwardhan et al., 2022). By validating traditional practices, ethnobotanical research enhances local participation in resource management and contributes to sustainable development (Gupta & Kumar, 2018).

Global Relevance of Ethnobotanical Knowledge

Ethnobotanical knowledge remains a cornerstone of conservation, healthcare and sustainable agricultural practices worldwide. Indigenous communities have long utilized plants for food, medicine and rituals passing down this knowledge through generations. Recent examples including studies on the Amazon Rainforest, where indigenous people continue to practice extensive pharmacopeia, contributing to modern medical advancements (Castãno et al., 2021). Similarly, research from Africa highlights how the San people's knowledge of *Hoodia gordonii* has been integrated into modern commercial practices for appetite suppression (Veldman et al., 2020).

Despite its value, ethnobotanical knowledge is threatened by deforestation, climate change and industrial agriculture. Recent research has documented how climate change affects the availability of key medicinal plants (Shaheen et al., 2021), which necessitates stronger conservation measures from global organizations like the United Nations and World Health Organization (WHO, 2023).

Ethnobotanical Studies in the Indian Context

India, with its diverse ecosystems, is a repository of ethnobotanical traditions. Indigenous communities like the Bhil, Gond and Santhal have preserved vast knowledge regarding the medicinal and agricultural uses of plants. Recent research has revealed new uses of common plants like *Azadirachta indica* (Neem) and *Emblica officinalis* (Amla) in modern medicine (Pandey et al., 2022).

In the Uttarakhand Himalayas, indigenous communities offer a wealth of knowledge that informs modern conservation and sustainable development strategies (Rawat & Nautiyal, 2020). This chapter focuses on the ethnobotanical studies in this region.

Ethnobotanical Studies in the Uttarakhand Himalayas

The Uttarakhand Himalayas, known for their biodiversity, are home to indigenous communities with intricate knowledge of local flora. In ethnomedicines, the highest percentage contribution is generally made by roots, followed by leaves, flowers and buds and others. (Fig 1). Recent studies (Table 1) have documented the medicinal and ecological importance of species such as:

Brahma Kamal (Saussurea obvallata**):** Revered in local culture, this sacred plant is used in religious ceremonies and traditional medicine. A 2022 study shows its potential for treating respiratory issues. (Negi et al., 2022).

Nardostachys jatamansi (Spikenard): Known for its anti-inflammatory properties, spikenard is used by the Bhotia and Jaunsari people in traditional remedies for neurological and inflammatory disorders. Recent studies affirm its medicinal efficacy (Tiwari & Uniyal, 2023). Picrorhiza kurroa **(Kutki):** Used extensively for treating liver disorders, kutki has been recognized for its hepatoprotective properties. However, its overharvesting for commercial purposes threatens its population (Bisht et al., 2023).

Berberis aristata (Daruhaldi): This plant is used to treat skin conditions due to its antimicrobial properties. Ethnobotanical studies document its growing popularity in both traditional and modern skincare (Sharma & Nautiyal, 2023).

Rhododendron arboreum **(Buransh):** Known for its antioxidant-rich flowers, which are used in traditional drinks. Recent ethnobotanical research suggests its potential in treating cardiovascular diseases (Rawat et al., 2023).

Aconitum heterophyllum (Atish) and Saussurea costus **(Kuth):** These critically endangered species are traditionally used by the Bhotia tribe for respiratory and digestive ailments. Despite their cultural importance, modern pressures like overharvesting threaten their survival (Negi et al., 2022).

Hippophae rhamnoides (Sea Buckthorn): Cultivated in harsh climates, sea buckthorn is critical for soil conservation and ecological restoration. It is traditionally used for its anti-inflammatory and antioxidant effects and its promotion aligns with sustainable development goals by providing economic benefits to local communities while preserving the environment (Singh et al., 2021).

Cordyceps spp: A medicinal fungus found at high altitudes, *Cordyceps* is traditionally used to enhance stamina and boost the immune system. The collection of *Cordyceps* is a major source of income for local communities but unsustainable harvesting practices pose a significant threat to its natural populations and the ecosystems it inhabits (Joshi et al., 2022).

The Role of Ethnobotanical Studies in Conservation

Ethnobotanical studies are critical for biodiversity conservation. They preserve indigenous knowledge and promote sustainable use of plant resources. Recent studies emphasize how aligning conservations efforts with traditional knowledge enhances the effectiveness of biodiversity conservation initiatives (Cunhingham et al., 2021).

Indigenous knowledge, especially related to the sustainable harvesting of medicinal plants, provides key insights into effective conservation strategies. The Bhotia tribe's practices of sustainable harvesting, which have been passed down through generations, are now at risk due to modern commercial pressures. Ensuring the continuity of these practices is critical for both ecological stability and the livelihoods of local communities (Bisht et al., 2022).

Sea Buckthorn cultivation and *Cordyceps* collection are prime examples of how ethnobotanical knowledge can contribute to sustainable development by offering both economic and ecological benefits. The promotion of sustainable harvesting practices for these species is crucial for maintaining their populations and the ecosystems their support.

The traditional Van Panchayat system in Uttarakhand, which involves community-led forest management, remains a relevant model for sustainable conservation, as recent studies indicate its success in preserving forest cover while maintaining local traditions (Bisht et al., 2022).

Challenges to Preserve Ethnobotanical Knowledge in Uttarakhand

Recent Studies have identified various challenges to preservation of ethnobotanical knowledge in the Uttarakhand Himalaya, such as modernization and migration, which result in younger generations relocating to urban areas and consequently diminishing traditional knowledge (Mehta et. al., 2023). Additionally, climate change poses a significant threat, as the changing climatic patterns jeopardize the habitats of numerous medicinal plants, thereby endangering traditional practices (Singh et al., 2022). Addressing these challenges requires to develop conservation strategies that blend modern scientific outlook with traditional knowledge.

Recommendations for Conservation

A comprehensive approach to preserve ethnobotanical knowledge in Uttarakhand involves systematic documentation, legal protection and community-based conservation. Systematic documentation is essential as traditional knowledge is increasingly at risk due to environmental degradation (Posey et al. 2023). Additionally, indigenous communities should have strong legal right over their traditional knowledge to guard against biopiracy and ensure fair use of resources (Shiva, 2022). Recent studies further emphasize the importance of community-led initiatives, such as Van Panchayats, which integrate traditional practices into conservation strategies, supporting sustainable biodiversity management (Bisht et al., 2022)

Prospects for Future Research:

Emerging research areas include the impact of climate change on medicinal plants, with initial studies exploring how climate change affects the distribution and sustainability of these species (Shaheen et al., 2023). Additionally, recent advances in phytochemical analysis highlights the potential for novel pharmaceutical applications derived from medicinal plants (Patwardhan et al. 2023). Furthermore, Collaborative research efforts that involve partnership among indigenous communities, scientists and policymakers are increasingly recognized as vital for the holistic ethnobotanical studies (Mehta et al., 2023)

Conclusion

Ethnobotanical studies play a crucial role in preserving traditional knowledge and promoting biodiversity conservation in the Uttarakhand Himalayas. By documenting indigenous practices and promoting sustainable harvesting, these studies contribute to the resilience of local communities and preservation of their cultural heritage. Integrating traditional knowledge with modern science ensures a sustainable future for both biodiversity and the people of Uttarakhand.

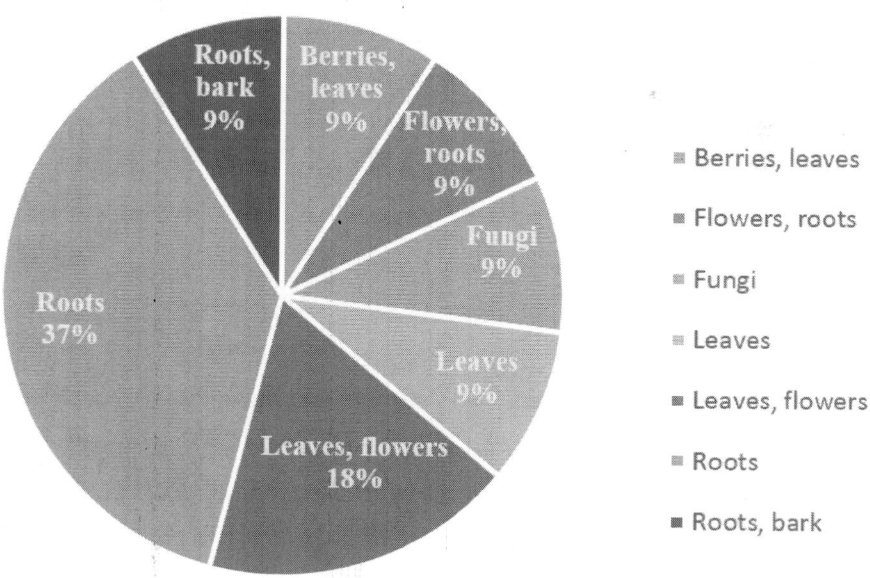

Figure 1 percentage contribution of various plant parts in ethnomedicinal preparations.

Table 1: Medicinal Plants Used by Indigenous Communities in Uttarakhand: Traditional Uses and Conservation Status

Scientific Name	Common Name	Traditional Uses	Parts Used	Citation
Saussurea obvallata	Brahm Kamal	Used for respiratory issues, digested disorders and as an anti-inflammatory; believed to have adaptogenic properties	Flowers, roots	Singh, A., & Sharma, D. (2014).
Picrorhiza kurrooa	Kutki	Known for hepatoprotective properties, used for liver disorders and gastrointestinal ailments; traditionally used in Ayurvedic formulations	Roots	Bhatt, L.D., & Bhatt, S. (2015).
Nardostachys jatamansi	Jatamansi	Valued for calming effects, often used in Ayurvedic formulations for enhancing cognitive function and alleviating stress	Roots	Sharma, P., & Kumar R. (2016).
Rhododendron spp.	Rhododendron	Consumed in various forms, known for antioxidant properties and traditional uses in treating fevers and respiratory ailments	Leaves, flowers	Negi, K.S., & Negi, S. (2018).
Hippophae rhamnoides	Seabuckthorn	Nutritionally rich, used to boost immunity, improve skin health and treat various ailments including wounds	Berries, leaves	Bhandari, U., & Kumar, S. (2019).
Cordyceps sinensis	Cordyceps	Enhances energy and stamina; traditionally used for improving vitality, athletic performance and as an adaptogen	Fungi	Ghosh, S., & Sharma, A. (2020).
Berberis aristata	Indian Barberry	Used for its anti-microbial and anti-inflammatory properties; traditionally employed for treating skin diseases and fever	Roots, bark	Kumar, P., & Kumari, S. (2021).
Glycyrrhiza glabra	Licorice	Used for digestive issues, respiratory ailments and skin problems noted for its sweet flavor and adaptogenic properties	Roots	Prakash, O., & Gupta, A. (2021).
Asparagus racemosus	Shatavari	Traditionally used as a galactagogue, to promote women's health and for its adaptogenic effects	Roots	Thakur, M., & Pankaj, B. (2020).
Ocimum sanctum	Holy Basil	Used in Ayurveda for respiratory ailments, stress relief and as an antioxidant also consumed as a culinary herb	Leaves	Sharma, R., & Kaur, M. (2019).

Scientific Name	Common Name	Traditional Uses	Parts Used	Citation
Thymus serpyllum	Wild Thyme	Traditionally used for digestive issues and respiratory problems; also used as a culinary herb	Leaves, flowers	Negi, P.S., & Kumari, R. (2018).

References

Agrawal, A. (2001). Common resources and institutional sustainability. World Development, 29(10), 1649-1672. https://doi.org/10.1016/S0305-750X(01)00079-5.

Berkes, F. (2012). *Sacred ecology*. Routledge.

Bhandari, U., & Kumar, S. (2019). The medicinal potential of seabuckthorn (*Hippophae rhamnoides* L.). *International Journal of Herbal Medicine,* 7(1), 11-15.

Bhatt, I. D., & Bhatt, S. (2015). Ethnobotanical exploration of medicinal plants in Uttarakhand. *Indian Journal of Traditional Knowledge,* 14(3), 463-472.

Bisht, D. S., Meena, P., and Maikhuri, R.K.(2022). Agroforestry and traditional knowledge of rural communities in the Himalayan region. *Jouranl of Himalayan Ecology,* 15(1), 45-59.

Bisht, R., Negi, P., & Joshi, A. (2023). Sustainable harvesting of medicinal plants in the Himalayas: Challenges and opportunities. *Journal of Himalayan Conservation,* 12(2), 105-125.

Brown, K., & White, J. (2017). *Ethnobotany: A global perspective.* Cambridge University Press.

Castaño, J., Gutiérrez, R., & López, M. (2021). Traditional knowledge of medicinal plants in the Amazon rainforest: The contribution of indigenous peoples to modern medicine. *Journal of Ethnopharmacology.* 274, 114057. https://doi.org/10.1016/j.jep.2021.114057.

Ghosh, S., & Sharma, A. (2020). Cordyceps in traditional medicine: A review. *Asian Pacific Journal of Tropical Medicine,* 13(4), 151-158. https://doi.org/10.4103/1995-7645.280934

Gupta, R., & Kumar, S. (2018). Traditional ecological knowledge and sustainable development. *Indian Journal of Ecology,* 45(4), 124-132.

Joshi, H., Rawat, S., and Pant, H.(2022).Role of traditional knowledge in ecosystem management in Uttarakhand. *Ecological Society Journal,* 24(2), 72-86.

Kala, C.P. (2005). Indigenous uses and conservation of ethnomedicinal plants among the tribal communities in Chhattisgarh, India. Journal of Ethnobiology and Ethnomedicine, 1(1), 20. https://doi.org/10.1186/1746-4269-1-20.

Kumar, P., & Kumari, S. (2021). Ethnobotanical studies on *Berberis aristata*: Conservation and traditional uses. *Journal of Medicinal Plants Research,* 15(8), 217-226.

Maundu, P., Kariuki, R., and Munguti, S. (2021). Ethnobotanical Knowledge and Sustainable use of Medicinal plants in Kenya. *African Journal of Ethnobotany*, 19(3), 55-70.

Mehta, S., Singh, D., & Rawat, A. (2023). Youth migration and the erosion of traditional knowledge in the Indian Himalayas. *Indian Journal of Ethnobotany*. 18(1), 88-110.

Molina, M., Leonti, M., & Pardo-de Santayana, M. (2020). Ethnobotanical knowledge and its relevance in a changing world. *Journal of Ethnobiology and Ethnomedicine*, 16, 22. https://doi.org/10.1186/s13002-020-00306-1.

Negi, G., Pant, S., & Tiwari, M. (2022). Medicinal plants of Uttarakhand: Conservation strategies for sustainable use. *Biodiversity Conservation Review*, 15(3), 229-245.

Negi, K.S., & Negi, S. (2018). Ethnobotanical and pharmacological properties of Rhododendron spp. in Uttarakhand. *Journal of Medicinal Plants Research*, 12(7), 101-108.

Negi, P.S., & Kumari, R. (2018). Ethnobotanical uses of *Thymus serphyllum* in Uttarakhand. *International Journal of Current Microbiology and Applied Sciences*, 7(6), 2032-2038.

Pandey, A., Kumar, A., & Mishra, P. (2022). New medicinal applications of *Azadirachta indica* and *Embilica officialis*: A review. *Asian Journal of Pharmaceutical Sciences*, 17(5), 689-698. https://doi.org/10.1016/j.ajps.2022.02.006.

Patwardhan, B., & Jadhav, A. (2023). Phytochemical analysis and medicinal applications of Himalayan plants. *Journal of Ethnopharmacology*, 291, 115256. https://doi.org/10.1016/j.jep.2022.115256.

Patwardhan, B., Jadhav, A., & Sinha, S. (2022). Ethnobotanical approaches to biodiversity conservation. Current Science. 123(1), 1-3. https://doi.org/10.18520/cs/v123/il/1-3

Posey, D. A., & Dutfield, G. (2023). Beyond intellectual property: Toward traditional resource rights for indigenous peoples. *WIPO Journal*, https://doi.org/10.2139/ssrn.4021584 14(2), 81-97.

Prakash, O., & Gupta, A. (2021). Glycyrrhiza glabra: A comprehensive review of its ethnobotanical and pharmacological aspects. *Journal of herbal medicine*, 27, 100418.

Rawat, A., & Nautiyal, S. (2020). Indigenous knowledge systems in the Uttarakhand Himalayas: A study of local communities. *Indian Journal of Ecology*, 47(2), 401-409.

Rawat, A., Singh, R., & Joshi, A. (2023). Ethnobotanical research on *Rhododendron arboreum* in Uttarakhand: A review of its traditional uses and conservation status. *Journal of Ethnopharmacology*, 295, 115353. https://doi.org/10.1016/j.jep.2022.115353.

Shaheen, H., Ahmad, M., & Singh, R. (2023). Impact of climate change on the distribution of medicinal plants in the Western Himalayas. *Environmental Change Journal*, 19(4), 301-315.

Shaheen, H., Khan, I., & Kumari, N. (2021). Effects of climate change on the availability of medicinal plants: A review. *Journal of Medicinal Plants Studies*, 9(5), 43-50.

Sharma, A., and Nautiyal, M.C. (2023). Impact of Climate Change on Traditional Agricultural practices in the Indian Himalayas. *Climate Change and Mountain Ecosystems*, 33(1), 112-128.

Sharma, P., & Kumar R. (2016). Jatamansi: A Review of Ethnobotanical and Pharmacological Studies. *International Journal of Pharmaceutical Sciences Review and Research*, 41(2), 151-157.

Sharma, R., & Kaur, M. (2019). Holy Basil (*Ocimum sanctum*): Ethnobotanical and pharmacological aspects. *Journal of Ethnopharmacology*, 245, 112-137.

Shiva, V. (2022). Protecting indigenous knowledge: The need for legal recognition. *World Journal of Agricultural Research*, 10(2), 37-42.

Singh, A., & Sharma, D. (2014). Ethnobotanical studies of medicinal plants in Uttarakhand. *Journal of Ethnopharmacology*, 154(3), 478-490.

Singh, D., Kumar, R., & Singh, R. (2021). The role of *Hippophae rhamnoides* in ecological restoration and sustainable development. *Journal of Sustainable Forestry*, 40(3), 227-241.

Singh, R., Sinha, P., & Singh, S. (2022). Climate change and its impact on traditional ecological knowledge in the Himalayas. *Global Ecology and Conservation*, 35, e02081. https://doi.org/10.1016/j.gecco.2022.002081.

Thakur, M., & Pankaj, B. (2020). Ethnobotanical aspects of *Asparagus racemosus* in Uttarakhand. *Indian Journal of Natural Products and Resources*, 11(3), 212-220.

Tiwari, S., & Uniyal, R. (2023). Medicinal efficacy of *Nardostachys jatamansi*: Traditional uses and pharmacological properties. *Journal of Ethnopharmacology*, 298, 115536. https://doi.org/10.1016/j.jep.2022.115536.

Veldman, A., Tredoux, A., and Nirenaber, A (2020). The integration of traditional knowledge and modern commercial practices: A case study of *Hoodia gordonii* and the San people. *Journal of Ethnopharmacology*, 260,113059.

World Health Organization. (2023). *Integrating traditional medicine in health systems: Strategies and practices*. World Health Organization Press.

CHAPTER 4

PHYTOCHEMICAL ANALYSIS OF MULETHI (GLYCYRRHIZA GLABRA): UNVEILING THE BIOACTIVE CONSTITUENTS AND THEIR POTENTIAL HEALTH BENEFITS

Pandey D.P.[1], S.K Kuriyal[2] and Pratibha Baluni[3]

1 Department of Chemistry, G.D.C Dehradun Shahar, Dehradun
2 Department of Botany, SDSUV Rishikesh Campus, Rishikesh
3 Department of Botany, G.D.C Dehradun Shahar, Dehradun

Abstract

Mulethi (*Glycyrrhiza glabra*), a perennial herb from the Fabaceae family, is valued for its extensive therapeutic properties in traditional medicine systems across the globe. This article provides a comprehensive review of its botanical characteristics, ethnomedicinal application, phytochemical profile, and pharmacological properties. The root of *Mulethi* contains bioactive compounds, primarily glycyrrhizin, which contributes to its sweetness and numerous health benefits, including anti-inflammatory, antiviral, hepatoprotective, and anticancer activities. The review highlights the significant roles of various phytochemicals, such as flavonoids and saponins, in promoting health and well-being. Additionally, it addresses sustainability concerns regarding the overharvesting of this valuable medicinal resource. As global interest in herbal medicine surges, understanding the intricate relationship between traditional uses and modern scientific validation becomes crucial.

Keywords: Mulethi; *Glycyrrhiza glabra*; Phytochemical analysis; Therapeutic properties; Ethnomedicine

Introduction

Licorice (*Glycyrrhiza glabra*), commonly known as Mulethi in India, is a perennial herb belonging to the fabaceae family. This medicinal plant has garnered attention in various traditional medicinal systems for its extensive therapeutic benefits. Native to parts of Europe, Asia, and the Mediterranean, licorice thrives in temperature and semi-arid climates, primarily in well-drained soils. Its roots are particularly esteemed for treating respiratory ailments, gastrointestinal disorders, and dermatological conditions. Recent pharmacological studies have corroborated its traditional uses by identifying numerous bioactive compounds, thus enhancing its reputation as a valuable medicinal resource (Das & Khanna, 2018; Sharma et al., 2019). This review aims to provide a comprehensive analysis of *Glycyrrhiza glabra*, focusing on its botanical characteristics, ethnomedicinal applications, phytochemical profile, and pharmacological properties. By bridging the gap between traditional knowledge and contemporary science, this study aims not only to recognize ancient wisdom but also to inspire innovative research directions.

Botanical description

Glycyrrhiza glabra can reach heights of 1 to 1.5 meters, characterized by its pinnate leaves, which consist of 9-17 ovate leaflets. The plant produces attractive purple to violet flowers that bloom in racemes. The root system, highly prized for its sweet flavour due to the presence of glycyrrhizin, contains yellowish internal tissues rich in this bioactive compound. This species thrives in well- drained soils with a temperate or semi-arid climate, making it suitable for cultivation in regions across Southern Europe, Central Asia, and India (li et al., 2020).

Ethnobotanical and traditional uses

Licorice has been utilized in traditional medicine for centuries. In Ayurveda, it is classified as a "Rasayana", a category of herbs that promote longevity, immunity, and homeostasis within the body. It is particularly revered for its efficacy in alleviating respiratory conditions such as coughs and bronchitis, as well as enhancing digestive health (Srivastava & Yadav, 2017). In Traditional Chinese Medicine (TCM), known as "Gan Cao", licorice is used to harmonize the effects of other herbs and is a common ingredient in formulations aimed at treating coughs, infections, and ulcers (Sharma et al., 2019). Unani medicine recognizes licorice for its anti- inflammatory, antioxidant, and hepatoprotective properties, further solidifying its importance across various cultural contexts. In regions across the Mediterranean and central Asia, licorice root has been employed as a remedy for sore throats, digestive issues, and skin ailments, reflecting its broad pharmacological applications (Das & Khanna, 2018). Mulethi's versatility across various cultures illustrates its profound impact on health practices, symbolizing a shared heritage of healing. These traditional applications present opportunities for future studies aimed at isolating active compounds and elucidating their mechanisms of action.

Phytochemical composition

The medicinal attributes of licorice are attributed to its rich phytochemical composition, which includes saponins, flavonoids, chalcones, and polysaccharides. The principal bioactive compound, glycyrrhizin, a triterpenoids saponin, is responsible for its characteristic sweetness and many of its therapeutic effects. Glycyrrhizin is metabolized in the body to glycyrrhetinic acid, which exhibits significant anti- inflammatory and antiviral properties (Li et al., 2020).

Key phytochemicals:

- **Saponins:** Glycyrrhizin and glycyrrhetinic acid are recognized for their anti-inflammatory, hepatoprotective, and antiviral properties (Sharma et al., 2019).
- **Flavonoids:** Compounds like liquiritin, isoliquiritin, and liquiritigenin demonstrate potent antioxidant capabilities and cardioprotective effects (Das & Khanna, 2018).
- **Chalcones:** Licochalcone A and glabridin possess estrogenic, antimicrobial, and anticancer properties (Li et al., 2020).
- **Polysaccharides:** These contribute to the immune-boosting properties of licorice by enhancing cytokine production (Sharma et al., 2019).

Pharmacological Activities

Anti- inflammatory and Immunomodulatory Effects

Licorice is well- documented for its remarkable anti- inflammatory properties. Glycyrrhizin inhibits the enzyme 11 - hydroxysteroid dehydrogenase, resulting in enhanced glucocorticoid activity and reduce inflammation (Srivastava & Yadav, 2017). Its immunomodulatory effects have been highlighted in studies where glycyrrhizin and liquiritigenin modulate cytokine production, particularly TNF- and IL-6, offering therapeutic potenitial for autoimmune and inflammatory diseases (Sharma et al., 2019).

Antiviral and antibacterial properties

Licorice exhibits antiviral activity against a variety of viruses, including Herpes simplex, Hepatitis C, and SARS- Cov-2. Glycyrrhizin interferes with viral replication, thereby limiting infection spread (LI et al., 2020). Additionally, it has demonstrated efficacy against Helicobacter pylori, a common gastric ulcer pathogen, due to its antimicrobial properties (Das & Khanna, 2018).

Hepatoprotective Properties

Historically, licorice has been utilized for liver protection. Glycyrrhizin has shown effectiveness in managing chronic liver conditions such as hepatitis and cirrhosis by reducing liver enzymes and oxidative stress. Studies in japan have incorporated glycyrrhizin into treatment protocols for chronic hepatitis, reinforcing its role in liver protection (Li et al., 2020).

Antidiabetic and Cardiovascular Benefits

Research indicates licorice's potential in diabetes management. The flavonoid liquiritigenin has been shown to enhance insulin sensitivity and lower blood glucose levels in experimental models. Furthermore, its antioxidant properties contribute to cardioprotective effects by reducing oxidative stress and improving vascular health (Sharma et al., 2019).

Anticancer potential

Compounds such as glabridin and licochalcone A have demonstrated promising anticancer properties. Preclinical studies indicate their ability to induce apoptosis in cancer cells and inhibit tumour growth, particularly in breast and prostate cancers, where they impede metastasis and angiogenesis (Das & Khanna, 2018). The promising anticancer properties of Mulethi-derived compounds open new avenues for development of adjunct therapies in oncology. Understanding these pharmacological activities highlight the need of holistic approach to health, integrating lifestyle diet and herbal remedies.

Production and consumption of Licorice

The global production of licorice root is estimated to be around **200,000 tons** annually, with primary producers including **China, Turkey, and Iran.** Notably, **China** accounts for approximately **70%** of the world's licorice root production, primarily sourced from the **Xinjiang region**. The expansion of global cultivation areas is driven by the increasing demand for natural sweeteners and medicinal products (Zhang et al., 2021; Research and Markets, 2023). In India, *Mulethi* is predominantly cultivated in the states of **Uttarakhand, Himachal Pradesh, and Punjab**, with an estimated annual production of **20,000 tons,** primarily for medicinal and cosmetic uses. The growing demand for herbal products has spurred increased interest in licorice cultivation within the country (Kaur et al., 2019; Kumar et al., 2022).

Globally, the demand for licorice is expected to grow at a **CAGR of 4.5%** from **2023 to 2030**, influenced by its applications in the food and beverage, pharmaceutical, and cosmetic industries. Major consumers of Licorice include the confectionery and beverage industries, which utilize its flavour and sweetness, with **Europe** and **North America** identified as the largest markets, benefiting from the rising interest in natural and organic products (Grand View Research, 2023; Mordor Intelligence, 2023). In India, the licorice market is anticipated to expand due to increasing awareness of natural remedies and herbal products. *Mulethi* is increasingly incorporated into Ayurvedic formulations and herbal teas, indicating a shift towards natural health products, with the primary consumers being health-conscious individuals and those seeking traditional remedies (Sharma et al., 2021; Nair et al., 2023).

Commercial Products and Applications

The therapeutic potential of Glycyrrhiza glabra has led to its incorporation into various commercial products, ranging from herbal medicines and dietary supplements to cosmetic formulations. Notable products include:

- **Digestive Health Supplements:** formulations that utilize licorice for its digestive benefits and to soothe gastrointestinal discomfort.
- **Cough Syrups and Lozenges:** Products aimed at alleviating coughs and throat irritations, leveraging the soothing properties of licorice.
- **Skin care Products:** licorice extracts are often included in formulations designed to reduce inflammation, hyperpigmentation, and improve skin texture.
- **Herbal Teas:** Infusions featuring licorice root for its flavour and health benefits, particularly for respiratory and digestive health.
- **Pharmaceutical formulations:** Glycyrrhizin is incorporated into various pharmaceuticals, particularly in formulations aimed at treating gastrointestinal disorders, respiratory conditions, and liver diseases.

As consumer interest in natural and herbal products grows, the market for licorice- based products is expected to expand. The global licorice market is projected to reach USD 1.5 billion by 2030, driven by the rising demand for natural sweeteners, herbal supplements and traditional remedies (Grand View Research, 2023). Establishing regulatory standards, transparency in sourcing and product formulation will be key to sustaining market growth.

Conservation and sustainability Concerns

The increasing demand for licorice in pharmaceutical and herbal industries raises concerns about the sustainability of *Glycyrrhiza glabra*. Unsustainable harvesting practices in regions such as Iran and Central Asia have led to declines in wild populations. Conservation strategies, including regulated cultivation and sustainable harvesting methods, are essential for preserving this valuable medicinal resource (Li et al., 2020).

Conclusion

Mulethi (*Glycyrrhiza glabra*) is a remarkable medicinal plant with a rich history in traditional medicine and an impressive profile of bioactive compounds. Its pharmacological properties, including anti-inflammatory, antiviral, hepatoprotective, and anticancer effects, underscore its potential for diverse health applications. However, the increasing demand for licorice products necessitates the implementation of sustainable practices to protect wild populations and ensure the availability of this valuable resource for future generations. Further research and collaboration among stakeholders will be essential to promote responsible cultivation, effective conservation, and the exploration of new applications for Glycyrrhiza glabra in modern medicine. Ultimately, the integration of traditional knowledge with scientific research can pave the way for the sustainable utilization of this medicinal treasure.

Table 1. Phytochemical Analysis of Mulethi (Glycyrrhiza glabra): Global findings and implications

Study	Phytochemical components	Key findings	Region
Das & Khanna (2018)	Glycyrrhizin, flavonoids (liquiritin, isoliquiritin), saponins	Highlighted the anti- inflammatory and hepatoprotective properties of glycyrrhizin.	India
Sharma et al. (2019)	Glycyrrhizin, glabiridin, liquiritigenin, licochalcone A	Explored the antioxidant, antimicrobial, and anticancer potential of various compounds.	India
Li et al. (2020)	Glycyrrhizin, flavonoids, polysaccharides	Discussed antiviral activity against viruses like HCV and the importance of sustainable harvesting.	China
Srivastava & Yadav (2017)	Glycyrrhizin, phenolic compounds	Documented the traditional uses and potential for modern therapeutic applications in respiratory diseases.	India
Badgujar et al. (2014)	Licochalcone A, glycyrrhizin	Evaluated the gastroprotective effects of licorice and its ability to enhance gastric mucosal defense.	India
Kim et al. (2018)	Licochalcone A, glycyrrhizin	Studied the neuroprotective effects of licorice components in neurodegenerative models.	Korea
Zhang et al. (2020)	Glycyrrhizin, flavonoids	Identified the role of licorice extracts in inhibiting TNF- and IL-6.	China

References:

Badgujar, B. P., Chavan, M. J., & Ghosh, S. (2014). Evaluation of gastroprotective effects of licorice (Glycyrrhiza glabra) extract on gastric mucosal defense in experimental models. *International Journal of Pharmacy and Pharmaceutical Sciences*, *6*(3), 381-386.

Das, S., & Khanna, A. (2018). Medicinal properties of licorice (Glycyrrhiza glabra): A review. *International Journal of Herbal Medicine*, 6(1), 34-41.

Kaur, G., Dhanjal, D. S., & Bansal, R. (2019). Current status and future prospects of licorice (Glycyrrhiza glabra) cultivation in India. *Journal of Medicinal Plants Research*, 13(8), 145-152.

Kim, H. J., Yoon, H. J., Kim, H. S., Kim, M. H., & Kim, K. K. (2018). Neuroprotective effects of licochalcone A and glycyrrhizin in neurodegenerative models. *Journal of Ethnopharmacology, 214*, 223-230. https://doi.org/10.1016/j.jep.2017.11.009.

Kumar, V., Raghav, P. K., & Kumar, A. (2022). Status of licorice cultivation and its medicinal applications: A review. *Agricultural Reviews*, 43(2), 104-110.

Li, C., Chen, F., & Hu, J. (2020). The pharmacological effects of licorice (Glycyrrhiza glabra) on human health: A review. *Frontiers in Pharmacology*, 11, Article 674. https://doi.org/10.3389/fphar.2020.00674

Mordor Intelligence. (2023). *Global licorice market - Growth, trends, COVID-19 impact, and forecasts (2023-2030)*. Retrieved from https://www.mordorintelligence.com

Nair, A. G., Nair, N. K., & Kumar, M. (2023). Herbal formulations of Glycyrrhiza glabra and their role in traditional medicine: A review. *Journal of Ethnopharmacology*, 305, Article 115123. https://doi.org/10.1016/j.jep.2022.115123

Research and Markets. (2023). *Licorice market - Growth, trends, COVID-19 impact, and forecasts (2023-2030)*. Retrieved from https://www.researchandmarkets.com

Sharma, D., Sharma, A., & Singh, H. (2019). Phytochemical profile and therapeutic potential of licorice (Glycyrrhiza glabra): A review. *Journal of Pharmacy & Bioallied Sciences*, 11(2), 100-106. https://doi.org/10.4103/jpbs.JPBS_221_18

Sharma, P., Singh, S., & Sharma, R. (2021). The medicinal significance of licorice (Glycyrrhiza glabra): A review. *Journal of Medicinal Plants Research*, 15(5), 75-85.

Srivastava, A., & Yadav, R. (2017). Ethnomedicinal uses and pharmacological activities of Glycyrrhiza glabra: A review. *Journal of Ethnopharmacology*, 197, 224-239. https://doi.org/10.1016/j.jep.2016.11.029

Zhang, Y., Zhang, M., & Wang, L. (2021). Global licorice market: Trends and opportunities. *Market Research Report*, 22(4), 68-74.

CHAPTER 5

ANTAGONISTIC EFFECT OF ROOT ENDOPHYTIC AQUATIC HYPHOMYCETES AGAINST PLANT PATHOGENIC FUNGI

Prabha Pant Upadhyay* and S. C. Sati

DSB Campus, Kumaun University, Nainital

Abstract

Two root endophytic aquatic hyphomycetous fungi *Campylospora purvula* Kuzuha and *Tetracladium setigerum* (Grove) Ingold isolated from riparian area of Nainital, Kumaun Himalaya were assessed for their antagonistic activity. Isolated fungi were evaluated against two plant pathogenic fungi to check their antifungal activity by dual culture and 'well- diffusion' method. In dual culture technique *C. parvula* inhibited the growth of pathogenic fungi up to 63.20% and 45.26% for *C. capsicii* and *F. oxysporum*, respectively. In case of *T. setigerum* it was observed 45% and 40.9% for *C. capsici* and *F. oxysporum* respectively. In 'well- diffusion' method the percentage of inhibition against the *C. capsici* and *F. oxysporum* were 32.14% and 26.66% by the crude extract of *C. parvula* while the crude extract of *T. setigerum* inhibited the growth of *C. capsicii* and *F. oxysporum* 19.35% and 23.33%, respectively.

Introduction

Natural products obtained from the microorganisms are potentially useful in agriculture, industry, and medicine (Demain 2014). Secondary metabolites derived from the various microbes have been the major source of antibiotics available in the market. Endophytic fungi represent an interesting group of fungi which remain in versatile interactions with the host plant and responsible for synthesizing a large number of bioactive secondary metabolites, including phenolics, flavonoids, and alkaloids (Liu et al. 2010).

Ingold in 1942 discovered some characteristic spores associated with decomposed leaf litter in a stream and named them as aquatic hyphomycetes. The diversity of aquatic hyphomycetes in the fresh water ecosystem has a main significance in the decomposition of submerged plant materials, nutrient release and productivity. However the endophytic nature of aquatic hyphomycetes reveals that these fungi also play an important role in plant health by producing various bioactive compounds. Some species of aquatic hyphomycetes are reported to have antagonistic activity and can be used against pathogenic bacterial and fungal strains (Gulis and Stephanovich, 1999, Arya and Sati 2012, Singh and Sati 2019, Pant and Sati 2021).

Keeping the above view in mind that there are a few reports on the antimicrobial potentiality of aquatic hyphomycetes, the present investigation was focused on the antifungal activity of two selected root endophytic aquatic hyphomycetes *Campylospora parvula* Kuzuha and *Tetracladium setigerum* (Grove) Ingold against plant pathogenic fungi.

Material and methods

Isolation of endophytic aquatic hyphomycetes

Roots of riparian plants were collected from the streams and their catchment areas of Nainital district of Kumaun Himalayas, Uttarakhand, India. Fresh and healthy roots from the living plants were sampled, kept in pre- sterilized bags, and brought to the laboratory for further processing. Root samples were thoroughly washed under running tap water (2–4 hours), then with sterile water (5–7 min), and finally with 2% sodium hypochlorite solution (1–2 min) to remove adhering free living microbes of the root surfaces. These root samples were then cut into small pieces (1–2 cm long) and placed into Petri dishes containing 20 mL of sterile water. The root segments in Petri dishes were incubated at 20 ± 2°C for 3–7 days and examined regularly. Observations were made with compound microscope to confirm mycelial growth and conidial liberation into the water. Floating conidia were picked up by a sterile needle and studied further by making semi-permanent slides. Pure cultures were initiated by transferring conidia onto 2% malt extract agar (MEA). Germinated conidia were allowed to grow into colonies and maintained on MEA slants at 4 °C (Sati and Belwal 2005).

The isolates were identified with the help of relevant monographs and literature. Pure cultures of the isolates have been submitted in Kumaun University Mycological Society (KUMS) collection of the Department of Botany, Kumaun University, Nainital, India.

Test microorganisms

Selected two root endophytic aquatic hyphomycetes (REAH) were evaluated against two plant pathogenic fungi *Colletotrichum capsici* (MTCC No. 2071) and *Fusarium oxysporum* (MTCC No. 284) to check their antifungal activity. Both the plant pathogenic fungi were borrowed from Microbial type culture collection (MTCC) Chandigarh.

Media and Plate preparation

Potato dextrose agar (PDA) media was used and autoclaved at 121 °C (15 Pascal pressure) for 20 minutes. After cooling the medium, 10 mg of streptomycin was added to it and mixed thoroughly to prevent bacterial contamination as suggested. 20 ml of medium was poured into Petri plates (90 mm) and after solidification of media these plates were used to check the antagonistic activity of used REAH.

Assessment of Antifungal activity

The antifungal activity of the endophytic fungi was tested using a 'dual culture' and 'well diffusion' method.

(i) Dual culture method- Fungal hyphae containing agar block of actively growing colonies of both the REAH (*C. parvula* and *T. setigerum*) and test fungi (*C. capsici* and *F. oxysporum*) were cut from the periphery of culture plates with the help of a sterile borer. These mycelial containing agar blocks (5 mm dia.) at a distance of 2.5 cm. were aseptically inoculated in Petri plates (90mm) containing 20 ml of Potato dextrose agar medium. After inoculation these plates were incubated at 25°C in a BOD incubator.

The antagonistic activity of root endophytic aquatic hyphomycetes was analyzed by measuring the radial growth of the test plant pathogenic fungi on 4th day of incubation in two directions: R_1 (radius in opposite direction) and R_2 (radius in direction of the used REAH). Percent inhibition of radial growth was calculated as suggested by (Sati and Arya 2010).

$$\text{Percent inhibition} = \frac{R_1 - R_2}{R_1} \times 100$$

Where, R_1 = Growth of pathogenic fungi in opposite direction of fungal extract of used REAH

R_2 = Growth of pathogenic fungi in direction of fungal extract of used REAH

(ii) Well diffusion method- 5 mm disc of fungal containing agar block was cut using sterile borer from the apical margin of 12-13 days old culture of selected root endophytic aquatic hyphomycetes and inoculated into 250 ml. of conical flask containing 100 ml of Malt extract broth. These flasks were incubated at 28°C for 10-12 days in a BOD incubator. After 10-12 days the cultures were filtered with the help of Whatman paper no. 1 and centrifuged at 5000 rpm for the separation of fungal mycelia. The mycelial mat and fungal broth were extracted by 100 ml of methanol and ethyl acetate respectively. These two extracts containing secondary metabolites were mixed, filtered and evaporated to dryness. The resulting crude extract was finally dissolved in DMSO and used for further experiment.

In solidified agar plates 2 wells (5 mm diameter) at a distance of 2.5 cm were made by using a sterile borer. One well was filled with 200 µl of fungal extract and another one was inoculated with mycelia disc of plant pathogenic fungi. The tests were performed in triplicates and these plates were incubated in BOD incubator at 25°C for 4-5 days. Observations were made after 4-5 days of incubation.

The antagonistic activity of root endophytic aquatic hyphomycetes was analyzed by measuring the radial growth of the test plant pathogenic fungi on 4th day of incubation in two directions: R_1 (radius in opposite direction) and R_2 (radius in direction of the used REAH). Percent inhibition of radial growth was calculated as suggested by (Sati and Arya 2010).

$$\text{Percent inhibition} = \frac{R_1 - R_2}{R_1} \times 100$$

Where, R_1 = Growth of pathogenic fungi in opposite direction of fungal extract of used REAH

R_2 = Growth of pathogenic fungi in direction of fungal extract of used REAH

Results

On the basis of morphological characters the isolates were identified as *Campylospora parvula* Kuzuha (KUMS 015) and *Tetracladium setigerum* (Grove) Ingold (KUMS 022) recovered from *Rubus elepticus* and *Lantana camara* respectively. These two isolates were evaluated for their antagonistic effect against plant pathogenic fungi.

Dual culture method-

Both the selected root endophytic aquatic hyphomycetes (*C. parvula* and *T. setigerum*) were found positive against the tested plant pathogenic fungi (Plate- 1). The results showed that both the used REAH inhibited the growth of pathogenic fungal strains by possessing inhibitory potential. The results of present study are summarized in table 1 and 3.

Well diffusion method-

Crude extract of used REAH were screened for antifungal activity against 2 plant pathogenic fungi and results are presented in table 2 and 4. As evident from the table both the used REAH are able to produce some antifungal substances which inhibit the growth of plant pathogenic fungi.

Antifungal activity of C. parvula Kuzuha

The used fungus isolated as an endophyte from the live roots of *Pilea scripta* was found as an antagonist against both the plant pathogenic fungi. In dual culture method the growth inhibition in terms of percentage were 63.20% and 45.26% for *Colletotrichum capsici* and *Fusarium oxysporum* respectively. However, in well diffusion method the percentage of inhibition against the *C. capsici* is 26.66% and *F. oxysporum* it was recorded 32.14% by the crude extract of *C. parvula* (Table 1 and 2).

Antifungal activity of T. setigerum (Grove) Ingold

Another used fungus recovered from the roots of *Barberris vulgaris* also found potent to inhibit the growth of test fungi. In dual culture assay the growth of *C. capsici* and *F. oxysporum* were inhibited 45% and 40.09% respectively by the mycelial disc of used aquatic hyphomycetes. The crude extract of *T. setigerum* inhibits the growth of *C. capsici* (23.33%) while against the *F. oxysporum* it was found (19.35%) in well diffusion assay (Table 3 and 4).

Table 1- Antifungal activity of *C. parvula* in dual culture method

S. No.	Fungal strains	Growth diameter		Percent inhibition
		R_1 (in mm)	R_2 (in mm)	
1.	*C. capsici*	21	12	63.20%
2.	*F. oxysporum*	24	9	45.26%

Table 2- Antifungal activity of *C. parvula* in well diffusion method

S. No.	Fungal strains	Growth diameter		Percent inhibition
		R_1 (in mm)	R_2 (in mm)	
1.	*C. capsici*	28	19	32.14%
2.	*F. oxysporum*	30	22	26.66%

Table 3- Antifungal activity of *T. setigerum* in dual culture method

S. No.	Fungal strains	Growth diameter		Percent inhibition
		R_1 (in mm)	R_2 (in mm)	
1.	*C. capsici*	20	11	45%
2.	*F. oxysporum*	22	13	40.09%

Table 4- Antifungal activity of *T. setigerum* in well diffusion method

S. No.	Tested pathogenic fungal strains	Growth diameter		Percent inhibition
		R_1 (in mm)	R_2 (in mm)	
1.	*C. capsici*	31	25	19.35%
2.	*F. oxysporum*	30	23	23.33%

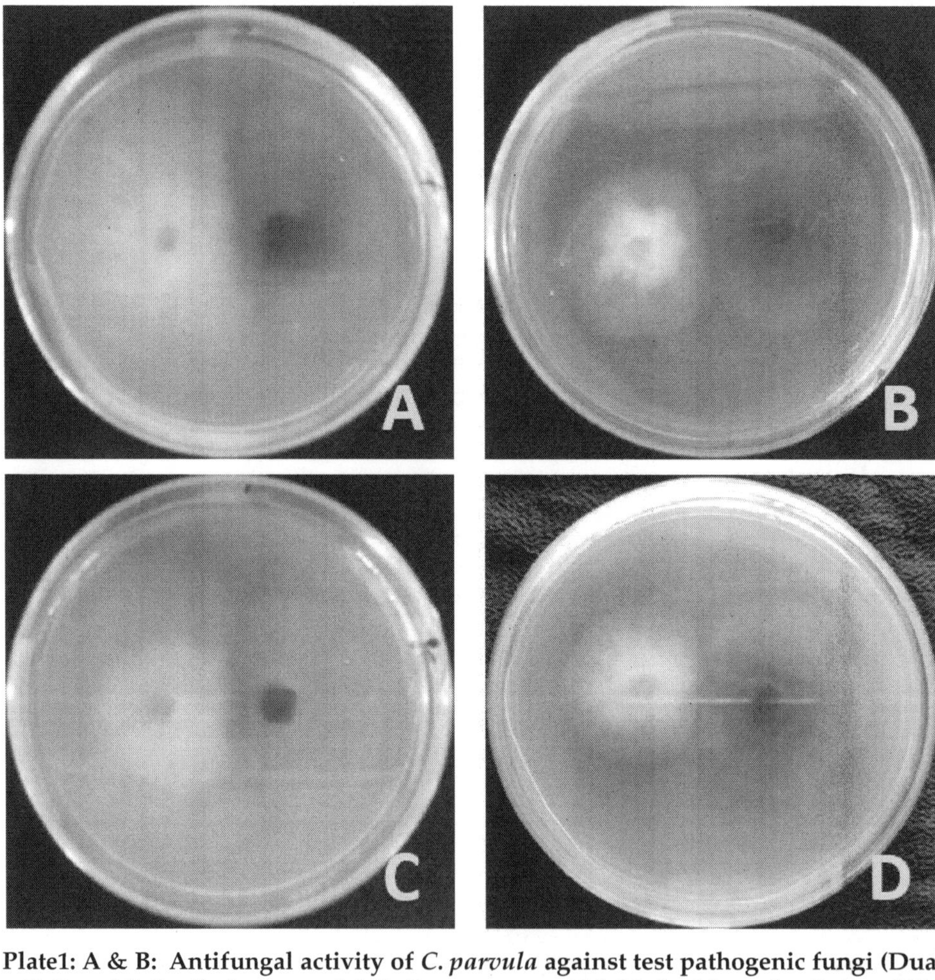

Plate1: A & B: Antifungal activity of *C. parvula* against test pathogenic fungi (Dual Culture Method): A- *C. capsici*: B- *F. oxysporum*; C & D: Antifungal activity of *T. setigerum* against test pathogenic fungi (Dual Culture Method): C- *C. capsici*: D- *F. oxysporum*

Discussion

Among the use of natural resources, fungi are getting significant interest as one of the important sources after the discovery of various antifungal drugs such as penicillin (Chain et al., 1940), cephalosporins (Abraham and Newton 1961), Anidulafungin (Cabello et al. 2001) micafungin (Vicente et al. 2016) etc from fungal sources. Endophytes generally exist asymptomatically, coexisting with their hosts and representing an underutilized group of microorganisms for the discovery of new compounds. Endophytic fungi are being considered as alternate sources of the bioactive metabolites mainly produced by plants (Deshmukh et al. 2019).

In the present study, endophytic aquatic hyphomycetes isolated from roots of riparian plants were screened for their antifungal activity against two plant pathogenic fungi using a dual culture and well diffusion techniques. As evident from the table 1-4 it is clear that both the used fungi (*Campylospora parvula* and *Tetracladium setigerum*) have definite antifungal potentiality against the test fungi. The formation of inhibition zone around the test fungi confirms the production of some antifungal substances by the aquatic hyphomycetes *C. parvula* and *T. setigerum*. Shearer and Zare-Maivan (1988) also reported some species of aquatic hyphomycetes as potent inhibiters for competitive species by the production of antibiotic substances.

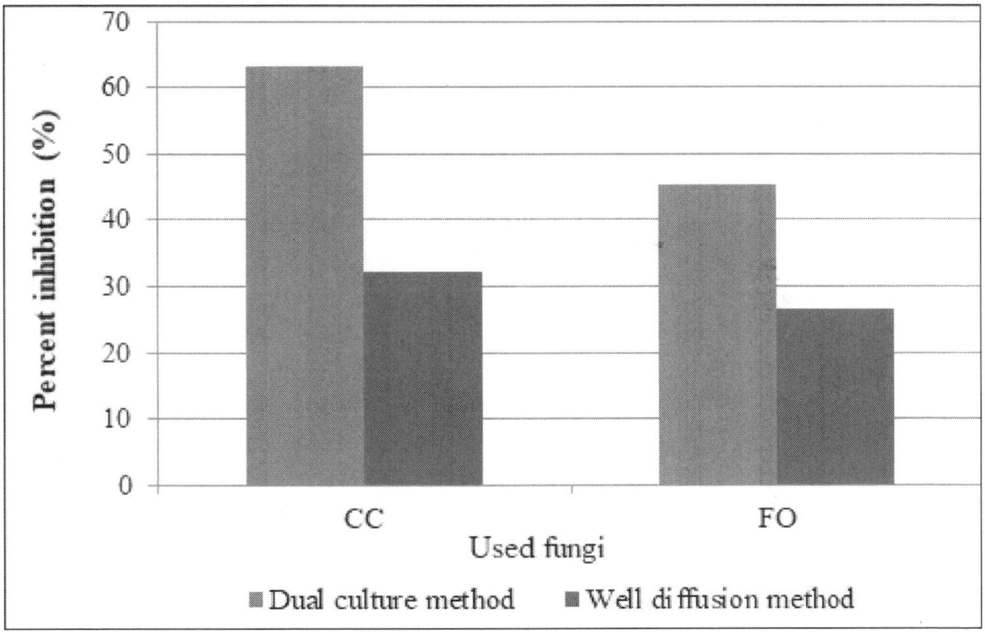

Figure 1: Antifungal activity of *C. parvula* against the plant pathogenic fungi (CC-*C. capsici*, FO- *F. oxysporum*)

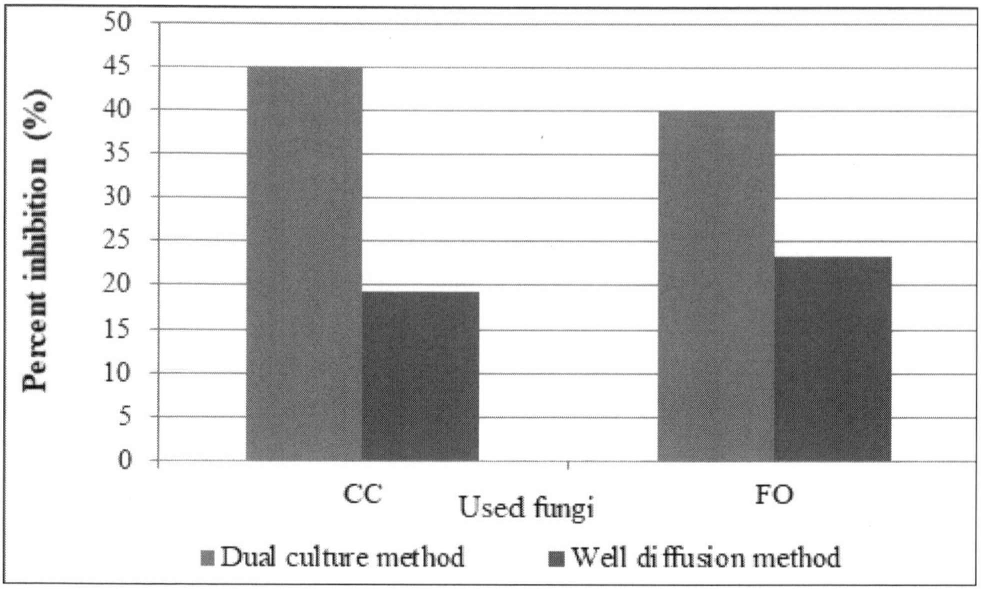

Figure 2: Antifungal activity of *T. setigerum* against the plant pathogenic fungi (CC- *C. capsici*, FO- *F. oxysporum*)

It is noteworthy that the fungal disc and crude extract of *C. parvula* showed similarity on the test pathogenic fungi. In the case of *C. parvula* fungal disc and crude extract maximum inhibition was found against *C. capsici* 63.20% and 45% respectively while the fungal disc and crude extract of *T. setigerum* worked differently. Antifungal activity of terrestrial endophytic fungi has been documented by some workers (Kumar et al. 2008, Vaz et al. 2009, Kumar and Kaushik 2013) but there are only few reports on the antifungal nature of aquatic hyphomycetes (Shearer and Zare-Maivan, 1988, Sati and Arya 2010, Singh and Sati 2019).

Earlier antifungal activity of five species of aquatic hyphomycetes, viz., *Heliscus lugdunensis*, *Tetrachaetum elegans*, *Tetracladium breve*, *T. marchalianum*, and *T. nainitalense*, against seven plant pathogenic fungi was studied using a dual culture technique by Sati and Arya (2010). They found a remarkable antagonistic activity of *T. elegans* against *Colletotrichum falcatum*, *Fusarium oxysporum*, *Pyricularia oryzae*, *Sclerotium sclerotiorum*, and *Tilletia indica*. Similarly *H. lugdunensis* showed antagonism against two plant pathogenic fungi, *Rhizoctonia solani* and *Colletotrichum falcatum* while *T. breve*, *T. marchalianum*, and *T. nainitalense* showed no response towards tested plant pathogenic fungi.

In an another study Singh and Sati (2019) also evaluated two root endophytic aquatic hyphomycetes (*Anguillospora longisima*) for their antifungal potentiality by dual culture and well diffusion method and found a significant antifungal activities against pathogenic fungi. Thus the present study supports the earlier findings and clearly indicates that aquatic hyphomycetous fungi have great potentiality to antagonize the pathogenic fungi. It could also be envisage that these fungi may have capability to synthesize some metabolites that inhibit the growth of plant pathogenic fungi. However, further screening of the compounds responsible for the antagonistic activity is required. This antagonistic property of aquatic hyphomycetes may also be exploited in future for the use of naturally occurring compounds, as fungicides.

Conclusion

The aquatic hyphomycetes constitute a major component of freshwater habitats but as endophyte they occupy a unique ecological niche and can influence the distribution, ecology, physiology and biochemistry of plants. Both the tested species of root endophtic aquatic hyphomycetes possess antifungal activity against the plant pathogenic fungi in dual culture and well diffusion assay. It is also interesting to note that this is the first report on the antifungal activity of *Campylospora parvula* and *Tertacladium setigerum*. On the basis of present investigation it can also be concluded that in future these species of aquatic hyphomycetes can be used as an alternative source of bio control agent against the plant diseases caused by fungi.

References

Abraham EP, Newton GGF (1961) The structure of cephalosporin. Biochemical Journal 9:377–393

Arya P, Sati SC (2012) Evaluation of endophytic hyphomycetes for their antagonistic activity against pathogenic bacteria International Journal of Microbiology 2(9):343–347

Cabello MA, Platas G, Collado J, Díez MT, Martín I, Vicente F, Meinz M, Onishi JC, Douglas C, Thompson J, Kurtz MB, Schwartz RE, Bills GF, Giacobbe RA, Abruzzo GK, Flattery AM, Kong L, Peláez F (2001) Arundifungin, a novel antifungal compound produced by fungi: biological activity and taxonomy of the producing organisms International Microbiology 4(2):93–102

Chain E, Florey W, Gardner AD, Heatley NG, Jennings MA, Orr-Ewing J, Sanders AG (1940) Penicillin as a chemotherapeutic agent Lancet 239: 226–228

Demain AL (2014) Importance of microbial natural products and the need to revitalize their discovery Journal of Industrial Microbiology and Biotechnology 41: 185-201

Gulis VI, Stephanovich AI (1999) Antibiotic effects of some aquatic hyphomycetes. Mycological Research 103: 111–115

Kumar S, Kaushik N, Edrada-Ebel R, Ebel R, Proksch P (2008) Endophytic fungi for pest and disease management. In: Integrated Management of Diseases caused by Fungi Phytoplasma and Bacteria. (Eds. Ciancio A, Mukerji KG). Springer, Berlin. pp. 365–387

Kumar S, Kaushik N (2013) Endophytic fungi isolated from oil-seed crop *Jatropha curcas* produces oil and exhibit antifungal activity PloS one 8(2):p.e56202

Liu X, Dong M, Chen X, Jiang M, Zhou J (2010). Antimicrobial activity of an endophytic *Xylaria* sp. YX-28 and identification of its antimicrobial compound 7- amino-4-methylcoumarin Applied Microbiology and Biotechnology 78: 241–247

Pant P, Sati SC (2021) Antibacterial activity of root endophytic aquatic hyphomycetes against plant pathogenic bacteria Vegetos 34: 785-789

Sati SC, Arya P (2010) Antagonism of some aquatic hyphomycetes against plant pathogenic fungi The Scientific World Journal 760–765

Shearer CA, Zare-Maivan H (1988) *In vitro* hyphal interactions among wood and leaf-inhabiting ascomycetes and fungi imperfecti from freshwater habitats Mycologia 80: 31–37

Singh L, Sati SC (2019) Antagonistic Activity of Isolated Root Endophytic Freshwater Fungus *Anguillospora Longissima* Against Pathogenic Fungi National Academic Science Letters https://doi.org/10.1007/s40009-019-00818-w

Vaz ABM, Mota RC, Bomfim MRQ, Vieira MLA, Zani CL (2009).Antimicrobial activity of endophytic fungi associated with Orchidaceae in Brazil Canadian Journal of Microbiology 55: 1381–1391

Vicente F, Reyes F, Genilloud O (2016) Fungal secondary metabolites as source of antifungal compounds. In: (Eds. Deshmukh SK, Misra JK, Tewari JP, Papp T), Fungi applications and management strategies. CRC Press, Taylor & Francis Group, Boca Raton, pp 80–116

CHAPTER 6

DIVERSITY ASSESSMENT OF LICHENS FROM HISTORICAL MONUMENTS OF GARHWAL HIMALAYA, UTTARAKHAND

Krishna Chandra[1], Shashi Upadhyay[2] & Yogesh Joshi[3]

Department of Botany,
1 Govt. P .G. College Gairsain 246428, Uttarakhand, India
2 Himmotthan Society- An initiative of TATA Trust Almora 263601, Uttarakhand, India
3 Lab. No. 15, Department of Botany, University of Rajasthan, Jaipur 302004, Rajasthan, India

Abstract

The present work was focused on diversity assessment of lichens from historical monuments of Uttarakhand. Investigation of 40 historical monuments revealed a total of 86 lichen species belonging to 50 genera and 23 families. Quantitative analysis showed the alpha diversity and Simpson's index (D') varies from 1.330 to 2.616 and 3.867 to 17.92 respectively while the value of concentration of dominance ranges from 0.089 to 0.278. Polynomial fits to the data suggest the lichen species richness shows the positive relationship with altitude of the studied monuments (r^2= 0.331, p= 0.0005).

Keywords: Diversity, lichens, monuments, SEM, Uttarakhand

Introduction

Among the different groups of cultural tourism, historical monuments and heritage sites are the unique element in tourism segments and are significantly involve in economic growth of a country. Stone-built historical monuments in an exposed environment are influenced by various environmental parameters as well as biological organisms. The colonization of biological organisms on historical monuments presents a significant challenge due to the combined chemical, physical and aesthetic damages, responsible for biodeterioration of monument.

Hence identification and determination of the biological agents involved in biodeterioration is a prerequisite for the study of the microbial ecology on historical monuments. Among various biological organisms lichens are one of them which invade monuments and performs irreversible changes to the substratum often called biodeterioration. The issue of biodeterioration is the resultant of complex interplay of climate, substrate and the organism itself. In order to determine the biodeterioration potential and the degree of damage caused by lichens, it is necessary to investigate their diversity, taxonomic details and ecological properties of each species growing on monuments.

Hence various studies were conducted on floristic account of lichens on historical monuments across the globe as well as in India. After the advancement of technical and multidisciplinary approaches in 1990's the complex processes of biodeterioration and remedial measures came into lime light. It has now been confirmed that the presence of some lichens on monument exhibit the inevitable transformations at lichen-rock interface (LRI) due to their physical and chemical activities. Evaluation of LRI by the different complementary techniques *viz.* scanning electron microscopy (SEM), transmission electron microscopy (TEM), Inductively Coupled Plasma Mass Spectrometry (ICPMS) and Energy Dispersive X-Ray Spectroscopy (EDS) showed the tremendous alteration and disintegration of minerals along with the changes in chemical composition in the LRI.

In context to India, little is known about biodeterioration effect of lichen on Indian monuments and requires such studies across the country. Hence considering all these the present work is focused on enumeration of lichens colonizing various historical monuments of Garhwal Himalaya, Uttarakhand.

Materials and methods

An extensive survey was conducted on 40 monuments of Uttarakhand (some are shown in Fig. 1.1 a-e), extending from 475 to 3600 m asl (Table 1.1). The foliose lichen samples colonizing monuments were collected with the help of snapper without damaging the monument. Samples of crustose lichens were collected in vicinity of monument. All lichen samples were identified following floristic/monographic keys and descriptions. Morphological examinations were carried out using a Stereozoom dissecting microscope (OLYMPUS SZ2-ILST) and anatomical studies of handmade sections were done using a Compound microscope (CX21iLEDFS1). Spot test reactions on thalli, medulla and fruiting bodies were performed with the standard reagents (K, C, KC and PD). Thin Layer Chromatography (TLC) was performed to chemical analysis.

Quantitative analysis

Alpha diversity was analyzed using Shannon-Wiener diversity index while concentration of dominance was analyzed using Simpson's index. Pearson's correlation coefficient was performed by using SPSS software 16.0 version for knowing the relationship between altitude and species richness, number of families, alpha diversity, Simpson's diversity and concentrations of dominance. Polynomial regression analysis was performed by using PAST software for knowing the relationship between altitude and species richness.

Results and Discussion

Lichens diversity

A total of 86 lichen species belonging to 50 genera and 23 families were recorded from 40 monuments of Uttarakhand (Table 1.2). Lichen family Parmeliaceae was dominating with 15 species while family Physciaceae was showing 12 of lichens on different monuments. Madmaheshwar temple shows the maximaum diversity of lichens (30 species). In contrast, minimum diversity was noticed in Laxminarayan temple Sumari (4 species) (Table 1.1).

Quantitative analysis shows that the alpha diversity and Simpson's index (D') varies from 1.330 (Nanda Devi Nandkeshri) to 2.616 (Madmaheshwar) and 3.867 (Nanda Devi Nandkeshri) to 17.92 (Madmaheshwar), respectively. However, the value of concentration of dominance ranges between 0.089 (Tungnath) to 0.278 (Nanda Devi Nandkeshri) (Table 1.1).

The lichen diversity in these monuments shows the scattered distribution. For the determination of similarity between monuments, PCA was done and it was confirmed that the dominant and co-dominant lichen species differed from site to site; indicating distinct lichen assemblages at the different sites and this is showed by PCA ordination plot. Out of 40 studied monuments most of the monuments were adjacent to each other in the ordination plot (Fig. 1.2). In the ordination plot, 80.92% of the variations in species composition were explained by the PCA axis 1, while PCA axis 2 explained 13.51% respectively. Both the axes explained 94.43% in the species composition.

PCA axis 1 was significantly related to altitude and species richness; indicating that altitude and species richness of lichens is the main factor concurrence of these different monuments and similarly, altitude was significantly related to PCA axis 2 (Table 1.3). Diversity variables of these monuments show the positive significant relationship with altitude. Species richness was significantly related to number of family, alpha diversity, Simpson's diversity index and concentration of dominance. All the diversity parameters were also significantly related to axis

1 scores. Diversity indices were also significantly related with each other in these monuments (Table 1.3). Polynomial fits to the data suggest that lichen species richness was linearly increased with altitude and shows the positive relationship with altitude of monuments (r^2= 0.331, p= 0.0005; Fig.1.3).

The significance of historical monuments devised by many scholars as their aesthetic, social, historical, political, educational, scientific, economic, cultural as well as spiritual. These wide dimensions of significance endorse their conservation to retain its historic information and values for the posterity. Hence IUCN World Conservation Congress held in 2008, accentuated and included the natural monuments in the category III of conservation list with the aim of protection and conservation of these values.

Since Uttarakhand is known as the abode of God, a number of ancient monuments are being protected by the Archeological Survey of India (ASI) for maintaining their antiquarian significance. These stable and undisturbed substrata occupies extensive lichen diversity and till now 46 monuments were surveyed for lichen studies from some monuments of Uttarakhand and enumerated a total of 71 lichens species belonging to 39 genera and 21 families.

This considerable amount of lichens on historical monuments is considered as the culprit by conservators and archaeologists because of their pedogenetic effect on stone artifacts. As the phenomenon of pedogenesis is a fundamental process and involves interaction between lithosphere, hydrosphere, atmosphere and biosphere and is responsible for supporting biodiversity; but this phenomenon has never been treated as commendable for historical monuments as they retain not only their intrinsic and instrumental values but also are the tangible historical evidence of that time period when these monuments were constructed.

This unfavorable process of pedogenesis on historical monuments via lichens either by physical or chemical activities shows the drastic alterations at both LRI and on the upper surface of rock; consequently, the deposition of mineral particles immediate below the thallus suggests their presence and detached from the same rock surface except for the idea of cross-contamination.

As for the physical deterioration, evidence reveals that the adhering organs penetrate the rock up to the depth of 2.0 cm forming micro and meso-pits, lead to the detachment of several millimeters of stone under the central part of the thallus. Since the intensity of damage depends on type of lichen thallus and rock property even the same stone shows the difference in depth of hyphal penetration. The significant depth of penetration was also observed in granite chlorite schist and quartzite rock samples of the previous study.

Apart from the physical deterioration, chemical deterioration is another challenge for the LRI through chelation property of lichen acids, and responsible for the weakening of the rock interface. Evaluation of lichen-rock-interface in Indian monuments by Singh and Dhawan showed the organic acid secretion and chelation activity by three lichen species. Lichens containing atranorin, salazinic acid, usnic acid, zeorin, divericatic acid, fumarprotocetraric acid, lecanoric acid and gyrophoric acid have chelation properties. Besides these lichen acids, evidence of metal-complex formation by norstictic acid was suggested by previous studies. In the present study most of the lichens recorded from these monuments are associated with the chelation property and responsible for chemical alteration of rock surface leading to deterioration of substratum.

Conclusion

1. Monuments of Uttarakhand are highly covered with lichen species of various growth forms.
2. Their presence on monuments is directly associated with the location, environmental condition and conservation status of monuments.
3. Monuments under the control of government authorities are reflecting less lichen diversity due to their regular maintenance and cleaning practices.
4. Each lichen species possess different biodeterioration potential; it is therefore imperative to quantify the floristic diversity growing on them.
5. Assessment of lichen-rock interface by using SEM and ICP-MS techniques indicates the deteriorative effect of lichens on monument stone.
6. Hence to prevent the encroachment of lichens on culturally rich artifacts and to combat the biodeterioration of precious stone an effective, risk free and environment friendly technique/measure should be used by conservators.

Acknowledgements

The authors are thankful to the Scientific and Engineering Research Board (SB/FT/LS-313/2012) for providing financial support.

References

Anagnostidis, K., Gehrmann, C. K., Gross, M., Krumbein, W. E., Lisi, S., Pantazidou, A. and Zagari, M. (1991). Biodeterioration of marbles of the Parthenon and Propylaea, Acropolis, Athens. associated organisms, decay and treatment suggestions. The Conservation of Monuments in the Mediterranean Basin, Decrouez D. Chamay J., and Zezza F. (eds.). 305-325.

Ascaso, C., and Wierzchos, J. (1994). Nuevas aplicaciones de las técnicas submicroscópicas en el estudio del biodeterioro producido por talos liquénicos. *Microbiologia*, **10**, 103-110.

Ascaso, C., and Wierzchos, J. (1994). Structural Aspects of the Lichen Rock Interface Using Back scattered Electron Imaging. *Botanica Acta*, 107(4), 251-256.

Ascaso, C., Wierzchos, J., Souza-Egipsy, V., De los Rıos, A., and Rodrigues, J. D. (2002). In situ evaluation of the biodeteriorating action of microorganisms and the effects of biocides on carbonate rock of the Jeronimos Monastery (Lisbon). *International Biodeterioration & Biodegradation*, 49(1), 1-12.

Awasthi, D. D. (1991). Key to the microlichens of India, Nepal and Sri Lanka. J. Cramer.

Awasthi, D.D. (2007) A Compendium of the Macrolichens from India, Nepal and Sri Lanka, Bishen Singh Mahendra Pal Singh Pub., Dehradun, India. 1-580.

Bajpai, R., and Upreti, D. K. (2014). Lichens on Indian monuments.

Chandra, K., Upadhyay, S., Bisht, K., and Joshi, Y. (2017) Diversity and distribution of lichenized fungi on some historical monuments of Kumaun Himalaya, Uttarakhand, *Austrian Journal of Mycology*, 26, 2017, 1-15.

Chandra, K., Joshi, Y. and Bajpai R. (2020) Lichen diversity and assessment of biodeterioration impact: a case study from shiva temple Gyupani, Almora, Uttarakhand. *International journal of research and analytical reviews*. 7(4), 612-620.

Csapo, J. (2012). The role and importance of cultural tourism in modern tourism industry. In strategies for tourism industry-micro and macro perspectives. (K. Murat ed.) 10, 201-212.

de Los Ríos, A., Cámara, B., Del Cura, M. A. G., Rico, V. J., Galván, V., and Ascaso, C. (2009). Deteriorating effects of lichen and microbial colonization of carbonate building rocks in the Romanesque churches of Segovia (Spain). *Science of the total environment*, 407(3), 1123-1134.

Divakar, P.K., and Upreti, D.K (2005) Parmelioid lichens in India, Bishen Singh Mahendra Pal Singh Pub., Dehradun, India. 488.

Dümcke, C., and Gnedovsky, M. (2013). The social and economic value of cultural heritage: literature review. EENC paper, 1, 101-114.

Hammer, Ø., and Harper, D. A. (2001). Past: paleontological statistics software package for educaton and data anlysis. *Palaeontologia electronica*, 4(1), 1.

Haas, J., R., and Purvis, O.W. (2006). Lichen biogeochemistry, Fungi in biogeochemical cycles, Gadd G.M. (ed), Cambridge university press, 344-376.

Kiurski, J. S., Ranogajec, J. G., Ujhelji, A. L., Radeka, M. M., and Bokorov, M. T. (2005). Evaluation of the effect of lichens on ceramic roofing tiles by scanning electron microscopy and energy dispersive spectroscopy analyses. *Scanning*, 27(3), 113-119.

Laso, B. M. (2001). Biodiversidad y colonización liquénica de algunos monumentos en la ciudad de Salamanca (España). *Botanica Complutensis*, (25), 93-102.

Mehta, P. (2013). Rock weathering and farmland formation: Its geology that determines biology, *Earth science India*, 6, 1-3.

Mohammadi, P., and Krumbein, W. E. (2008). Biodeterioration of ancient stone materials from the Persepolis monuments (Iran). *Aerobiologia*, 24(1), 27-33.

Orange, A., James, P. W., and White, F. J. (2001). Microchemical methods for the identification of lichens. Twayne Publishers.

Piervittori, R., Favero Longo, S. E., and Gazzano, C. (2009). Lichens and biodeterioration of stonework: a review. *Chimicaoggi-Chemistry today*, 27, 8-11.

Purvis, O. W., Elix, J. A., Broomheadj, J. A., and Jones, G. C. (1987). The occurrence of copper norstictic acid in lichens from cupriferous substrata. *The Lichenologist*, 19(2), 193-203.

Puy-Alquiza, M. J., Gómez Peralta, M., Miranda-Avilés, R., Reyes-Zamudio, V., Salazar-Hernández, M., and Ordaz Zubia, V. Y. (2015). The role of lichen communities in superficial deterioration of their rock substrates: studies of the lichen-rock interface of two historical buildings in the city of Guanajuato, Mexico. *Acta universitaria*, 25(4), 35-47.

Rigamonti, M. (2008). Lichens and monuments: searching for the balance between the necessity of restoration and preservation, and the improvement of the naturalistic-environmental aspects of the historical-architectural heritage. *Scientifica Acta*, 2(2), 93-96.

Robert, M., Galsomies, L., Tessier, D., Jaunet, A. M., Iñigo, A. C., and Elsass, F. (2003). Degradation of historical monuments. Role of the different mechanisms, processes and microsystems. Applied Study of Cultural Heritage and Clays, 13, 17.

Rosato, V. G. (2006). Diversity and distribution of lichens on mortar and concrete in Buenos Aires province, Argentina. *Darwiniana*, 44(1), 89-97.

Salvadori, O., Pinna, D., & Grillini, G. C. (1990). Deterioration induced by lichens on volcanite of the Vulsini complex (Central Italy), Proceedings of the International Meeting Lavas and Volcanic Tuffs, Charola, A.E., Koestler, R.J. and G. Lombardi (eds)). Easter Island Roma, 215-225.

Saxena, S., Upreti, D. K., Singh, A., & Singh, K. P. (2004). Observations on lichens growing on artifacts in the Indian subcontinent. Biodeterioration of Stone Surfaces: Lichens and Biofilms as Weathering Agents of Rocks and Cultural Heritage, 181-193.

Seaward, M. R. D. (1988). Lichen damage to ancient monuments: a case study. *The Lichenologist*, 20(3), 291-294.

Shannon, C.E. and Weaver, W. (1963). The mathematical theory of communication. University of Illinois Press.

Shukla, V., Upreti, D. K., and Bajpai, R. (2014). Lichens to biomonitor the environment. New Delhi, India: Springer India. 97-170.

SIMPON, E. (1949). Measurement of diversity. *Nature*, 688, 163.

Singh, A., and Dhawan, S. (1991). Interesting observations on stone weathering of an Indian monument by lichens. *Geophytology*, 21, 119-123.

Sohrabi, M., Favero-Longo, S. E., Pérez-Ortega, S., Ascaso, C., Haghighat, Z., Talebian, M. H. and de los Ríos, A. (2017). Lichen colonization and associated deterioration processes in Pasargadae, UNESCO world heritage site, Iran. *International Biodeterioration & Biodegradation*, 117, 171-182.

Tiano, P. (2002). Biodegradation of cultural heritage: decay mechanisms and control methods. In Seminar article, new university of Lisbon, Department of Conservation and Restoration, 7-12.

Tripathi, M., Joshi, S., & Jinnah, Z., (2013) Lichens Growing on Monuments of Almora, Uttarakhand, Biodeterioration of Cultural Property-7, Dhawan, S., Abduraheem, K. and Nath V. (eds), 65-73.

Upreti, D. K. (2002). Lichens of Khajuraho temple and nearby area of Mahoba and Chattarpur district. In Proceedings of National Seminar on biodeterioration of materials, 123-127.

Upreti, D. K., Bajpai, R., and Nayaka, S. (2009). Indian monuments need lichen biodeterioration study. *New Horizon*, 7, 64-69.

Upreti, D. K., Nayaka, S., and Joshi, Y. 2004. Lichen activity over rock shelter of Bhimbetka world Heritage Zone, Madhya Pradesh. In Rock Art research: Changing Paradigms. Rock Art Society of India, 30-31.

Fig.1.

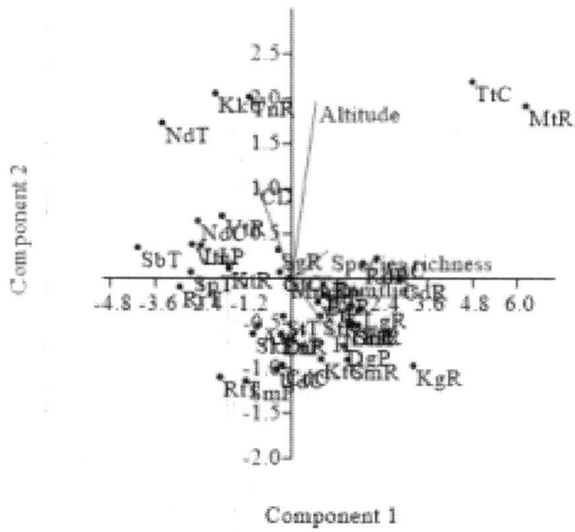
TtC
MtR
NdT
Altitude
SbT
Species richness
KgR
Component 2
Component 1

Fig. 2

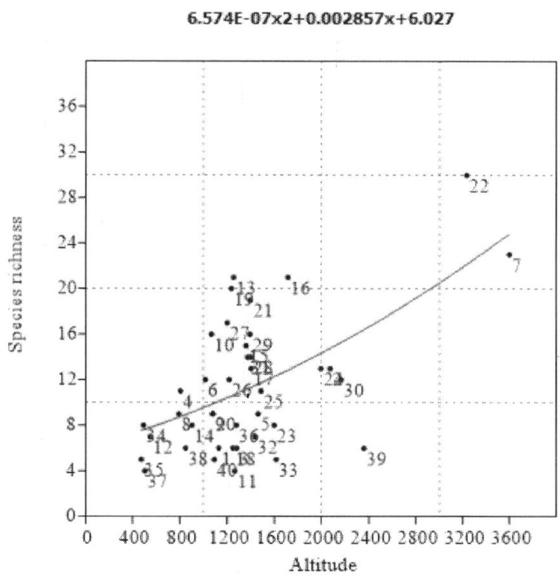
6.574E-07x2+0.002857x+6.027
Species richness
Altitude

Fig. 3

Legend for figures

- Fig. 1. Some monuments of Garhwal Himalaya, Uttarakhand: A, Laxmi Narayan group; B), Adi Badri; C, Tungnath; D, Kulsari temple; E, Silla temple
- Fig. 2. Principal Component Analysis ordination plot of different monuments of Garhwal Himalaya.
- Fig. 3. Polynomial regression analysis between altitude and species richness of lichens (Y = 6.574E-07x2+0.002857x+6.027).

Table 1.1 Qualitative and quantitative characteristics of studied monuments. [SR- number of lichen species or species richness, H'- Shannon-Wiener Index or alpha diversity, D'- Simpson's Diversity index, CD - Concentration of dominance (Simpson's index)].

S N.	Name of Monument.	Latitude (N)	Longitude (E)	Altitude (m asl)	SR	H'	D'	CD
1	Aadi badri	30°43.676′	78°26.358′	1132	6	1.792	6	0.167
2	Anushuya Devi	3028.792′	79°17.022′	2074	13	2.311	10.96	0.112
3	Nanda Devi Nandkeshri	30°02.441′	79°34.048′	1282	6	1.330	3.867	0.278
4	Chandika Devi Simli	30°13.974′	79°15.463′	806	11	1.894	6.927	0.157
5	Gopinath temple	30°24.833′	79°18.963′	1463	9	1.831	6.917	0.185
6	Kulsari temple	30°43.676′	78°26.358′	1018	12	2.095	8.939	0.139
7	Tungnath	30°29.22′	79°12.55′	3600	23	2.525	13.95	0.089
8	Uma Devi temple	30°15.754′	79°13.044′	792	9	1.889	6.944	0.160
9	Danari Group of temple	30°10.721′	78°49.561′	1080	9	1.889	6.944	0.160
10	Devalgarh	30°13.232′	78°51.622′	1070	16	2.133	9.917	0.141
11	Laxminarayan temple Sumari	30°12.059′	78°48.847′	1265	4	1.386	4	0.25
12	Shankar Math	30°13.417′	78°46.110′	551	7	1.748	5.952	0.184
13	Shiva temple Paithani	30°08.895′	79°59.212′	1255	21	1.898	7.871	0.170
14	Shivalaya Kukhargaun	30°11.695′	78°44.788′	903	8	1.733	5.929	0.188
15	Vaishnava temple	30°11.695′	78°44.788′	1361	15	1.991	7.914	0.147
16	Chandika temple Bhanaj	30°25.499′	78°08.395′	1715	21	2.306	11.92	0.120
17	Damyanti temple	30°32.340′	79°04.910′	1403	13	2.205	9.949	0.124
18	Kalimath	30°35.256′	79°08.497′	1249	6	1.561	4.933	0.222
19	Kandara group	30°26.118′	79°05.167′	1235	20	2.484	12.95	0.09

S N.	Name of Monument.	Latitude (N)	Longitude (E)	Altitude (m asl)	SR	H′	D′	CD
20	Laxmi Narayan temple Bairangana	30°12.505′	78°57.399′	1082	9	1.889	6.944	0.160
21	Laxmi Narayan group	30°38.484′	78°58.615′	1397	19	2.160	9.918	0.130
22	Madmaheshwar	30°38.123′	79°13.296′	3234	30	2.616	17.92	0.107
23	Makku math	29°39.722′	79°35.135′	1598	8	1.906	6.964	0.156
24	Rakeshwari Devi	30°35.257′	79°08.497′	1997	13	2.245	9.962	0.112
25	Shivalaya Gahatara	30°32.166′	79°04.815′	1487	11	1.768	6.873	0.207
26	Shivalaya Mahad	30°12.600′	78°57.948′	1218	12	2.254	9.970	0.111
27	Silla temple	30°24.244′	79°01.443′	1200	17	1.987	7.904	0.149
28	Shivalaya Naratankoti	30°32.678′	79°04.363′	1399	14	2.206	9.945	0.122
29	Tripura sundari	30°32.678′	79°04.367′	1394	16	2.220	9.942	0.117
30	Triyuginarayan temple	29°39.007′	80°01.753′	2166	12	1.474	5.758	0.306
31	Bashukedar group of temple	30°33.526′	79°05.289′	1373	14	2.008	8.901	0.163
32	Vishwanath temple	29°34.520′	80°14.401′	1436	7	1.475	4.857	0.265
33	Nanda devi Bajinga	30°32.340′	77°04.908′	1615	5	1.055	2.8	0.36
34	Raj Rajeshwari Ranihat	30°13.400′	78°46.527′	495	8	1.255	3.786	0.313
35	Raghunath temple	30°08.461′	78°35.800′	475	5	1.609	5	0.2
36	Shiva temple Bagi	30°14.076′	78°35.287′	1279	8	1.906	6.964	0.156
37	Shivalaya Bagwan	30°13.305′	78°40.527′	506	4	1.040	2.833	0.375
38	Sun temple Palethi	30°13.262′	78°40.800′	850	6	1.330	3.867	0.278
39	Kalpkedar	31°02.258′	78°46.939′	2358	6	1.330	3.867	0.278
40	Vishwanath	30°40.483′	78°26.939′	1093	5	1.332	3.9	0.28

Table 1. 2. Lichen diversity on historical monuments of Garhwal Himalaya.

Lichen species	1	2	3	4	5	6	7	8	9	10	11	12	13	14	15	16	17	18	19	20	21	22	23	24	25	26	27	28	29	30	31	32	33	34	35	36	37	38	39	40
Acarospora badiofusca (Nyl.)Th. Fr.																		+							+															
Amandinea subduplicata (Vain.) Marbach													+			+													+											
A. submontana Marbach															+		+									+														
Anema decipiens (Massal.) Forssell																						+																		
Bacidia **sp.**							+			+						+	+																							
Blennothallia crispa (Huds.) Otálora, P.M. Jørg. &Wedin							+	+													+				+															
Buellia **sp. 1**							+						+															+												
Buellia **sp. 2**				+								+																												
Bulbothrix isidiza (Nyl.) Hale																													+		+									

Lichen species	1	2	3	4	5	6	7	8	9	10	11	12	13	14	15	16	17	18	19	20	21	22	23	24	25	26	27	28	29	30	31	32	33	34	35	36	37	38	39	40
Brownliella cinnabarina (Ach.) S.Y. Kondr. *et al*						+							+		+						+										+		+	+			+	+		
Caloplaca ochroplaca Poelt & Hinter.															+	+			+					+																
C. subbassiae Y. Joshi & Upreti			+		+				+											+									+					+						
Candelaria concolor (Dicks.) Stein		+			+	+										+			+	+	+	+	+			+	+			+									+	
Candelariella **sp.**												+		+	+	+	+		+			+		+			+		+		+					+		+		
Canoparmelia texana (Tuck.) Elix & Hale				+									+																		+									
Catillaria **sp.**																						+																		
Circinaria calcarea (L.) A. Nordin, Savi & Tibell							+	+								+			+			+																		
Cladonia **sp.**		+					+															+								+										
Chrysothrix chlorina (Ach.) J.R. Laundon																	+													+										

Lichen species	1	2	3	4	5	6	7	8	9	10	11	12	13	14	15	16	17	18	19	20	21	22	23	24	25	26	27	28	29	30	31	32	33	34	35	36	37	38	39	40
Coccocarpia erythroxyli (Spreng.) Swinscow & Krog																						+																		
Cratiria obscurior (Stirt.) Marbach & Kalb							+						+																											
Dermatocarpon miniatum (L.) W. Mann.						+	+								+			+										+				+								
D. vellereum Zschacke	+	+		+				+		+											+				+		+		+	+				+	+		+			+
Dimelaena oreina (Ach.) Norman							+																																	
D. tenuis **(Müll.Arg.)** H. Mayhofer & Wipple															+																									
Diploschistes actinostomus (Pers. ex Ach.) Zahlbr.						+	+						+			+	+		+	+	+	+	+		+	+		+	+		+				+	+				
Endocarpon rosettum Ajay Singh & Upreti							+		+						+	+	+							+			+					+	+	+			+			

Lichen species	1	2	3	4	5	6	7	8	9	10	11	12	13	14	15	16	17	18	19	20	21	22	23	24	25	26	27	28	29	30	31	32	33	34	35	36	37	38	39	40
E. subrosettum Ajay Singh & Upreti				+				+		+				+				+		+		+			+															+
Gyalolechia flavovirescens (Wulfen) Søchting, Frödén & Arup				+	+		+	+			+	+	+				+	+			+	+																		
Heterodermia **sp.**																						+																	+	
Heterodermia diademata (Taylor) D.D. Awasthi		+	+																+										+	+										
H. himalayana Y. Joshi, K. Chandra & M. Tripathi													+												+					+										
H. japonica (M. Satô) Swinscow & Krog																														+										
H. pseudospeciosa (Kurok.) W.L. Culb.			+			+																											+							
H.speciosa (Wulf.) Trevis.															+						+		+	+			+			+	+									

Lichen species	1	2	3	4	5	6	7	8	9	10	11	12	13	14	15	16	17	18	19	20	21	22	23	24	25	26	27	28	29	30	31	32	33	34	35	36	37	38	39	40
Ioplaca pindarensis (Räsänen) Poelt & Hinter.																						+																		
Lecanora **sp.**				+			+						+		+	+					+			+		+	+	+	+		+				+					
Lecanora campestris (Schaer.) Hue						+						+					+					+																		
L.formosula Lumbsch																+															+		+							
L. frutulosa (Dicks.) Ach.																						+		+																
L. fuscococcin-eaNyl.															+				+			+																		
L. leuteomarginata Nayaka & Upreti																						+																		
L. pseudistera Nyl.							+		+		+					+			+				+						+											
L. wilsonii Müll. Arg.																						+																		
Lecidea **sp.**						+										+					+							+												
Lecidella carpathica Körb.						+							+								+	+				+		+	+		+					+				
Lecidella stigmatea (Ach.) Hertel & Leuckert	+			+	+	+									+	+	+		+		+	+	+				+													

Lichen species	1	2	3	4	5	6	7	8	9	10	11	12	13	14	15	16	17	18	19	20	21	22	23	24	25	26	27	28	29	30	31	32	33	34	35	36	37	38	39	40
Lepraria **sp.**	+		+	+	+			+	+	+			+		+				+		+			+		+	+	+	+		+	+					+	+		+
Leptogium **sp.**		+					+													+		+					+													
Montanelia sorediata (Ach.) Divakar, A. Crespo, Wedin & Essl.																						+																		
Mycobilimbia hunana (Zahlbr.) D.D. Awasthi																			+							+														
Myriospora smaragdula (Wahlenb.) Nägeli ex Uloth		+					+								+	+						+		+																+
Parmotrema andinum (Müll. Arg.) Hale																												+												
P. crinitum (Ach.) M. Choisy												+	+														+													
P. grayanum (Hue) Hale									+																															
P. hababianum (Gyeln.) Hale		+																																						
P. mesotropum (Müll. Arg.) Hale													+																											

Lichen species	1	2	3	4	5	6	7	8	9	10	11	12	13	14	15	16	17	18	19	20	21	22	23	24	25	26	27	28	29	30	31	32	33	34	35	36	37	38	39	40
P. praesorediosum (Nyl.) Hale		+				+			+				+			+	+		+						+				+	+	+									
P. reticulatum (Taylor) M. Choisy														+																										
P. tinctorum (Despr. ex Nyl.) Hale																										+	+				+					+				
Pertusaria indica Preeti Srivast. & D.D. Awasthi																			+				+	+		+		+	+		+					+				
P. leucosora Nyl.		+																				+																		
Phaeophyscia ciliata (Hoffm.) Moberg																					+				+		+			+										
P. hispidula (Ach.) Moberg	+			+	+		+	+	+	+	+	+	+	+	+	+		+		+	+	+		+	+	+	+	+		+		+		+					+	
P. orbicularis (Neck.) Moberg																			+									+												
Physcia dubia (Hoffm.) Lettau														+		+																								
Pleopsidium flavum (Bell.) Körb.							+															+																		
Porpidia albocoerulescens (Wulfen) Hertel & Knoph							+															+																		

Lichen species	1	2	3	4	5	6	7	8	9	10	11	12	13	14	15	16	17	18	19	20	21	22	23	24	25	26	27	28	29	30	31	32	33	34	35	36	37	38	39	40
P. macrocarpa (DC.) Hertel & Knoph		+																				+																		
Protoparmeliopsis muralis (Schreber) M. Choisy							+									+						+																	+	
Punctelia borreri (Sm.) Krog																								+																
P. subrudecta (Nyl.) Krog		+																												+										
Pyxine berteriana **var. himalaica** D.D. Awasthi										+				+							+						+				+									
P. cocoes (Sw.) Nyl.	+		+			+				+			+				+			+				+										+	+			+		
P. sorediata (Ach.) Mont.				+	+			+	+	+	+		+			+	+	+	+		+	+	+		+		+	+	+			+						+		+
Rhizoplaca chrysoleuca (Sm.) Zopf																						+																		
Rinodina oxydata (A. Massal.) A. Massal.												+	+			+					+				+				+											

Lichen species	1	2	3	4	5	6	7	8	9	10	11	12	13	14	15	16	17	18	19	20	21	22	23	24	25	26	27	28	29	30	31	32	33	34	35	36	37	38	39	40
Rusavskia elegans (Link) S.Y. Kondr. & Kärnefelt																																							+	
Squamulea subsoluta (Nyl.) Arup, Søchting & Frödén	+		+	+	+	+	+	+	+	+			+	+	+	+			+	+			+	+		+		+	+			+	+	+	+	+		+		
Staurothele fissa (Taylor) Zwackh							+			+									+		+						+	+				+	+	+		+				
Tephromela khatiensis (Räsänen) Lumbsch		+															+																							
Trapelia coarctata (Turner ex Sm.) M. Choisy		+			+		+			+									+																					
Umbilicaria badia Frey							+															+																		
Verrucaria coerulea (Ramond) DC.													+						+	+	+						+	+								+				
Xanthoparmelia antleriformis (Elix) Elix & J. Johnst.																			+																					
X. mexicana (Gyeln.) Hale													+	+												+														

Name of Monuments: 1) Aadi badri, 2) Anushuya Devi, 3) Nanda Devi Nandkeshri, 4) Chandika Devi Simli, 5) Gopinath temple, 6) Kulsari temple, 7) Tungnath, 8) Uma Devi temple, 9) Danari Group of temple, 10) Devalgarh, 11) Laxminarayan temple Sumari, 12) Shankar Math, 13) Shiva temple Paithani, 14) Shivalaya Kukhargaun, 15) Vaishnava temple, 16) Chandika temple Bhanaj, 17) Damyanti temple, 18) Kalimath, 19) Kandara group, 20) Laxmi Narayan temple Bairangana, 21) Laxmi Narayan group, 22) Madmaheshwar, 23) Makku math, 24) Rakeshwari Devi, 25) Shivalaya Gahatara, 26) Shivalaya Mahad, 27) Silla temple, 28) Shivalaya Naratankoti, 29) Tripura sundari, 30) Triyuginarayan temple, 31) Bashukedar group of temple, 32) Vishwanath temple, 33) Nanda devi Bajinga, 34) Raj Rajeshwari Ranihat, 35) Raghunath temple, 36) Shiva temple Bagi, 37) Shivalaya Bagwan, 38) Sun temple Palethi, 39) Kalpkedar, 40) Vishwanath.

Table 1.3. Pearson's Correlation Coefficient of Uttarakhand monuments. H'- Shannon-Wiener Index or alpha diversity, D'- Simpson's Diversity index, CD - Concentration of dominance, PC – Principal Component 1 & 2.

	PC1	PC2	Altitude	Species richness	Family	H'	D'	CD
PC1	1							
PC2	0.000	1						
Altitude	0.609**	0.760**	1					
Species richness	0.912**	0.120	0.571**	1				
Family	0.990**	0.007	0.587**	0.911**	1			
H'	0.966**	-0.230	0.437**	0.813**	0.946**	1		
D'	0.990**	-0.002	0.583**	0.906**	1.000**	0.950**	1	
CD	-0.872**	0.406**	-0.282	-0.673**	-0.829**	-0.961**	-0.836**	1

** Correlation is significant at the 0.01 level (2-tailed).

CHAPTER 7

REVOLUTIONIZING AGRICULTURE: UNLOCKING THE POWER OF CRISPR CAS9

Adhyatm Rawat[1], Niki Nautiyal[2] and Kartik Uniyal[3]

Sardar Bhagwan Singh University, Balawala, Dehradun -248001 (Uttarakhand)

Abstract

The term genome editing refers to the modifications such as insertions, deletions or substitutions in the DNA sequence of the living organism. Gene editing is performed using specialized technologies and the most widely accepted technologies nowadays is CRISPR-Cas9 which stands for Clustered Regularly Interspaced Short Palindromic Repeat and Cas9 Protein. CRISPR-Cas9 technology has its widespread application in the field of medicine, agriculture and basic life sciences. In the field of agriculture it is used to enhance the crop yield, quality, prolonging shelf life and developing tolerance to the various biotic and abiotic stresses. CRISPR-Cas9 is more accepted than other gene editing technologies because it is easier to design and implement with a higher rate of success. It bridges a gap between conventional techniques and the molecular biology researches and has eventually affected the production and sustainability of crops. However some concerns remains but CRISPR-Cas9 can be a promising technology that has a power to revolutionize and can lead to the benefit of growing human population.

Keywords – CRISPR, ZFN, TALENs, Gene/Genome Editing, Nuclease

Introduction

What is Genome Editing

Genome editing, the ability to make very specific changes to the DNA sequence of a living organism, significantly altering its genetic makeup. Gene editing uses enzymes, specifically nucleases, that are engineered to target a specific DNA sequence, where they add nicks to DNA strands that allow the existing DNA to be removed and replacement DNA to be inserted. Gene editing technology makes it possible to add, delete or change genetic material at specific locations in the genome. Genome editing technology changes physical characteristics such as eye color and disease risk. Researchers use different techniques for this. These techniques work like scissors. The first genome editing techniques were developed in the late 20th century. Recently, a new genome editing tool, CRISPR, invented in 2009, has made DNA editing easier than ever. CRISPR is easier, faster, cheaper and more accurate than older genome editing methods. Many genome editing researchers are now using CRISPR.

Genome Editing Tools

Genome editing is the process of changing an organism's genetic code. There are several tools available for gene editing, all of which use enzymes called nucleases that act on DNA. Different editing tools work on the same basic principle: the enzyme is guided to a specific target site in the genome either by a control sequence or by specific DNA-binding domains in the nuclease itself. Once the nuclease recognizes and binds to the target DNA, it can be used for editing in a number of ways.

Restriction Enzymes: The original genome editors

The possibility of gene editing became a reality when restriction enzymes were discovered in the 1970s. Restriction enzymes recognize specific patterns of nucleotide sequences and cut at that location, allowing DNA material to be inserted at that location. Restriction enzymes are not commonly used for new gene editing today because they are limited by their nucleotide patterns detect but are still widely used for molecular cloning. In addition, certain classes of restriction enzymes play a key role in DNA mapping, epigenome mapping, and DNA library construction.

Zinc finger nucleases: increased identification potential

Over time, the need for precision in genome editing has become more apparent. Researchers needed a gene-editing technique that would identify the site to be altered, because off-target effects can be harmful. Zinc finger nucleases (ZFN) are artificial restriction enzymes created by fusing the DNA-binding domain to a zinc finger DNA-binding domain. Zinc finger domains can be engineered to target specific desired DNA sequences and this allows zinc finger nucleases to target unique sequences in complex genomes. ZFNs are more efficient and reliable in creating double-strand breaks (DSBs) (1). ZFN was used as the first genome manipulation method developed using engineered nucleases. One of the main obstacles to using ZFN was that the 3 base pair requirement made the design more complex. Guanine-rich target sites appeared to be more effective than guanine-free sites. Furthermore, because ZFN interaction with DNA is adjustable (i.e. each ZF interacts with DNA independently), editing efficiency was also compromised. That's why researchers had to address these questions if they wanted more efficient genome editing.

Transcription Activator-Like Effector Nucleases (TALENs)

Transcription activator-like effector (TALE) nucleases (TALENs) have as of late risen as a progressive genome altering apparatus in numerous distinctive living beings and cell sorts. The site-specific chromosomal double-strand breaks presented by TALENs essentially increment the effectiveness of genomic modification. TALENs have been broadly utilized for numerous a long time. TALENs were created by uniting the FokI slice domain to the DNA-binding domain of the TALE proteins. These TALEs contain complex duplications of 34 amino acids for the compelling change of a single base pair (2). TALEN innovation was created approximately 15yrs a long time after the portrayal of ZFNs. Comparative to ZFNs, TALENs are chimeric proteins comprising of a modular DNA-binding domain combined to a FokI nuclease that produces a DSB at a particular site. Transcription activator-like effectors (TALENs) are proteins discharged by *Xanthomonas* microscopic organisms to help the disease of plant species. TALEs help contaminations by binding to particular DNA groupings and enacting the expression of host genes.

Base Editing: Single Nucleotide Substitutions

Base altering (BE) speaks to a unused dimension of CRISPR/Cas intervened precise altering to produce single-nucleotide changes in DNA or RNA freely of DSBs and homology-directed repair. It is a modern genome-editing procedure that creates transformations at single-base determination. All four transition changes, C to T, G to A, A to G and T to C, can be delivered in the genome with the accessible CRISPR/Cas9 base editors (3). Base altering frameworks, created by David Liu's lab at the Wide Organized, not at all like conventional CRISPR framework, base editors do not initiate duplex stranded breaks in the genome. Base altering devices holds incredible potential for the treatment of various hereditary maladies, through transitory RNA or changeless DNA base alterations. Cytidine deaminases which initiate C to T substitutions, are actually happening in microscopic organisms whereas Adenine deaminases which actuate A to G substitutions were built from bacterial proteins for base altering. Combining dCas9 to either a cytidine deaminases (CBE) or an adenine deaminases (ABEs) and giving a sgRNA to coordinate it to the target arrangement permits analysts to present modern substitutions in the DNA.

Prime Editing

Prime altering is a 'search-and-replace' genome-editing innovation that presents all base-to-base transformations, as well as little additions and deletions, without the requirement for double-strand breaks (DSBs) or benefactor DNA templates. The nonattendance of DSB profoundly diminishes the chances of unforeseen changes and/or off-target mutant sites. Prime Editing straightforwardly presents exogenous DNA molecules arrangements into the target location utilizing a preliminary altering gRNA (pegRNA) that incorporates the arrangement to be embedded or combined with a catalytically disabled Cas9-nickase and a Designed reverse transcriptase (RT). Prime altering permits an sgRNA and a benefactor layout for repair, prime altering utilizes a single design developed known as a prime altering guide RNA (pegRNA), which is made up of the preliminary binding site(PBS) arrangement and an arrangement containing the wanted edit. After it finds the target, the Cas9 nicklase makes a nick in one strand of the DNA and the reverse transcriptase utilizes the pegRNA as a layout for reverse transcription, joining the comparing nucleotides to the incised DNA end. This effective device combines the DNA filtering and arrange identification capabilities of the CRISPR Cas9 framework with a reverse transcriptase enzyme which utilizes an RNA format to synthesize a modern single strand DNA (ssDNA) arrangement and place it into the DNA.

CRISPR Cas9

Working of CRISPR Cas9

CRISPR or Clustered Regularly Interspaced Short Palindromic repeat and the associated protein Cas9 is one of the most effective and accurate tool for editing the genome of the living cells. There are numerous associated proteins and the guide RNAs that are the necessary factors of the CRISPR Cas9 system. This working of the CRISPR Cas9 takes place in three way Recognition, Cleavage and Repair, before performing these way CRIPSR associated Cas9 proteins are being selected with the help of the Proto spacer-adjacent motif (PAM) sequence. The Cas9 protein introduces nicks in the DNA strand for the process of genome editing. The Cas9 and Cas12a are one of the main enzymes used in this process, Cas9 was the first enzyme used in this trial. The guide RNA attendants the CAS enzyme towards the target point, hence the guide RNA (sgRNA) should be designed impeccably. The guide RNA recognizes the target point and the Cas9 nuclease introduces the double stranded break(DSB) at the point 3 upstream to the proto spacer-adjacent motif sequence(PAM), after the double stranded break the sequence suffer non-homologous end joining(NHEJ) and homologous directed repair(HDR). The non-homologous end joining allows the elimination of the sequence for the study of the gene function whereas, the homologous directed repair not only requires Ribonucleoprotein but also needed a donor template for the repair of the sequence, it's used to exclude out different types of the mutations. The illustration has been mentioned below. This technology is used for applying certain variations, CRISPR Cas9 has come significant tool in molecular biology but it still demands some of the advancements. CRISPR/ Cas9 genome editing fashion is defined as clustered regularly interspaced short palindromic repeat, short, repeating variants of inheritable material and set up in utmost archaea as well as in numerous bacterial species. CRISPR/ Cas9 and its affiliated proteins develop a veritably strong defensive system that safeguards plants against foreign agents like contagions, bacteria, and other rudiments. CRISPR/ Cas9 is frequently used to change the targeted genes within the system (4). CRISPR gene editing fashion is a inheritable engineering fashion in molecular biology by which the genomes of living organisms may be modified by delivering the Cas9 nuclease along with a synthetic guide RNA (sgRNA) into a cell, the cell genome can be cut at a asked locations, allowing genes to be removed or new one to be added. It involves two essential factors a guide RNA to match a asked target gene and Cas9(CRISPR–associated protein 9) an endonuclease which causes a double-stranded DNA break, allowing variations to the genome. The introductory way to use Cas9 involves the actuality of a Protospacer-adjacent motif (PAM) sequence near and directly refocused to a given target. Numerous spacer sequences are necessary for the operation of Cas9 as a target; thus, CRISPR/ Cas9 is quick, effective, cost-effective, and adaptable. Parts of CRISPR/ Cas9, either in DNA or RNA, are administered to plant cells for the CRISPR/ Cas9-intermediated manipulation of genomes to cut plant DNA in a destined series. In this manner, plant cells initiate processes to "patch" the break to maintain genome integrity, and this hints at the generation of multitudinous forms of differences in the directed sequence.

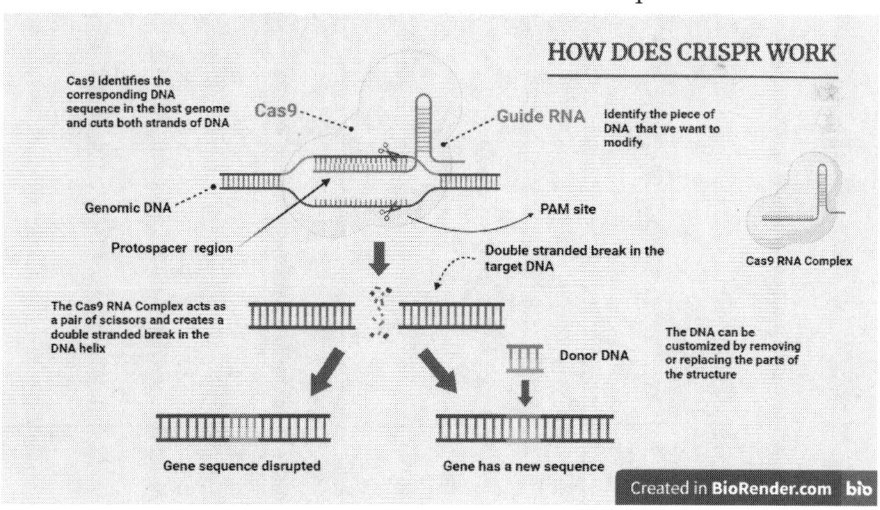

Fig 1 Working of the CRISPR Cas9 based genome editing

When the break is fixed via NHEJ/ homology- directed repair (HDR), small elisions or insertions do that can lead to the mutation of genes (5). In plants, CRISPR/ Cas9- grounded gene- editing consists of multiple steps including the selection of target spots, designing and synthesis of sgRNA, delivery of metamorphosis carrier or Ribonucleoprotein (RNP) in plant cells, metamorphosis, and screening of gene-edited plants. At present, the plant CRISPR/ Cas9 and its derived system have shown genome-editing capability, similar as gene knock-in, knockout, knockdown, and expression activation as well (6).

Genome editing in plants with CRISPR CAS9

Till date, multiple efforts have been successfully employed for targeted gene editing in model as well as in major crop plants via the CRISPR/ Cas9- grounded gene editing toolbox. Numerous factors have been reported that affect the editing capability of the CRISPR/ Cas9 system including targeted DNA, GC contents, Cas9 codons, sgRNA structure, and expression of cas9 and sgRNA. All these factors must be largely optimized to achieve lesser effectiveness of CRISPR/ Cas9 system (7). CRISPR cas9 have opened numerous openings to accelerate plant breeding and to bridge the gap between conventional breeding and the knowledge acquired through plant molecular biology to study and ameliorate complex traits. CRISPR Cas9 intermediated genome editing enables veritably precise and effective targeted variations in utmost crops, therefore large proliferation in speed of crop enhancement compared to conventional breeding. Plant Biotechnology is entering a new period with the preface of genome editing technologies that enables the precise mutations. Plant CRISPR Cas9 are intended for agrobacterium intermediated plant metamorphosis. The product grounded on type IIA CRISPR Cas9 deduced from *Streptococcus pyogenes*.

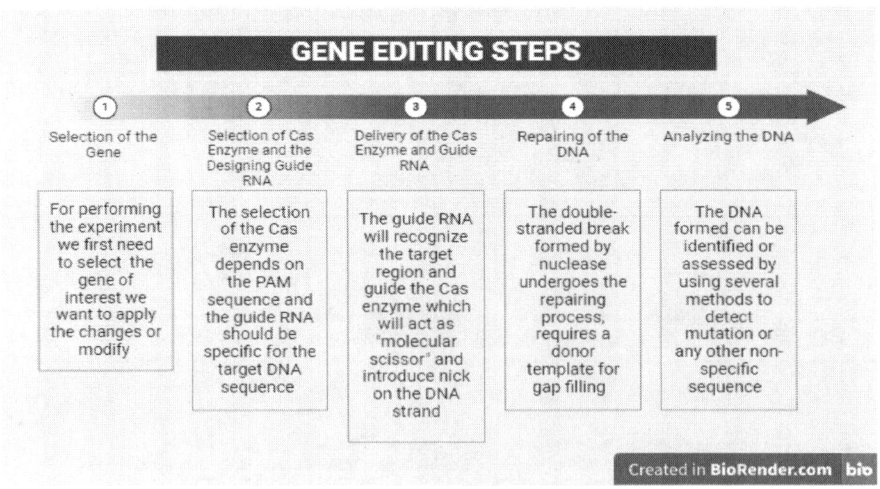

Fig 2 Steps involved in the process of gene editing

The most common way to produce genome-edited plants is to make transgenic plants using Cas9 and sgRNA genes in their genomes. Expression of these genes leads to the accumulation of Cas9 and sgRNA in plants cells and to the editing of target sequences in the genomic DNA. Generally, transgenic plants are developed by *Agrobacterium*-intermediated metamorphosis (8). Plant tissue fractions – explants – are co-cultivated with a suspense of Agrobacterium tume-faciens cells. A tume-faciens can deliver the T- DNA (transfer DNA) into the nucleus of a plant cell, where it's fitted into genomic DNA. Whole T0 generation transgenic plants are regenerated from converted cells in vitro on special culture media. However, their expression will induce the editing of target sequences in the genomes of the T0 plants, if the T- DNA contains the Cas9 and sgRNA genes. In numerous cases, this approach has been relatively effective. Occasionally, indeed T0 plants contain only edited variants of the target sequences (9). In contrast to the transgenic approach, which leads to arbitrary insertions and veritably frequently arbitrary phenotypes, genome editing styles produce defined mutants, therefore getting a potent tool in functional genomics and crop breeding. Genome edited crops have a fresh advantage over transgenic plants since they carry their edited DNA for the asked trait, T-DNA-free plants with homozygous mutations can be fluently named. Although plant metamorphosis and rejuvenescence in vitro using tissue culture ways are extensively used, this approach has some disadvantages. These downsides are substantially defined by the demand for rejuvenescence of transgenic plants from the somatic cells. This process is demanded because it's insolvable to gain invariant metamorphosis of every cell in the whole plant still, to force somatic plant cells to return to active cell division, several conditions must be met, generally taking treatment with exogenous growth controllers (8). The design for effective metamorphosis and rejuvenescence procedures is laborious and time-consuming and should be optimized not only for different plant species but also for different ecotypes or cultivars of the same species. In addition, the rejuvenescence process promotes inheritable and epigenetic changes in regenerated plant, called soma clonal variation (10, 11, and 12)

Applications Of CRISPR CAS 9

Crop Yield and Quality Improvements via CRISPR/Cas9

Arising attestations are showing rapid-fire expansion of CRISPR/Cas9 in a variety of field including agriculture. This section updates and highlights the operation of CRISPR/Cas9 technology in enhancement of tropical crops. Genome editing at target spots by specific nucleases has shown great pledge for enhancement of crop plant to meet the growing global demands of food. This technology offers an occasion for biotechnologists to develop a sustainable and productive agricultural system in terms of perfecting yield, abiotic stress bearance, enhancing resistance to diseases and pests, and modification of plants for product quality. In the history, crop plants have been modified by conventional and modified plant breeding styles. These ways are falling out of favor due to the demand of longer time, problems of comity among the species, and the ever dwindling inheritable variation of plants (13). Crop quality has played a vital part in determining the market value of crops. In general, crop quality is determined by external and internal traits. The external quality attributes include physical and aesthetic characteristics, similar as size, color, texture, and scent. In discrepancy, the internal quality factors include nutrients (like protein, starch, lipids etc.) and bioactive composites (similar as carotenoids, lycopene, -amino butyric acid, flavonoid and so on). CRISPR/Cas9- intermediated crop quality enhancement concentrated on the physical appearance, comestible quality, fruit texture and nutritive value. In this regard, crop enhancement requires genome editing with CRISPR/Cas9 technology to modify plant traits or add new characteristics within a short period of time (14). Presently, CRISPR/Cas9 is also being used to produce point specific mutagenesis or targeted transcriptional regulation in several tropical crops. Meanwhile, CRISPR/Cas9 technology have been used in numerous crop plants cultivated in the tropical regions. The trends of operation of the CRISPR/as9 technology for enhancement of crop plans in tropical climates within the last 5 times are phenomenal, and it indicates a rapid-fire large-scale operation of it in addressing arising challenges in product of crops in tropical climates. In addition to increase yield, CRISPR/Cas9 has also been used to ameliorate quality of crops cultivated in tropical climates. Carotenoids play an important part in numerous physiological processes in plants and the phytoene desaturase genes render important enzymes in the carotenoid biosynthesis pathway. Phytoene desaturase genes, RAS-PDS1 and RAS-PDS2 were lately shifted by the operation of CRISPR/Cas9 with a 59% success rate in bananas (15). This finding indicates that CRISPR/Cas9 can be used for the modification of quality of bananas, and a farther enhancement may just be on the horizon. Cassava is one of the most studied tropical crops and is an essential source of energy in tropical regions. Using CRISPR/Cas9 technology, MePDS mutants were generated in cassava, which displayed albino or partial albino at cotyledonary-stage physical embryos (16). Therefore, CRISPR/Cas9 could be applied for the targeted enhancement of this important tropical food crop.

Improving the Crop Physical Appearance

Modification of Shape and Size

CRISPR/Cas9 technology has been operated to optimize the shape and size of the crops consisting to consumer preferences. Several genes quantitative attribute loci (QTLs) accountable for crop appearance quality have been proposed. The most knowledge on fruit shape and size regulation was disclosed in rice and tomato. GS3 (GRAIN SIZE 3), the first QTL linked in controlling grain length, has been successfully knocked out in five japonica rice breeds. The length of the T1 lines has been accelerated compared to wild type in all the different inherited background (17, 18). Shape of the grain not only affects the quality but also the weight of the grain(GW), for illustration, rice GW has been increased by disturbance of multiple grain weight negative regulators, GW2, GW5, and GW6 (19).

Modification in Color

Plant pigments, which are made up of carotenoids, anthocyanins, and polyphenols, are what give plants their color. The color of the fruit, leaves, and flower buds, in particular, influences the consumer's decision when it comes to plant edible parts. For example, Asian customers prioritize pink tomatoes, but Europeans and Americans prefer red tomatoes (20). Research has shown that the lack of flavonoid pigments in the peel was the cause of the pink phenotype. Therefore, using CRISPR/Cas9 to delete genes involved in the pigment manufacturing pathway, it is possible to manipulate the color of fruits. As a transcription factor involved in the flavonoid biosynthesis pathway, MYB12 influences the accumulation of flavonoids and controls the phenotypic of pink skin. It has been successful to create pink-fruited tomatoes by eliminating SlMYB12 (21). Additionally, scientists produced purple and yellow tomatoes.

Prolonging Shelf Life

Another important factor in the commercial cultivation of crops is fruit texture. Breeders constantly strive to alter textural features for increased shelf life. The technology known as CRISPR/Cas9 has the potential to significantly increase the shelf life of tomatoes and bananas. Numerous naturally occurring mutant genes, including Nr, alc, rin, nor, and Cnr, have the ability to increase shelf life. According to one study, alc mutation preserved the fruit's color and scent in addition to extending shelf life (22). Tomato ALC gene mutations have been created via HDR-mediated gene replacement, and the intended alc homozygous mutants in the T1 generation showed good storage performance (23) Similar to this, fruit phenotypes with extended shelf lives were seen in SlPL gene knockout mutations based on CRISPR/Cas9

Table 1 Some Uses of CRISPR/Cas9 for quality and yield enhancement in crops

Gene	Crops	Trait	Technique	Reference
CLE	Maize	Grain yield	CRISPR/Cas9	(68)
SLIAA9	Tomato	Seedless Fruit	CRISPR/Cas9	(69)
RVE7	Chickpea	Grain quality	CRISPR/Cas9	(70)
OsTB1	Rice	High grain yield	CRIPSR/Cas9	(71)
IncRNA1459	Tomato	Prolonging shelf life	CRISPR/Cas9	(72)

Development of herbicide resistance varieties using CRISPR Cas9

Traditional plant breeding methods have been extensively employed in the past to enhance a variety of crops. To generate commercial varieties, however, involves a labor-intensive process that takes time to move from screening genotypes and phenotypes to crosses. Afterwards, by transferring a gene or genes into elite kinds, genetic modification (GM) technology has been employed to create value-added crops with beneficial features (24). Weeds pose a serious danger to agricultural output because they compete with crops for resources including light, water, space, and nutrients (25). Consequently, the main technique for controlling weeds in agronomic crops is the extensive use of herbicides. In order to give farmers affordable tools for managing weeds, great efforts have been made to generate crop types with herbicide resistance or tolerance (26). The definitions of herbicide resistance and herbicide tolerance. Herbicide resistance is the inherited capacity of a plant to live and procreate after exposure to an amount of herbicide that would typically be fatal to the wild type, according to the Weed Science Society of America (WSSA). Plants can develop resistance either spontaneously or by the use of methods like mutagenesis, tissue culture, genetic engineering, or selection of variations. Herbicide tolerance, on the other hand, is the natural capacity of a species to endure and procreate following herbicide treatment. This indicates that no choice was made. The need for HR crops rises along with the use of herbicides in agriculture. Recent developments in plant genome editing technology have created a new avenue for significantly accelerating the development of HR crops in this regard. Adenine deaminase or cytidine deaminase combined with a catalytically impaired CRISPR/Cas9 mutant (i.e., Cas9 nickase, called nCas9) results in A-to-G or C-to-T substitution without the introduction of a DSB, respectively (27, 28). This creates programmable base-editing methods. These base editors facilitate deamination on the non-complementary strand, which has been effectively employed in a number of crops, by guiding the nCas9-gRNA complex to the target locus. As previously mentioned, the majority of herbicide-insensitive enzymes, including 2mEPSPS (T102I/P106S) and SuRB/ALS (P196A/W573L), are produced by amino acid substitutions brought about by base alterations. As a result, a precise and effective technique for developing herbicide-resistant crops is to use base editors. To target ALS in developing HR plants, Cas9-mediated gene editing has been performed to maize and soybeans (29, 30). Sulfonylurea herbicide resistance in maize is conferred by a single unique amino acid change in the acetolactate synthase (ALS) gene. A Cas9 gene and sgRNA gene that target the ALS2 gene near the crucial amino acid were thrown at maize embryos. In order to verify that herbicide-resistant plants were produced through template repair, DNA repair templates in the form of either 127-mer oligonucleotides or a 794 bp DNA fragment were also co-introduced. Each template encoded numerous extra alterations in addition to the proper alteration at the crucial amino acid. Two to four repair events were seen for every thousand processed embryos when callus was cultured on conditions containing the sulfonylurea chlorosulfuron. T1 plants were proven to exhibit resistance to chlorosulfuron during tissue culture. A very similar experiment was conducted in soybeans using a 1 kb repair sequence of the ALS1 gene, which codes for resistance to chlorosulfuron. Once again, ALS1 was damaged using Cas9 and a sgRNA, and an error-free sequence exchange event was restored (29). Additionally, HDR in the rice ALS gene has been obtained (31). These investigations show how effective this strategy may be when a repair template's particular sequence modification is known to provide a desirable phenotype.

Table 2 some of the herbicide-resistant plants generated by CRISPR/Cas9-mediated gene editing.

Gene	Mutation Site	Herbicide Resistance Plant	Repair Pathway	Reference
ALS	Ala96	Rice	CBE	(73, 74)
ALS	Pro165	Maize	HDR	(75)
ALS	Pro178	Soybean	HDR	(76)
ALS	Pro165	Maize	CBE	(77)
ALS	Pro174	Wheat	CBE	(78, 79)

CRISPR/Cas9 System for Plant Disease Resistance

The main causes of biotic stresses and decreased agricultural output are viruses, bacteria, fungus, nematodes, and insects. Furthermore, the ongoing emergence of multiple novel strains of deadly pests makes the fight against these infections extremely difficult (32). Microorganisms and plants coexist and evolve in the natural world. The relationship between bacteria and plants is essential to their existence, and it can be classified as either useful or harmful depending on how it affects the host, particularly plants (33). Several chemicals have been applied in the past as fertilizers, pesticides, or insecticides to reduce yield loss, enhance plant health, and help plants become resistant to diseases. These chemical products endanger not only the host and other creatures but also contaminate soil and water (34, 35). However because of these compounds' detrimental effects, researchers are currently using another method. Plant composition can be changed by using novel genomic techniques like genetic engineering (GE) and new breeding techniques (NBT), which can also increase a plant's resilience to microbial infections (36). The application of gene editing by biotechnologists results in a more stable product than traditional breeding methods. The ability to exchange genes between species is another benefit of genetic engineering, which makes it a useful tool for creating disease-resistant plants to fulfill the world's growing food need. More recently, techniques based on CRISPR-Cas9 (clustered regularly interspaced short palindromic repeats)-Cas9 (CRISPR-associated protein) have revolutionized agricultural science by demonstrating the ability to modify plant species' genomes and opening up new avenues for research (37). When compared to other GE methods like transcription activator-like effector nuclease (TALEN) and zinc-finger nucleases (ZFNs), the CRISPR-Cas9 approach is more affordable and user-friendly, making it a popular technology (38). Gene editing (GE) utilizing CRISPR/Cas9 can be used directly to sabotage "S-genes," or genes that cause illness, and create crops that are immune to it. For instance, the CRISPR/Cas9 technology was utilized to create targeted knockout plants for the ethylene-responsive gene OsERF922, which demonstrated less blast lesions and enhanced resistance against *Magnaporthe oryzae* -caused rice blast (39). *Citrus sinensis's* resistance to *Xanthomonas citri* is enhanced by editing effector binding elements (EBEs) in the CsLOB1 gene promoter region using the CRISPR/Cas9 system, according to research by Peng and colleagues (40). Shan et al. employed the CRISPR/Cas9 method in wheat protoplasts to modify the TaMLO gene and create wheat lines resistant to *Blumeria graminis* f. sp. Tritici-caused powdery mildew (41). According to estimates, pathogenic viruses are the cause of over half of plant illnesses, which severely reduce crop yields worldwide (42). Consequently, CRISPR/Cas9-mediated GE is a superb tool for enhancing genetic makeup and gives them the ability to fight off different infections. Here are several examples of CRISPR Cas9-developed disease-resistant crops.

Table 3 some of the disease-resistant crops developed via CRISPR/Cas9

Crop	Target gene	Pathogen	Gene Function	Trait Improvement	Editing Result	Repair Pathway	Delivery Technique	Reference
Oryza sativa	eIF4G	*Rice tungro spherical virus*	Translation initiation factor	Resistance against *Rice tungro spherical virus*	Knock-out	NHEJ	*Agrobacterium*-mediated transformation	(80)
Vitis vinifera	VvWRKY52	*Botrytis cinera*	Transcription factor	Increased resistance against *Botrytis cinera*	Knock-out	NHEJ	*Agrobacterium* mediated transformation	(81)
Gossypium hirsutum	Gh14-3-3d	*Verticillium dahlia*	Negative-regulator disease resistance	Resistance to cotton verticillium wilt	Knock-out	NHEJ	*Agrobacterium* mediated transformation	(82)
Solanum lycopersicum	SIJAZ2	*Pseudomonas syringae*	Co-receptor of coronatine	Bacterial speck resistant	Knock-out	NHEJ	*Agrobacterium* mediated transformation	(83)
Solanum lycopersicum	CP and Rep sequences	Tomato yellow leaf curl virus	Negative regulator of viral resistance	Improved resistance against Tomato yellow leaf curl virus	Knock-out	NHEJ	Agrobacterium-mediated transformation	(84)

Development of Climate Smart Crops

Cas9 has been used to address a number of abiotic stressors in important crops, including rice, wheat, maize, cotton, and potatoes. By creating crops that are resistant to abiotic stressors or climate change, plant breeding has updated plant breeding. Any sequence can be changed, and any character can be introduced into crops using CRISPR/Cas9. Numerous genes linked to abiotic stress tolerance were found by molecular breeders, who then incorporated them into crops (43). Using CRISPR/Cas9, two genes—TaDREB3 and TaDREB2—linked to abiotic stress resistance in wheat protoplasts have been investigated. Slmapk3 protein gene mutants increased defense response to drought stress in tomatoes (44). In almost 70% of the transfected wheat protoplasts, mutant gene expression was detected using the T7 endonuclease test. When compared to plants of the wild type, all mutant plants demonstrated drought tolerance (45). The CRISPR/Cas9 method has been used to modify three rice genes: betaine aldehyde dehydrogenase (OsBADH2), phytoene desaturase (OsPDS), and mitogen-activated protein kinase (OsMPK2). For transformation of CRISPR/Cas9 machinery, particle bombardment and protoplast transformation methods were applied and found that these genes are responsible for regulating several abiotic stresses (46). The function of the rice OsAnn3 gene, which was altered to withstand cold stress, was investigated in plants with altered genomes (47). In a similar vein, the rice stress tolerance mechanism was investigated by mutating the gene SAPK2. The findings showed that saline and drought stress conditions increased the expression level of SAPK2 (48). Targeting the ARGOS8 gene to produce novel variations has led to the development of drought tolerance, and this is incredibly significant for maize (49). Likewise, the Cas-based mutation of two genes, Drb2a and Drb2b, was investigated; these genes regulated soybean salt and drought tolerance. Numerous genes regulate a wide range of significant properties, including crop productivity and stress tolerance. Numerous investigations have been conducted to identify the quantitative trait loci (QTLs) governing significant attributes in crop development initiatives. In order to create variants that perform better. On the other hand, CRISPR/Cas technology can be an intriguing method for producing and analyzing targeted mutagenesis). Therefore, the research mentioned above demonstrated the enormous potential of CRISPR/Cas9-based GE for the generation of crops resistant to climate change (50).

Table 4 some of the climate smart crops developed using CRIPSR Cas9

Crop	Target Gene	Trait Study	Editing Result	Repair Mechanism	Delivery Technique	Reference
Oryza sativa	OsNAC041	Salinity tolerance	Knockout	NHEJ	*Agrobacterium*-mediated transformation	(85)
Oryza sativa	OsOTS1	Salinity tolerance	Knockout	NHEJ	*Agrobacterium*-mediated transformation	(86)

Crop	Target Gene	Trait Study	Editing Result	Repair Mechanism	Delivery Technique	Reference
Solanum lycopersicum	SlNPR1	Drought tolerance	Knockout	NHEJ	*Agrobacterium*-mediated transformation	(87)
Glycine max	Drb2a, Drb2b	Drought and salt tolerance	Knockout	NHEJ	*Agrobacterium-rhizogenes*	(88)
Solanum lycopersicum	SlCBF1	Cold tolerance	Knockout	NHEJ	*Agrobacterium*-mediated transformation	(89)

Pros and Cons of CRISPR CAS9

Researchers have access to numerous gene editing methods, including CRISPR-Cas9. Although it can be a very labor-intensive method of altering genetic material, it has a number of benefits and drawbacks that should be considered when planning an experiment, just like any other gene editing application.

Pros

- Quick, adaptable design — CAS nuclease and a guide RNA are the two main components needed for CRISPR-Cas9 gene editing. Researchers sometimes choose to employ a sgRNA and other times the 2-part crRNA-tracrRNA strategy. The resultant ribonucleoprotein (RNP) complex is formed when endonuclease and guide RNA bind to one another. The target sequence can be bound by the guide RNA and editing can take place within this complex.
- Multiplexed gene editing: When sgRNA are deployed and targeted to distinct genomic loci, CRISPR-Cas9 can edit multiple genes at once (51). Because of this, CRISPR-Cas9 gene editing is a compelling and effective method for modifying various genomic regions.
- Cost-effectiveness - Compared to alternative gene editing techniques like zinc finger nucleases (ZFNs) and transcription activator-like effector nucleases (TALENs), which need protein engineering, CRISPR-Cas9 can be incredibly cost-effective due to the previously mentioned factors. Researchers may begin editing rapidly and make changes with ease thanks to the simplicity of CRISPR-Cas9, which mostly relies on two molecules to induce alterations, and the flexibility afforded in constructing sgRNA to bind to any/numerous target(s).

Cons

- Delivery constraints: In order to cause the required modifications, CRISPR-Cas9 needs to be properly delivered into cells. Delivery of the Cas9/gRNA complex into a sufficient number of cells can be difficult, as with many other gene editing applications, because each component needs to be supplied into cells at the exact concentration and at the proper time of the cell cycle. The distribution of CRISPR reagents can be achieved using lipofection, electroporation, microinjection, nanoparticles, and viral plasmids.

- Efficiency restrictions - One further significant restriction to take into account when employing CRISPR-Cas9 gene editing is the possibility that the gene editing activity itself may not take place when the Cas9/gRNA combination is absorbed by cells. This is particularly true when the objective is to introduce or knock out a gene, as these operations depend on homology directed repair (HDR). Nonetheless, HDR efficiency rates have improved significantly thanks to CRISPR research. Check out the DECODED article, improving efficiency of homology-directed repair (HDR), to learn more about enhancing the effectiveness of CRISPR-Cas9 gene editing.
- Off-target effects: Unwanted changes might arise when the Cas9 nuclease modifies a region of the genome that is not intended for editing. For CRISPR-Cas9 research, off-target consequences are a big worry since they can be hard to foresee. There has been work done to enhance Cas9 nucleases to lessen off-target effects and to enhance in silico off-target prediction techniques (52, 53).

Ethical considerations for the CRISPR-Cas9 system

Biotechnology has advanced quickly during the past 20 years, allowing us to create scientific instruments that were previously only seen in science fiction. Because genome editing, and the CRISPR-Cas9 system in particular, allow us to structurally modify the genetic background of live creatures, it presents one of the best examples of this kind of biotechnology advancement. Although the CRISPR-Cas9 system enables efficient manipulation of genomic DNA.

Numerous moral, ethical, political, and scientific issues could result in a moratorium that would prevent further research in this area. Consequently, it is crucial to recognize the issues and resolve them (54, 55). The maxim that the reward should always outweigh the risk is a crucial ethical consideration in research. Given that the hazards have the potential to harm both the environment and living things, more focus should be paid to them. Because CRISPR Cas9 can result in unintended and perhaps harmful alterations, using it carries some risk. Numerous studies have documented the effectiveness of the CRISPR-Cas9 system in creating transgenic plants that are resistant to disease. CRISPR-edited tomato plants were recently reported (56, 57). This was achieved by regularly expressing cas9, which targets the viral coat protein and replicase genes. Up to the third generation, these transgenic plants remained stable and exhibited an exceptionally low virus burden. Nevertheless, the use of such continuously expressing nuclease crops falls under the category of genetically modified organisms (GMOs), which are subject to regulatory requirements specific to GMOs, are not permitted in many countries, and have a high risk of off-target mutation. Therefore, it is preferable to create transgene-free processes that can simultaneously use many pathogen effectors and resistant genes and produce foreign crops that are exempt from GMO laws. Therefore, it is anticipated that we will be able to overcome these constraints and that RNP, a DNA-free CRISPR system, will prove useful to us in the future. The adoption of CRISPR-edited crops is subject to varying legislative frameworks, cultural perspectives, and governance across the globe (58, 59 and 60). The world is still battling over whether or not CRISPR crops need to be controlled (as GMOs). Because GMO licensing and testing are difficult processes that cost millions of dollars, the USA and Canada recently exempted CRISPR from being classified as a GMO under regulatory systems and permitted its production and sale without a GMO label. Crops modified by CRISPR-Cas9 that are now free from GMO laws; a white button mushroom that resists browning by having the polyphenol oxidase (PPO) gene deleted; 2. Waxy corn, which lacks the wx1 gene that gives it its waxy look; 3. Green bristle grass that blooms later; 4. Camelina that has a higher oil content; and 5. Soybeans could withstand the effects of drought. Plant kinds that are (61, 62) resistant to biotic and environmental stressors and can give civilization an adequate supply of food. Other crops with more desired qualities, like increased yield or therapeutic capabilities (in the form of an edible vaccine), can be developed in addition to the CRISPR modified crops that are already established and waiting to pass regulatory hurdles. As a potent yet adaptable gene editing and control instrument, CRISPR-Cas9 technology is ultimately propelling every field of study and emerging as the foundation for sustainable agriculture. In the years to come, we think its wide-ranging applications in the study of plant and microbial biology will greatly expand our understanding of both basic biology and disease resistance.

Future Prospects

Speaking of which, due to its widespread acceptance and uses, genome editing has become a well-known process. Because the world's population is growing rapidly and there is a greater need for food, this can be used to increase the generation of food in the years to come. However, abrupt changes in the environment and global warming have a major impact on food production. Therefore, it is necessary to reduce these effects in order to improve the agrarian generation. Both the atomic and conventional breeding techniques are used to increase the fruitfulness of the following generation; nevertheless, discussing both of these techniques or strategies takes time, as conventional breeding or the ordinary strategy calls for work. In order to ensure food security in the future, breeders are now focusing their attention on CRISPR for the development of desired traits or features through genome control. Consequently, CRISPR has proven to be a very labor-saving and time-efficient method. On the other hand, there have also been reports of some unanticipated consequences of utilizing CRISPR. One such consequence is off-target mutation, which arises when there is a base mismatch between the sgRNA and the non-targeted sequence. The beginning of an unidentified mutation while a mistake is being repaired is obviously not something that can be ignored. (63, 64 and 65). Before proceeding with its therapeutic uses, off-targeted mutations must be detected and improved upon since they may have major repercussions. Scientists have already identified a few CRISPR versions that have minimal unintended consequences and maintain the enzyme's effectiveness. Methods like GUIDE-Seq and whole-genome sequencing have been developed to identify the existence of off-target effects (66, 67). In summary, the biggest obstacle to the widespread use of CRISPR-based genome editing is off-target consequences. As a result, it's possible that by using Cas9 to change sgRNA, unknown mutations will be prevented from occurring.

Conclusion

In conclusion, the science of plant biotechnology has a strong and unconventional tool called genome editing that makes it possible to accurately tweak or alter the genetic makeup of living things. The methods for genome editing have developed further, from the use of restriction enzymes to the sophisticated CRISPR/Cas9 system, enabling increased precision, efficiency, and a wider range of applications. Researchers can now achieve single-nucleotide alterations, insertions, deletions, and base-to-base conversions with previously unheard-of accuracy thanks to the development of zinc finger nucleases (ZFNs), transcription activator-like effector nucleases (TALENs), base editing, and prime editing. Furthermore, the potential for accurate and effective targeted alterations in crops has greatly altered agriculture with the use of CRISPR/Cas9 in plant genome editing. By bridging the gap between traditional breeding and molecular biology research, this method has sped up crop improvement and eventually affected food production and sustainability. It is crucial to think about the ethical and legal ramifications of using genome editing technologies as they develop. Furthermore, there is hope that this field's continued research may help treat genetic illnesses, advance biotechnology, and advance other scientific disciplines. This ever-changing field of genome editing instruments highlights the tremendous advancements in genetic engineering, their significant influence on biological research, and its potential to make significant improvements to human health and welfare. Enhancing tropical crops through the use of CRISPR/Cas9 technology is a novel way to meet the world's food production needs. There are a lot of interesting opportunities for sustainable agriculture because of its potential to fulfill customer demands, resist diseases, improve stress tolerance, and raise crop output. Moreover, the advancement of climate-smart and herbicide-resistant cultivars highlights its importance in influencing the trajectory of global food security.

It is becoming increasingly evident that CRISPR/Cas9 holds the key to transforming agricultural methods and providing a sustainable path towards resilient and high-quality food systems to fulfill the demands of an expanding population as we observe its amazing growth and potential. The quick and adaptable design of CRISPR-Cas9 genome editing, its cost-effectiveness, and its capacity to edit numerous genes at once are only a few of its benefits. But there are additional restrictions to take into account, like the difficulties in introducing the Cas9/gRNA complex into cells, the limitations on the efficacy of gene editing activity, and the possibility of off-target effects. There is a lot of promise for genetic modification with the CRISPR-Cas9 system, especially when creating transgenic plants that are resistant to disease. To guarantee that the advantages outweigh the hazards, ethical issues, legal restrictions, and possible hazards must be properly addressed. This will pave the way for future developments in science and sustainable agriculture.

Refrences

1. Durai, S.; Mani, M.; Kandavelou, K.; Wu, J.; Porteus, M.H.; Chandrasegaran, S. Zinc finger nucleases: Custom-designed molecular scissors for genome engineering of plant and mammalian cells. Nucleic Acids Res. 2005, 33, 5978–5990. [CrossRef]
2. Zhang, H.-X.; Zhang, Y.; Yin, H. Genome editing with mRNA encoding ZFN, TALEN, and Cas9. Mol. Ther. 2019, 27, 735–746. [Cross Ref]
3. Molla, K.A. and Yang, Y. (2019) CRISPR/Cas-mediated base editing: technical considerations and practical applications. Trends Biotechnol.. https://doi.org/ 10.1016/j.tibtech.2019.03.008
4. Liu, Q.; Yang, F.; Zhang, J.; Liu, H.; Rahman, S.; Islam, S.; Ma, W.; She, M. Application of CRISPR/Cas9 in Crop quality improvement. Int. J. Mol. Sci. 2021, 22, 4206. [CrossRef] [PubMed]
5. Symington, L.S.; Gautier, J. Double-strand break end resection and repair pathway choice. Ann. Rev. Genet. 2011, 45, 247–271. [CrossRef]
6. Ricroch, A.; Clairand, P.; Harwood, W. Use of CRISPR systems in plant genome editing: Toward new opportunities in agriculture. Emerg. Top. Life Sci. 2017, 1, 169–182. [PubMed]
7. Ma, X.; Zhang, Q.; Zhu, Q.; Liu, W.; Chen, Y.; Qiu, R.; Wang, B.; Yang, Z.; Li, H.; Lin, Y.; et al. A Robust CRISPR/Cas9 System for Convenient, High-Efficiency Multiplex Genome Editing in Monocot and Dicot Plants. Mol. Plant 2015, 8, 1274–1284. [CrossRef]
8. Anami S, Njuguna E, Coussens G, et al. Higher plant transformation: principles and molecular tools. Int J Dev Biol. 2013;57(6–8):483–494.
9. Luo M, Gilbert B, Ayliffe M. Applications of CRISPR/ Cas9 technology for targeted mutagenesis, gene replacement and stacking of genes in higher plants. Plant Cell Rep. 2016;35(7):1439–1450
10. Bent AF. Arabidopsis in planta transformation. Uses, mechanisms, and prospects for transformation of other species. Plant Physiol. 2000;124(4):1540–1547.
11. Krishna H, Alizadeh M, Singh D, et al. Somaclonal variations and their applications in horticultural crops improvement. 3 Biotech. 2016;6(1):54.

12. Wang QM, Wang L. An evolutionary view of plant tissue culture: somaclonal variation and selection. Plant Cell Rep. 2012;31(9):1535–1547

13. Chen, K., and Gao, C. (2014). Targeted genome modification technologies and their applications in crop improvements. Plant Cell Rep. 33, 575–583. doi: 10.1007/s00299-013-1539-6

14. Zhang, L., and Zhou, Q. (2014). CRISPR/Cas technology: a revolutionary approach for genome engineering. Sci. China Life Sci. 57, 639–640. doi: 10.1007/s11427-014-4670-x

15. Kaur, N., Alok, A., Shivani, Kaur, N., Pandey, P., Awasthi, P., et al. (2018). CRISPR/Cas9-mediated efficient editing in phytoene desaturase (PDS) demonstrates precise manipulation in banana cv. rasthali genome. Funct. Integr. Genomics 18, 89–99. doi: 10.1007/s10142-017-0 577-5

16. Odipio, J., Alieai, T., Ingelbrecht, I., Nusinow, D. A., Bart, R., and Taylor, N. J. (2017). Efficient CRISPR/Cas9 genome editing of phyteone desaturase in cassava. Front. Plant Sci. 8:1780. doi: 10.3389/fpls.2017.01780

17. Shen, L.; Wang, C.; Fu, Y.; Wang, J.; Liu, Q.; Zhang, X.; Yan, C.; Qian, Q.; Wang, K. QTL editing confers opposing yield performance in different rice varieties. J. Integr. Plant Biol. 2018, 60, 89–93. [CrossRef]

18. Yuyu, C.; Aike, Z.; Pao, X.; Xiaoxia, W.; Yongrun, C.; Beifang, W.; Yue, Z.; Liaqat, S.; Shihua, C.; Liyong, C.; et al. Effects of GS3 and GL3.1 for grain size editing by CRISPR/Cas9 in Rice. Rice Sci. 2020, 27, 405–413. [CrossRef]

19. Xu, R.; Yang, Y.; Qin, R.; Li, H.; Qiu, C.; Li, L.; Wei, P.; Yang, J. Rapid improvement of grain weight via highly efficient CRISPR/Cas9-mediated multiplex genome editing in rice. J. Genet. Genom. 2016, 43, 529. [CrossRef] [PubMed]

20. Velasco, C.; Wan, X.; Salgado-Montejo, A.; Woods, A.; Oñate, G.A.; Mu, B.; Spence, C. The context of colour–flavour associations in crisps packaging: A cross-cultural study comparing Chinese, Colombian and British consumers. Food Qual. Prefer. 2014, 38, 49–57

21. Yang, T.; Deng, L.; Zhao, W.; Zhang, R.; Jiang, H.; Ye, Z.; Li, C.-B.; Li, C. Rapid breeding of pink-fruited tomato hybrids using the CRISPR/Cas9 system. J. Genet. Genom. 2019, 46, 505. [CrossRef]

22. Wang, R.; Lammers, M.; Tikunov, Y.; Bovy, A.G.; Angenent, G.C.; de Maagd, R.A. The rin, nor and Cnr spontaneous mutations inhibit tomato fruit ripening in additive and epistatic manners. Plant Sci. 2020, 294, 110436. [CrossRef] [PubMed]

23. Casals, J.; Cebolla-Cornejo, J.; Roselló, S.; Beltrán, J.; Casañas, F.; Nuez, F. Long-term postharvest aroma evolution of tomatoes with the alcobaça (alc) mutation. Eur. Food Res. Technol. 2011, 233, 331–342. [CrossRef]

24. Prado JR, Segers G, Voelker T, Carson D, Dobert R, Phillips J, Cook K, Cornejo C, Monken J, Grapes L, Reynolds T, Martino-Catt S (2014) Genetically engineered crops: from idea to product. Annu Rev Plant Biol 65:769–790

25. Delye C, Jasieniuk M, Le Corre V (2013) Deciphering the evolution of herbicide resistance in weeds. Trends Genet 29:649–658

26. Duke SO (2015) Perspectives on transgenic, herbicide-resistant crops in the United States almost 20 years after introduction. Pest Manag Sci 71:652–657

27. Yin K, Gao C, Qiu JL (2017) Progress and prospects in plant genome editing. Nat Plants 3:17107

28. Zhang Y, Malzahn AA, Sretenovic S, Qi Y (2019b) The emerging and uncultivated potential of CRISPR technology in plant science. Nat Plants 5:778–794

29. Li Z, Liu ZB, Xing A, Moon BP, Koellhofer JP, Huang L, Ward RT, Clifton E, Falco SC, Cigan AM (2015) Cas9-guide RNA directed genome editing in soybean. Plant Physiol 169:960–970

30. Svitashev S, Young JK, Schwartz C, Gao H, Falco SC, Cigan AM (2015) Targeted mutagenesis, precise gene editing, and site-specifc gene insertion in maize using Cas9 and guide RNA. Plant Physiol 169:931–945

31. Sun Y, Zhang X, Wu C, He Y, Ma Y, Hou H, Guo X, Du W, Zhao Y, Xia L (2016) Engineering herbicide resistant rice plants through CRISPR/Cas9-mediated homologous recombination of the acetolactate synthase. Mol Plant. doi:10.1016/j.molp.2016.01.0

32. Al-Sadi, A.M.; Al-Moqbali, H.S.; Al-Yahyai, R.A.; Al-Said, F.A. AFLP data suggest a potential role for the low genetic diversity of acid lime (Citrus aurantifolia Swingle) in Oman in the outbreak of witches' broom disease of lime. Euphytica 2012, 188, 285–297. [CrossRef]

33. . Shelake RM, Pramanik D, Kim J-Y. Exploration of plant-microbe interactions for sustainable agriculture in CRISPR era. Microorganisms. 2019;7(8):269.

34. Tyagi S, Mulla SI, Lee K-J, Chae J-C, Shukla P. VOCsmediated hormonal signaling and crosstalk with plant growth promoting microbes. Crit Rev Biotechnol. 2018;38(8):1277–96. doi:10.1080/ 07388551.2018.1472551.

35. Vannier N, Agler M, Hacquard S. Microbiota-mediated disease resistance in plants. PLoS Pathog. 2019;15(6): e1007740–e1007740. doi:10.1371/journal.ppat.1007740.

36. Dong OX, Ronald PC. Genetic engineering for disease resistance in plants: recent progress and future perspectives. Plant Physiol. 2019;180:26–38.

37. Yin K, Qiu JL. Genome editing for plant disease resistance: applications and perspectives. Philos Trans R Soc Lond B Biol Sci. 2019;374(1767):20180322. doi:10.1098/ rstb.2018.0322.

38. Bedell VM, Wang Y, Campbell JM, Poshusta TL, Starker CG, Krug II RG, Tan W, Penheiter SG, Ma AC, Leung AYH, et al. In vivo genome editing using a high-efficiency TALEN system. Nature. 2012;491(7422):114–18. doi:10.1038/nature11537.

39. Wang, F.; Wang, C.; Liu, P.; Lei, C.; Hao, W.; Gao, Y.; Liu, Y.G.; Zhao, K. Enhanced rice blast resistance by CRISPR/ Cas9-Targeted mutagenesis of the ERF transcription factor gene OsERF922. PLoS ONE 2016, 11, e0154027. [CrossRef]

40. Peng, A.; Chen, S.; Lei, T.; Xu, L.; He, Y.; Wu, L.; Yao, L.; Zou, X. Engineering canker-resistant plants through CRISPR/Cas9-targeted editing of the susceptibility gene CsLOB1 promoter in citrus. Plant Biotechnol. J. 2017, 15, 1509–1519. [CrossRef]

41. Zhang, H.; Zhang, J.; Wei, P.; Zhang, B.; Gou, F.; Feng, Z.; Mao, Y.; Yang, L.; Zhang, H.; Xu, N.; et al. The CRISPR/Cas9 system produces specific and homozygous targeted gene editing in rice in one generation. Plant Biotechnol. J. 2014, 12, 797–807. [CrossRef]

42. Shan, Q.; Zhang, Y.; Chen, K.; Zhang, K.; Gao, C. Creation of fragrant rice by targeted knockout of the OsBADH2 gene using TALEN technology. Plant Biotechnol. J. 2015, 13, 791–800. [CrossRef]

43. Zaidi, S.S.A.; Tashkandi, M.; Mansoor, S.; Mahfouz, M.M. Engineering plant immunity: Using crispr/cas9 to generate virus resistance. Front. Plant Sci. 2016, 7, 1673. [CrossRef]

44. Razzaq, M.K.; Aleem, M.; Mansoor, S.; Khan, M.A.; Rauf, S.; Iqbal, S.; Siddique, K.H. Omics and CRISPR-Cas9 approaches for molecular insight, functional gene analysis, and stress tolerance development in crops. Int. J. Mol. Sci. 2021, 22, 1292. [CrossRef]

45. Wang, L.; Chen, L.; Li, R.; Zhao, R.; Yang, M.; Sheng, J.; Shen, L. Reduced drought tolerance by CRISPR/Cas9-mediated SlMAPK3 mutagenesis in tomato plants. J. Agric. Food Chem. 2017, 65, 8674–8682. [CrossRef] [PubMed]

46. Kim, D.; Alptekin, B.; Budak, H. CRISPR/Cas9 genome editing in wheat. Funct. Integr. Genom. 2018, 18, 31–41. [CrossRef] [PubMed]

47. Shen, C.; Que, Z.; Xia, Y.; Tang, N.; Li, D.; He, R.; Cao, M. Knock out of the annexin gene OsAnn3 via CRISPR/Cas9-mediated genome editing decreased cold tolerance in rice. J. Plant Biol. 2017, 60, 539–547. [CrossRef]

48. Lou, D.; Wang, H.; Liang, G.; Yu, D. OsSAPK2 confers abscisic acid sensitivity and tolerance to drought stress in rice. Front. Plant Sci. 2017, 8, 993. [CrossRef]

49. Shi, J.; Gao, H.; Wang, H.; Lafitte, H.R.; Archibald, R.L.; Yang, M.; Hakimi, S.M.; Mo, H.; Habben, J.E. ARGOS 8 variants generated by CRISPR-Cas9 improve maize grain yield under field drought stress conditions. Plant Biotech. J. 2017, 15, 207–216. [CrossRef]

50. Curtin, S.J.; Xiong, Y.; Michno, J.M.; Campbell, B.W.; Stec, A.O.; Cerm ˇ ák, T.; Starker, C.; Voytas, D.F.; Eamens, A.L.; Stupar, R.M. Crispr/cas9 and talen s generate heritable mutations for genes involved in small rna processing of glycine max and medicagotruncatula. Plant Biotech. J. 2018, 16, 1125–1137. [CrossRef] [PubMed]

51. Khan FJ, Yuen G, Luo J. Multiplexed CRISPR/Cas9 gene knockout with simple crRNA:tracrRNA co-transfection. Cell Biosci. 2019;9:41.

52. Guo C, Ma X, Gao F, *et al*. Off-target effects in CRISPR/Cas9 gene editing. *Front Bioeng Biotechnol*. 2023;11:1143157.

53. Vakulskas CA, Dever DP, Rettig GR, *et al*. A high-fidelity Cas9 mutant delivered as a ribonucleoprotein complex enables efficient gene editing in human hematopoietic stem and progenitor cells. Nat Med. 2018;24(8):1216-1224.

54. de Lecuona, I., Casado, M., Marfany, G., Baroni, M.L. and Escarrabill, M. (2017) Focus: genome editing: gene editing in humans: towards a global and inclusive debate for responsible research. *Yale J. Biol. Med.* **90**, 673.

55. Lyon, J. (2017) Bioethics panels open door slightly to germline gene editing. *JAMA*, **318**, 1639–1640.

56. Pyott DE, Sheehan E, Molnar A. Engineering of CRISPR/Cas9-mediated potyvirus resistance in transgene-free Arabidopsis plants. Mol Plant Pathol. 2016;17(8):1276–88. doi:10.1111/mpp.12417.

57. Macovei A, Sevilla NR, Cantos C, Jonson GB, SlametLoedin I, ermák T, Voytas DF, Choi IR, ChadhaMohanty P. Novel alleles of rice eIF4G generated by CRISPR/Cas9-targeted mutagenesis confer resistance to Rice tungro spherical virus. Plant Biotechnol J. 2018;16 (11):1918–27.

58. Globus R, Qimron U. A technological and regulatory outlook on CRISPR crop editing. J Cell Biochem. 2018;119(2):1291–98. doi:10.1002/jcb.26303.

59. Friedrichs S, Takasu Y, Kearns P, Dagallier B, Oshima R, Schofield J, Moreddu C. An overview of regulatory approaches to genome editing in agriculture. Biotechnol Res Innov. 2019;3(2):208–20.

60. Kawall K, Cotter J, Then C. Broadening the GMO risk assessment in the EU for genome editing technologies in agriculture. Environ Sci Eur. 2020;32:1–24.

61. Waltz E. With a free pass, CRISPR-edited plants reach market in record time. Nat Biotechnol. 2018;36(1):6–7. doi:10.1038/nbt0118-6b.

62. Wang F, Wang C, Liu P, Lei C, Hao W, Gao Y, Liu Y-G, Zhao K. Enhanced rice blast resistance by CRISPR/ Cas9-targeted mutagenesis of the ERF transcription factor gene OsERF922. PLoS One. 2016;11(4): e0154027. doi:10.1371/journal.pone.0154027.

63. Xu, C. F. et al. Rational designs of in vivo CRISPR-Cas delivery systems. *Adv. Drug. Deliv. Rev.* **168**, 3–29 (2021).

64. Zhang, S., Shen, J., Li, D. & Cheng, Y. Strategies in the delivery of Cas9 ribonucleoprotein for CRISPR/Cas9 genome editing. *Theranostics* **11**, 614–648 (2021).

65. Fu, Y. et al. High-frequency off-target mutagenesis induced by CRISPR-Cas nucleases in human cells. *Nat. Biotechnol.* **31**, 822–826 (2013).

66. Zischewski, J., Fischer, R. & Bortesi, L. Detection of on-target and off-target mutations generated by CRISPR/Cas9 and other sequence-specific nucleases. *Biotechnol. Adv.* **35**, 95–104 (2017).

67. Wienert, B. et al. Unbiased detection of CRISPR off-targets in vivo using DISCOVER-Seq. *Science* **364**, 286–289 (2019).

68. Liu, L.; Gallagher, J.; Arevalo, E.D.; Chen, R.; Skopelitis, T.; Wu, Q.; Bartlett, M.; Jackson, D. Enhancing grain-yield-related traits by CRISPR–Cas9 promoter editing of maize CLE genes. Nat. Plants 2021, 7, 287–294. [CrossRef]

69. Ueta, R.; Abe, C.; Watanabe, T.; Sugano, S.S.; Ishihara, R.; Ezura, H.; Osakabe, Y.; Osakabe, K. Rapid breeding of parthenocarpic tomato plants using CRISPR/Cas9. Sci. Rep. 2017, 7, 507. [CrossRef]

70. Badhan, S.; Ball, A.S.; Mantri, N. First report of CRISPR/Cas9 mediated DNA-free editing of 4CL and RVE7 genes in chickpea protoplasts. Int. J. Mol. Sci. 2021, 22, 396. [CrossRef]

71. Wang, S.; Yang, Y.; Guo, M.; Zhong, C.; Yan, C.; Sun, S. Targeted mutagenesis of amino acid transporter genes for rice quality improvement using the CRISPR/Cas9 system. Crop J. 2020, 8, 457–464. [CrossRef]

72. Li, R.; Fu, D.; Zhu, B.; Luo, Y.; Zhu, H. CRISPR/Cas9-mediated mutagenesis of lncRNA1459 alters tomato fruit ripening. Plant J. 2018, 94, 513–524. [CrossRef]

73. Shimatani Z., Fujikura U., Ishii H., Matsui Y., Suzuki M., Ueke Y., Taoka K.I., Terada R., Nishida K., Kondo A., et al. Inheritance of co-edited genes by CRISPR-based targeted nucleotide substitutions in rice. *Plant Physiol. Biochchem.* 2018;131:78–83. doi: 10.1016/j.plaphy.2018.04.028. [PubMed] [CrossRef] [Google Scholar] [Ref list]

74. Li C., Zong Y., Wang Y., Jin S., Zhang D., Song Q., Zhang R., Gao C. Expanded base editing in rice and wheat using a Cas9-adenosine deaminase fusion. *Genome Biol.* 2018;19:59. doi: 10.1186/s13059-018-1443-z. [PMC free article] [PubMed] [CrossRef] [Google Scholar] [Ref list]

75. Endo M., Mikami M., Toki S. Biallelic gene targeting in rice. *Plant Physiol.* 2016;170:667–677. doi: 10.1104/pp.15.01663. [PMC free article] [PubMed] [CrossRef] [Google Scholar] [Ref list]

76. Sun Y., Zhang X., Wu C., He Y., Ma Y., Hou H., Guo X., Du W., Zhao Y., Xia L., et al. Engineering herbicide-resistant rice plants through CRISPR/Cas9-mediated homologous recombination of acetolactate synthase. *Mol. Plant.* 2016;9:628–631.

doi: 10.1016/j.molp.2016.01.001. [PubMed] [CrossRef] [Google Scholar] [Ref list]

77. Zhang R., Liu J., Chai Z., Chen S., Bai Y., Zong Y., Chen K., Li J., Jiang L., Gao C., et al. Generation of herbicide tolerance traits and a new selectable marker in wheat using base editing. *Nat. Plants.* 2019;5:480–485. doi: 10.1038/s41477-019-0405-0. [PubMed] [CrossRef] [Google Scholar] [Ref list]

78. Zong Y., Song Q., Li C., Jin S., Zhang D., Wang Y., Qiu J., Gao C. Efficient C-to-T base editing in plants using a fusion of nCas9 and human APOBEC3A. *Nat. Biotechnol.* 2018;36:950–953. doi: 10.1038/nbt.4261. [PubMed] [CrossRef] [Google Scholar] [Ref list]

79. Tian S., Jiang L., Cui X., Zhang J., Guo S., Li M., Zhang H., Ren Y., Gong G., Zong M., et al. Engineering herbicide-resistant watermelon variety through CRISPR/Cas9-mediated base-editing. *Plant Cell. Rep.* 2018;37:1353–1356. doi: 10.1007/s00299-018-2299-0. [PubMed] [CrossRef] [Google Scholar] [Ref list]

80. Macovei, A.; Sevilla, N.R.; Cantos, C.; Jonson, G.B.; Slamet-Loedin, I.; Cerm ˇ ák, T.; Voytas, D.F.; Choi, I.R.; Chadha-Mohanty, P. Novel alleles of rice eIF4G generated by CRISPR/Cas9-targeted mutagenesis confer resistance to Rice tungro spherical virus. Plant Biotechnol. J. 2018, 16, 1918–1927. [CrossRef]

81. Wang, X.; Tu, M.; Wang, D.; Liu, J.; Li, Y.; Li, Z.; Wang, Y.; Wang, X. CRISPR/Cas9-mediated efficient targeted mutagenesis in grape in the first generation. Plant Biotechnol. J. 2018, 16, 844–855. [CrossRef]

82. Zhang, Z.; Ge, X.; Luo, X.; Wang, P.; Fan, Q.; Hu, G.; Xiao, J.; Li, F.; Wu, J. Simultaneous Editing of Two Copies of Gh14-3-3d Confers Enhanced Transgene-Clean Plant Defense Against Verticillium dahliae in Allotetraploid Upland Cotton. Front. Plant Sci. 2018, 9, 842. [CrossRef]

83. Ortigosa, A.; Gimenez-Ibanez, S.; Leonhardt, N.; Solano, R. Design of a bacterial speck resistant tomato by CRISPR/Cas9-mediated editing of SlJAZ2. Plant Biotechnol. J. 2018, 17, 665–673. [CrossRef]

84. Tashkandi, M.; Ali, Z.; Aljedaani, F.; Shami, A.; Mahfouz, M.M. Engineering resistance against Tomato yellow leaf curl virus via the CRISPR/Cas9 system in tomato. Plant Signal. Behav. 2018, 13. [CrossRef]

85. Bo, W.; Zhaohui, Z.; Huanhuan, Z.; Xia, W.; Binglin, L.; Lijia, Y.; Xiangyan, H.; Deshui, Y.; Xuelian, Z.; Chunguo, W.; et al. Targeted Mutagenesis of NAC Transcription Factor Gene, OsNAC041, Leading to Salt Sensitivity in Rice. Rice Sci. 2019, 26, 98–108. [CrossRef]

86. Sadanandom, A.; Srivastava, A.K.; Zhang, C. Targeted mutagenesis of the SUMO protease, Overly Tolerant to Salt1 in rice through CRISPR/Cas9-mediated genome editing reveals a major role of this SUMO protease in salt tolerance. BioRxiv 2019, 555706. [CrossRef]

87. Li, R.; Liu, C.; Zhao, R.; Wang, L.; Chen, L.; Yu, W.; Zhang, S.; Sheng, J.; Shen, L. CRISPR/Cas9-Mediated SlNPR1 mutagenesis reduces tomato plant drought tolerance. BMC Plant Biol. 2019, 19, 38. [CrossRef]

88. Curtin, S.J.; Xiong, Y.; Michno, J.M.; Campbell, B.W.; Stec, A.O.; Cerm ˇ ák, T.; Starker, C.; Voytas, D.F.; Eamens, A.L.; Stupar, R.M. CRISPR/Cas9 and TALENs generate heritable mutations for genes involved in small RNA processing of Glycine max and Medicago truncatula. Plant Biotechnol. J. 2018, 16, 1125–1137. [CrossRef]

89. Li, R.; Zhang, L.; Wang, L.; Chen, L.; Zhao, R.; Sheng, J.; Shen, L. Reduction of Tomato-Plant Chilling Tolerance by CRISPR-Cas9-Mediated SlCBF1 Mutagenesis. J. Agric. Food Chem. 2018, 66, 9042–9051. [CrossRef]

CHAPTER 8

SUSTAINABLE DEVELOPMENT THROUGH EVERGREEN REVOLUTION: FUTURE PROSPECTIVE

Sandeep Sirohi*[1], Ritu Vishnoi[1], Sanjay Kumar[1] and Surya Prakash D.V[2]

1 Department of Botany, Hariom Saraswati P.G. College, Dhanauri, Haridwar, Uttarakhand,

2 Department of Biotechnology, Meerut Institute of Engineering and Technology, Meerut

---

Abstract

The present study focuses on the paradigm shift in agricultural practices aimed at achieving sustainable food production while maintaining ecological balance. The Evergreen Revolution is a response to the shortcomings of the earlier Green Revolution, which significantly increasing food production. It advocates for the use of eco-friendly farming practices, biotechnology, organic methods, and modern agricultural techniques like of Permaculture to boost productivity while maintaining requirements of future generation. It presents a holistic approach to agriculture that not only enhances yields but also preserves the environment by promoting practices like intercropping, agroforestry, and organic farming, ensuring economic and social sustainability for rural communities. Evergreen Revolution integrates innovative technologies; improve crop verities and holistic farming techniques to enhance productivity without compromising environmental health, climate change and economic pressures.

Keywords: Evergreen Revolution, Sustainable Agriculture, Food Security, Eco-friendly Farming, Premaculture, Precision Farming.

Introduction

The concept of the Evergreen Revolution aims to address the global challenge of ensuring healthy and sustainable food for all by combining the successes of the Green Revolution with modern environmental sustainability. The Green Revolution (in the 1960s and 1970s) dramatically increased food production in developing countries through high-yield crop varieties, synthetic fertilizers, and pesticides. While it helped to alleviate hunger for many, it also contributed to environmental degradation, soil depletion, and inequality among small-scale farmers.

Evergreen revolution is the advanced terminology of green revolution which leads to food security to nutrition security under the aegis of technological intervention because the population is increasing day by day enormously, but the existing agricultural land area is not increasing, so the present challenge is to overcome through an evergreen revolution for sustainable agricultural production.

In the present scenario, the healthy food in the approach for everyone is a very big issue because in our country the production of grains especially wheat is big enough as Wheat production is estimated to increase to a record 108.75 million tones in 2020-21 from 107.86 million tones in the previous year, while the output of coarse cereals is likely to increase to 49.66 million tones, from 47.75 million tonnes a year ago but production of pulses and oilseeds are not big enough, while the last decades were completely dedicated for pulses production.

According to the Government of India, Ministry of Agriculture, food grain production is estimated to rise 2.66 percent to a new record of 305.43 million tonnes in the current crop year 2020-21, on the better output of rice, wheat, and pulses while the year 2019-2020 at 297.5 million tones.

The production of oilseeds is estimated at 36.56 million tonnes in 2020-21 as against 33.21 million tonnes in the previous year. But this much production is not supposed to fulfill the requirement of Humankind.

Source: Adobe stock

According to the recent data around 95 crores, or one-fourth of the world's population, get less than 2000 calories per day and face a severe starvation condition that leads to chronically undernourished and other health-related issues.

We all are aware that in the present pandemic situation the sufferers are mainly those who are having weak immune systems because of so many factors one, the key factor is the unavailability of healthy food. Vaccination is not a permanent solution but eating the right type of food that contains fats, protein, carbohydrates, and vitamins in adequate amounts.

The Evergreen Revolution takes a more balanced approach by focusing on sustainable agriculture that increases food production while preserving the environment, improving soil health, reducing the overuse of chemicals, and ensuring equitable access to resources for all farmers. This approach is essential for creating a resilient food system that can feed the world's growing population without compromising future generations.

Some key challenges and solutions in achieving this goal include:

Sustainability in Food Production: Shifting to environmentally friendly farming practices such as organic farming, conservation agriculture, agroecology, and precision farming to maintain soil health, water conservation, and biodiversity. This will ensure that agricultural practices can continue to provide for future populations.

Healthy and Nutritious Food: Ensuring that food systems not only produce enough calories but also provide the essential nutrients needed for a balanced diet. This involves promoting the cultivation of diverse crops, including fruits, vegetables, legumes, and biofortified crops.

Climate Resilience: Developing agricultural systems that can withstand the effects of climate change, such as droughts, floods, and temperature variations. This requires breeding climate-resilient crop varieties and implementing sustainable water management techniques.

Equity and Access: Addressing the disparities in food access, especially in rural and marginalized communities. This involves improving infrastructure, providing education, and ensuring that smallholder farmers have access to technology, markets, and financial support.

Reduction of Food Waste: Reducing food loss during production, storage, and transportation, as well as decreasing consumer-level food waste. This could have a significant impact on global food availability.

Policy and Governance: Governments and international organizations need to create policies that support sustainable farming practices, incentivize environmentally friendly innovations, and address global food security through comprehensive programs.

Sustainable farming practices aim to maintain the health of ecosystems while producing food and other agricultural products. These practices focus on minimizing environmental impact, conserving resources, and ensuring long-term productivity. Here are some examples of sustainable farming:

Agroforestry

Agroforestry integrates trees and shrubs into agricultural landscapes, combining agriculture and forestry. There are number of benefits like trees improve soil health, provide shade, reduce wind erosion, and increase biodiversity. They also act as carbon sinks, helping mitigate climate change. Examples: In tropical regions, farmers grow coffee or cocoa under the shade of trees, which protects the crops from extreme temperatures and helps retain soil moisture.

Crop Rotation

Crop rotation involves alternating the types of crops grown on a piece of land each season to maintain soil fertility and disrupt pest cycles. This practice prevents soil degradation, reduces the need for synthetic fertilizers, and naturally controls pests and diseases by breaking their life cycles. A common rotation includes growing nitrogen-fixing legumes like beans or peas one year, followed by nutrient-demanding crops like corn or wheat the next.

Cover Cropping

Cover crops, such as clover or rye, are grown in between main crop cycles to cover the soil. They prevent soil erosion, improve soil structure, enhance water retention, and add organic matter to the soil. Cover crops can also suppress weeds and fix nitrogen. **Example**: Farmers plant winter rye after harvesting summer crops to protect and enrich the soil during the off-season.

Organic Farming

Organic farming avoids synthetic pesticides, fertilizers, and GMOs, instead relying on natural inputs like compost, manure, and biological pest control. This practice maintains soil health, promotes biodiversity, and reduces water and chemical pollution.

Organic fruit and vegetable farms use compost and mulch for soil fertility and introduce beneficial insects like ladybugs to control pests.

Permaculture

Permaculture is a design system for sustainable agriculture that mimics natural ecosystems. It integrates water management, crop diversity, and animal husbandry in a holistic approach.

Permaculture promotes biodiversity, resource efficiency, and resilience against climate shocks. It often includes no-till farming, rainwater harvesting, and multi-layered crop systems. In permaculture gardens, crops are often interplanted with fruit trees, shrubs, and herbs, maximizing space and mimicking a natural forest structure.

No-Till Farming

No-till farming involves planting crops without plowing the soil. Instead, seeds are directly sown into undisturbed soil with minimal disruption. No-till farming reduces soil erosion, improves water retention, and promotes healthier soils by preserving microbial life and organic matter. It also reduces carbon emissions by limiting tractor use.

Farmers use no-till techniques in grain production, allowing crop residues like straw or stalks to remain on the field, which improves soil health.

Integrated Pest Management (IPM)

IPM combines biological, cultural, and chemical practices to manage pest populations in an environmentally friendly way. It emphasizes the use of natural predators, crop rotation, and targeted pesticide application.

IPM reduces the need for synthetic chemicals, protects beneficial organisms, and minimizes environmental damage. Farmers might introduce ladybugs or other beneficial insects to control aphids in vegetable crops, and only apply pesticides when absolutely necessary, based on pest monitoring.

Aquaponics and Hydroponics

These are soilless farming techniques where crops are grown in nutrient-rich water (hydroponics) or in combination with fish farming (aquaponics). These systems use less water than traditional farming, allow for year-round production, and can be set up in urban areas. Aquaponics provides an added benefit by cycling nutrients between fish and plants, reducing waste. Urban farms use aquaponics to grow vegetables like lettuce, herbs, and tomatoes while raising fish like tilapia in a closed-loop system.

Conservation Agriculture

Conservation agriculture includes practices such as minimal soil disturbance (no-till), crop rotation, and maintaining soil cover year-round. It improves soil fertility, reduces erosion, conserves water, and increases biodiversity. A farmer growing maize may leave crop residues on the field after harvest to maintain soil cover and plant legumes in the next cycle to fix nitrogen.

Biological Pest Control

Biological control involves using natural predators, parasites, or pathogens to control pest populations. This reduces the need for chemical pesticides and encourages a balanced ecosystem. Farmers introduce predatory beetles, birds, or spiders to control insect pests in orchards or vineyards.

Rotational Grazing

Rotational grazing involves moving livestock between different pastures to prevent overgrazing and allow vegetation recovery. This improves pasture health, increases biodiversity, and enhances soil fertility through natural nutrient cycling. Livestock farmers rotate cattle across several fields, allowing grass in grazed areas to recover before the animals return.

Water Conservation Techniques

Sustainable water management involves using techniques like drip irrigation, rainwater harvesting, and efficient water use systems to reduce water waste in farming. These methods conserve water, reduce soil erosion, and minimize the strain on freshwater resources. Drip irrigation delivers water directly to the plant roots, reducing evaporation and maximizing water use efficiency in crops like tomatoes and grapes.

Biodynamic Farming

Biodynamic farming is a holistic approach that considers farms as interconnected ecosystems. It includes crop rotation, composting, and the use of natural preparations to enhance soil health. It fosters long-term soil health, biodiversity, and ecological balance.

Biodynamic farms often grow a variety of crops alongside raising livestock, using animal manure to fertilize fields and composting organic matter.

Reducing food waste is critical for improving food security, conserving resources, and minimizing environmental impact. Here are some effective strategies to reduce food waste at various stages of the food supply chain:

At the Production Level

- **Improved Harvesting Techniques**: Train farmers to use better harvesting methods and tools to reduce crop damage and post-harvest losses.
- **Efficient Storage Facilities**: Invest in proper storage facilities like silos, warehouses, and refrigeration to prevent spoilage. In developing countries, lack of infrastructure often leads to significant waste.
- **Improved Transportation**: Use better transportation methods to prevent damage to perishable goods. Cold chain logistics (temperature-controlled transportation) are crucial for fruits, vegetables, dairy, and meat.

- **Precision Agriculture**: Use technology to apply the right amounts of water, fertilizers, and pesticides to crops, ensuring optimal growth and reducing surplus production.

At the Processing and Retail Level

- **Smarter Packaging**: Improve packaging to extend shelf life and reduce spoilage. Innovations such as biodegradable, moisture-absorbing, or resealable packaging can help.
- **More Accurate Forecasting**: Retailers can use data analytics to improve inventory management, reducing overstock and minimizing unsold goods.
- **Discounting Near-Expiry Products**: Encourage retailers to sell near-expiry products at a discount, encouraging consumers to purchase them before they go to waste.
- **Upcycling Food Waste**: Use by-products from food processing for other purposes (e.g., making animal feed, biofuels, or compost).

At the Consumer Level

- **Meal Planning**: Encourage consumers to plan meals in advance and buy only what is needed. This helps avoid over-purchasing and wasting food at home.
- **Understanding Labels**: Educate consumers about food labeling, especially the difference between "use by" and "best before" dates. "Best before" often indicates quality rather than safety.
- **Portion Control**: Serving appropriate portion sizes can reduce food waste in households and restaurants. Leftovers can be stored for future meals.
- **Composting**: Teach households and communities to compost food scraps rather than sending them to landfills. Composting recycles nutrients back into the soil, improving future food production.
- **Freezing and Storing**: Encourage freezing excess food or ingredients to extend their shelf life. Proper storage of fruits, vegetables, and dairy helps avoid spoilage.

At the Restaurant and Hospitality Level

- **Menu Design**: Restaurants can design menus with flexibility to repurpose ingredients and avoid surplus. For example, using ingredients in multiple dishes reduces the risk of waste.
- **Smaller Portions**: Offering smaller portion sizes or half portions gives consumers a choice and reduces food waste from uneaten food.
- **Donating Excess Food**: Establish partnerships with food banks and charities to donate excess food instead of discarding it. Many organizations can repurpose surplus food to feed those in need.

Government and Policy Actions

- **Food Waste Policies**: Governments can implement policies to incentivize food donation, such as tax breaks for companies that donate food to charities.
- **National Campaigns**: Launch public awareness campaigns to educate citizens about the importance of reducing food waste and ways to do so.
- **Food Waste Audits**: Encourage businesses, especially in the hospitality sector, to conduct regular food waste audits to identify areas for improvement and track progress in reducing waste.

Technology and Innovation

- **Food Rescue Apps**: Develop and promote apps that connect restaurants, stores, and individuals with excess food to organizations that can use it. Examples include Too Good To Go and OLIO.
- **Artificial Intelligence**: AI can help retailers and restaurants predict demand, ensuring they stock the right amounts of food and reducing over-purchasing.
- **Smart Fridges**: Smart refrigerators can alert users when food is about to expire, helping consumers use items before they spoil.

Cultural Shifts

- **Shift Perceptions of Imperfect Produce**: Promote the acceptance of "ugly" fruits and vegetables that may not meet aesthetic standards but are still perfectly edible. Campaigns like "ugly produce" initiatives encourage people to value all food regardless of appearance.
- **Sharing Economy for Food**: Encourage food-sharing platforms and community kitchens where people can share excess food instead of letting it go to waste.

Educational Initiatives

- **School Programs**: Educate children about the value of food, the impact of food waste, and how to minimize it. Building awareness early can help create lifelong habits.
- **Workshops and Community Gardens**: Community-based workshops and garden initiatives can teach people how to grow, store, and preserve their food sustainably.

Reducing food waste requires collaboration between producers, retailers, policymakers, and consumers. By adopting these strategies, food systems can become more efficient, equitable, and sustainable, helping to address food insecurity while minimizing environmental harm.

Precision farming, also known as **precision agriculture**, is a farming management approach that uses technology and data analysis to optimize crop production and farming practices. By applying precise amounts of inputs like water, fertilizers, and pesticides based on the specific needs of crops in different parts of a field, precision farming aims to maximize yields, reduce waste, conserve resources, and minimize environmental impact.

Key Features of Precision Farming:

Data-Driven Decision Making: Precision farming relies on data from various sources such as satellite imagery, GPS, sensors, drones, and field mapping to monitor crops and soil conditions. This data helps farmers make informed decisions on when, where, and how to apply resources.

Site-Specific Crop Management (SSCM): Instead of treating the entire field uniformly, SSCM allows farmers to manage specific areas within the field based on their unique conditions. This can include adjusting planting density, irrigation, fertilization, and pest control in different zones of the same field.

Use of Technology:

- **GPS (Global Positioning System)**: GPS technology helps farmers map fields, track the movement of machinery, and apply inputs precisely where needed.
- **Drones and Remote Sensing**: Drones equipped with cameras and sensors can capture real-time images and data on crop health, soil moisture, and other factors, providing detailed insights for targeted action.
- **Soil and Crop Sensors**: These sensors measure variables such as soil moisture, nutrient levels, and plant health. This data helps farmers optimize irrigation, fertilization, and pest management.
- **Variable Rate Technology (VRT)**: VRT allows the application of inputs (e.g., water, fertilizers, seeds) in varying amounts across different areas of the field, based on specific needs determined by data analysis.

Geographic Information Systems (GIS): GIS integrates various data sources (like soil, climate, topography) to create detailed maps and models that guide farmers in field management. This technology allows farmers to analyze spatial data to identify patterns and make predictions about crop performance.

Yield Monitoring and Mapping: Sensors installed on harvesters measure and record crop yields in real-time, creating yield maps that show variations across the field. This information helps farmers understand which areas of the field are most productive and why.

Benefits of Precision Farming:

- **Increased Efficiency**: Precision farming optimizes the use of inputs, ensuring that crops get what they need when they need it. This reduces waste and cuts costs for farmers.

- **Higher Yields**: By providing targeted care for crops, precision farming can improve plant health and maximize yield potential in each part of the field.
- **Environmental Sustainability**: By minimizing overuse of water, fertilizers, and pesticides, precision farming reduces runoff, soil degradation, and pollution, promoting sustainable farming practices.
- **Resource Conservation**: Precision irrigation, for example, ensures that water is used efficiently, conserving valuable freshwater resources and reducing water consumption.
- **Climate Resilience**: Precision agriculture helps farmers adapt to changing climate conditions by allowing them to respond quickly to variations in temperature, rainfall, and other environmental factors.
- **Cost Savings**: By applying inputs more efficiently, farmers can reduce production costs associated with fertilizers, pesticides, and water, while still maintaining or improving yields.

Examples of Precision Farming Applications:

- **Precision Irrigation**: Using soil moisture sensors and real-time data to apply water only where and when it's needed. This method reduces water waste and prevents overwatering, which can lead to root diseases or soil erosion.
- **Variable Rate Fertilization**: Soil tests and sensors determine which areas of a field need more or less fertilizer. VRT equipment then applies the exact amount of fertilizer needed for each specific zone.
- **Automated Machinery**: GPS-guided tractors and harvesters can follow precise paths to minimize overlaps and gaps, improving efficiency and reducing fuel use.
- **Crop Health Monitoring**: Drones or satellite imagery can detect early signs of plant stress (due to pests, disease, or nutrient deficiency) by analyzing color patterns in the crops. This allows for timely interventions before problems become widespread.

Challenges of Precision Farming:

- **High Initial Costs**: The investment in advanced technology, such as GPS systems, drones, sensors, and data software, can be expensive, especially for small-scale farmers.
- **Technical Expertise**: Farmers need training and technical knowledge to interpret data and use precision farming tools effectively.
- **Data Management**: Managing and analyzing large amounts of data from various sources can be complex and time-consuming.

- **Access to Technology**: Precision farming technologies may not be accessible or affordable for farmers in developing countries or remote areas.

Permaculture is a holistic approach to designing sustainable agricultural systems that mimic natural ecosystems. It integrates principles of ecology, biology, and sustainable design to create self-sustaining environments that minimize waste, conserve resources, and enhance biodiversity. Unlike conventional agriculture, permaculture focuses on long-term sustainability and resilience by considering how plants, animals, people, and the environment interact in a mutually beneficial way.

Here's a breakdown of how permaculture works:

Principles of Permaculture

Permaculture is guided by three main ethics and several design principles:

Ethics:

- **Care for the Earth**: Protecting and restoring natural ecosystems, soil health, and biodiversity.
- **Care for People**: Ensuring that human needs (food, water, shelter, etc.) are met in ways that are equitable and sustainable.
- **Fair Share (Return of Surplus)**: Distributing surplus (resources, food, and energy) to benefit the community and future generations, and reinvesting in the ecosystem.

Design Principles (among others):

- **Observe and Interact**: Before making changes to the land, observe natural processes and interact with the environment to understand its dynamics, such as water flow, sunlight patterns, and wildlife.
- **Use and Value Renewable Resources**: Rely on natural processes, like solar energy, wind power, and composting, rather than non-renewable resources.
- **Produce No Waste**: Close loops in systems, such as recycling nutrients back into the soil through composting, or reusing water through rainwater harvesting.
- **Design for Diversity**: Encourage biodiversity by integrating a variety of plants, animals, and microorganisms, leading to a resilient ecosystem.
- **Integrate Rather than Segregate**: Ensure that all elements in the system work together, for instance, integrating animals to help with pest control and nutrient cycling.
- **Use Small, Slow Solutions**: Focus on gradual, sustainable improvements rather than quick fixes that could have long-term negative impacts.

Key Elements of Permaculture Design

Zoning and Planning:

- Permaculture systems are often designed with **zones** that dictate where different elements (e.g., crops, animals, energy sources) should be located based on how frequently they need human attention. The zones range from 0 (the home) to 5 (wilderness).
- **Zone 0 (Home)**: The living area where energy efficiency and resource management (like water and waste recycling) are key.
- **Zone 1**: Areas closest to the home, like vegetable gardens and herb beds, that require daily attention.
- **Zone 2**: Slightly farther from the home, used for crops that require less care, such as fruit trees or poultry enclosures.
- **Zone 3**: Includes larger fields for staple crops or livestock that need occasional attention.
- **Zone 4**: Used for forestry, foraging, and wild-growing plants, requiring minimal human intervention.
- **Zone 5 (Wilderness)**: An untouched area left for wildlife to maintain biodiversity and ecological balance.

Polyculture and Crop Integration:

- **Polyculture** involves growing multiple plant species together in the same space, unlike monoculture farming, where only one type of crop is grown. Plants are chosen to complement each other through mutual benefits like pest control, nutrient sharing, and providing shade.
  - **Example**: The "Three Sisters" method—growing corn, beans, and squash together—is a traditional permaculture practice. The corn provides a natural trellis for the beans, the beans fix nitrogen into the soil, and the squash shades the soil to retain moisture.

Agroforestry and Food Forests:

- **Food forests** are designed to mimic the structure of a natural forest, with multiple layers of plants, including canopy trees (fruit/nut trees), understory plants (shrubs), ground cover crops, and root vegetables. This creates a self-sustaining system that requires minimal maintenance and provides diverse yields.
  - **Example**: In a food forest, nut trees (overstory) provide shade, berry bushes (mid-layer) grow underneath, while herbs and vegetables (ground layer) flourish below. Each layer supports the others in terms of nutrients, shade, or pest protection.

Water Management:

- **Permaculture designs prioritize efficient water use and conservation** by harvesting, storing, and directing water where it's most needed.
  - » **Swales**: These are shallow, water-harvesting ditches dug along contour lines to capture and hold rainwater, allowing it to slowly percolate into the ground, thereby reducing runoff and erosion.
  - » **Rainwater Harvesting**: Collecting rainwater from roofs and directing it into ponds, tanks, or underground storage for irrigation.
  - » **Keyline Design**: A technique used to guide water to dry areas of land, making the most of rainfall and preventing soil erosion.

Soil Health:

- **Composting and Mulching**: Organic matter like food scraps, plant waste, and animal manure are composted to improve soil fertility. Mulching with leaves, straw, or other organic materials helps retain moisture, suppress weeds, and add nutrients to the soil as it decomposes.
- **No-Till Farming**: Soil is left undisturbed to maintain its structure, minimize erosion, and promote the health of soil organisms like worms and beneficial bacteria.

Animal Integration:

- **Animals play a key role** in permaculture systems by providing natural pest control, fertilization, and waste management. For example, chickens can be used to control insect pests in gardens, while their manure fertilizes the soil.
  - » **Example**: A "chicken tractor" is a movable coop that allows chickens to graze in different parts of the garden, controlling weeds and pests while fertilizing the ground.

Energy Efficiency:

Permaculture incorporates renewable energy sources such as solar panels, wind turbines, and biogas systems to meet energy needs. In homes and farms, passive solar design is used to heat buildings or greenhouses with minimal energy use.

Social and Economic Sustainability:

Permaculture also promotes building **resilient communities** by encouraging local food production, community gardens, and cooperative economies. It emphasizes reducing reliance on external inputs and fostering systems that can support themselves and contribute to the local economy.

Closed-Loop Systems:

Permaculture strives to create systems where outputs from one element are inputs for another, mimicking nature's cycles. For example, waste from animals (manure) is used as fertilizer for crops, while plant leftovers can be composted and returned to the soil to close the nutrient loop.

Example of Permaculture in Action:

In a typical permaculture farm:

- Chickens provide eggs, manure, and pest control while eating food scraps.
- Fruit trees provide shade for the garden, while their leaves fall and decompose into mulch.
- Vegetables are interplanted with nitrogen-fixing legumes to enrich the soil naturally.
- Rainwater is collected from the roof, stored in tanks, and used to irrigate the garden.

Conclusion

The Evergreen Revolution is a long-term solution to feed the world sustainably and ensure that healthy food is accessible to all, without depleting the earth's resources. By adopting these sustainable farming practices, farmers can maintain productivity while protecting the environment, conserving resources, and fostering biodiversity. Sustainable farming is key to achieving long-term food security and mitigating the impact of climate change. Precision farming represents a leap forward in agriculture by leveraging technology and data to improve efficiency, productivity, and sustainability. As the global population grows and climate change poses new challenges, precision agriculture offers a way to produce more food with fewer resources, while reducing environmental impacts and promoting long-term farm viability while the Permaculture works by designing agricultural systems that mimic the natural processes of ecosystems, ensuring long-term sustainability. It emphasizes using resources wisely, integrating diverse elements (plants, animals, water, and energy) in mutually beneficial ways, and minimizing waste. Through these practices, creates resilient, low-maintenance systems that support both people and the environment.

Reference

Aguilar-Rivera N, de Jesús Debernardi-Vázquez T (**2018**) Sustainable development for farmers transforming agro-industrial wastes into profitable green products. In: Sustainable Development Research and Practice in Mexico and selected Latin American Countries. Springer, Cham, pp 53–75

Bandara JMRS, Wijewardena HVP, Liyanege J, Upul MA, Bandara JMUA (**2010**) Chronic renal failure in Sri Lanka caused by elevated dietary cadmium: Trojan horse of the green revolution. Toxicol Lett 198(1):33–39

Borges CD, Carvalho JLN, Kölln OT, Sanches GM, Silva MJ, Castro SG, … Vargas VP (**2019**) Can alternative N-fertilization methods influence GHG emissions and biomass production in sugarcane fields. Biomass Bioenergy 120:21–27

Cardoso TF, Watanabe MD, Souza A, Chagas MF, Cavalett O, Morais ER, … Bonomi A (**2018**) Economic, environmental, and social impacts of different sugarcane production systems. Biofuels Bioprod Biorefin 12(1):68–82

Carvalho JLN, Nogueirol RC, Menandro LMS, Bordonal RDO, Borges CD, Cantarella H, Franco HCJ (**2017**) Agronomic and environmental implications of sugarcane straw removal: a major review. GCB Bioenergy 9(7):1181–1195

Flores-Jiménez DE, Algara-Siller M, Aguilar-Rivera N, Carbajal N, Aldama-Aguilera C, Ávila-Galarza A, Álvarez-Fuentes G (**2016**) Influence of sugarcane burning on soil carbon and nitrogen release under drought and evapotranspiration conditions in a Mexican sugarcane supply zone. Revista Internacional de Contaminación Ambiental 32(2):177–189

Herrera-Solano A, Milanés-Ramos N, Ordóñez-Barahona P, Castillo-Morán A, Enríquez-Ruvalcaba V, Heredia-Espejo C (**2011**) Raw material cleaner at Ingenio San Miguelito containers with the use of the crop of sugar cane (*Saccharum* spp.). Cultivos Tropicales 32:49–53

Pimentel D, Acquay H, Biltonen M, Rice P, Silva M, Nelson J, … D'amore M (**1992**) Environmental and economic costs of pesticide use. Bioscience 42(10):750–760

Pretty J, Bharucha ZP (**2014**) Sustainable intensification in agricultural systems. Ann Bot 114(8):1571–1596

Rockström J, Williams J, Daily G, Noble A, Matthews N, Gordon L, de Fraiture C (**2017**) Sustainable intensification of agriculture for human prosperity and global sustainability. Ambio 46(1):4–17

Singh, S., J.S. Sidhu, N. Huang, Y. Vikal, Z. Li, D.S. Brar, H.S. Dhaliwal and G.S. Khush. (**2001**). Pyramiding three bacterial blight resistance genes (xa5, xa13 and Xa21) using marker-assisted selection into indica rice cultivar PR106. *Theor. Appl. Genet.* **102**: 1011–1015

CHAPTER 9

DISTRIBUTION PATTERNS OF RINGAL (HILL-BAMBOO) IN DIFFERENT STUDY SITES IN RUDRAPRAYAG GARHWAL HIMALAYA.

Dr. Subodh Kumar Kandari

Department of Botany,
O.N.S. Govt. Degree College Devprayag, Tehri Garhwal -249301 Uttarakhand, India.

Abstract

Forests are the most important natural resource in Uttarakhand providing firewood, fodder, biomass, and other major and minor forest produce. Natural fiber, medicinal, aromatic plants, lichens, and moss are one among the potential resources available in different villages and adjoining forests in this state. Several communities are involved in bamboo and natural fiber crafts work for their survival. *Ringal* (hill-bamboo) play an important role for economic progress in the region. Many people are depending upon the *ringal* (hill-bamboo) for their economic wealth.

Gradually *ringal* (hill-bamboo) work acquired a low status and it became an occupation identified with the Rudiya community. They made and supply *ringal* (hill-bamboo) articles like Supa, Kanddi, Dhaloa, Konda, Wodga, Dalona, Mathi, Hathkandi, Swalta and Breadbaskets etc. Mostly people belonging to this casts in

Garhwal and Beda in Kumaon are engaged in *ringal* (hill-bamboo) craft. In some places, other casts such as Badhani are also involved. The present study was made to define patterns of distribution in some sites of Rudraprayag, Grahwal Himalaya, in winter, rainy and summer session, which may be helpful to the explorers, scholars, and naturalist fòr various aspects or purposes. A total number of 4 species of *ringal* (hill-bamboo) like *Thamnocalamus falconeri, Thamnocalamus spethiflora, Thamnocalamus jaunsarensis* and *Sinarundineria falcate* were found common to all five sites.

Keywords: fiber, *Ringal* (hill-bamboo), economic, Rudiya, articles.

Introduction

The structure of vegetation of an area largely influenced by the quantity and quality of floristic elements in it. The total number of species in any vegetational aspect reflects the adaptational strength of any vegetational unit (Community) in which transformation, accumulation and flow of energy are involved (Odum, 1963). The functioning of this system is closely related to the component of the community. In case the diversity of species in a community shows that the adaptational Potential is greater to changing condition of an environment. The fluctuations indicate influences by population of each Species to the incoming heat, moisture and light as modified by vegetation itself.

The present study was made to define structure, composition, and patterns of distribution in some sites of Rudraprayag, Grahwal Himalaya, which may be helpful to the explorers, scholars, and naturalist for various aspects or purposes.

Review of the literature

Grahwal Himalaya has been a Centre of floristic as well as ecological studies in last eighty years. Duthei, (1906) revised and supplemented the catalogue of the plants of Kumaun and Grahwal regions based on the collection made by Strachey and Winter bottom during the years 1846-1849. Duthei, (1903 - 1929) is also well known for this classical work "Flora of Upper Gangetic Plain and of the Adjacent Siwalik and Sub- Himalayan Tracts" which includes many species of outer Himalaya. With the inclusion of collections of many other workers. One of the works on arborescent vegetation of West Himalaya was published as 'A Forest Flora for Kumaun' by Osmaston, (1927). Holdsworth and Smythe collect the plant species from valley of Flower in 1931 and enumerated them in Smythe's book 'Kamet Conquered' (1932). Smythe once again came to Bhyundar Valley and collected more than 250 species from the valley and enumerated them in his book "The Valley of Flowers" (1938). After establishment of northern circle of Botanical Survey of India at Dehradun in 1956, there has been a spurt of botanical exploration in North-West and West Himalaya and several plants collectors at this regional center collected plants from remote localities of Grahwal. Ghildiyal, (1957) collected plants from Valley of Flower and Gupta, (1957) enumerated the plants from Bhilangana Valley, and enumerated 738 species of flowering plants and ferns. Bor, (1960) in his voluminous work "Grasses of Burma, Ceylon, India and Pakistan" have referred large number of grass species from Grahwal. Dey et al., (1969) have enumerated flowering plants of Bhilangna Valley of Tehri district. Rau, (1975) published "High Altitude Flowering Plants of West Himalaya". Balapure, (1985) recorded flowering plants of Madhmaheswar area of Grahwal Himalaya. Some other contributions were made by Bhattacharyya and Malhotra, (1982); Joshi (1982); Mehrotra and Aswal, (1985); Mishra and Joshi, (1974) and Hajra and Balodi, (1995) have published "Plant Wealth of Nanda Devi Biosphere Reserve", enumerating flora of this area. An account of the occurrence, floristic composition, and soil characteristics of the Karangas forest in Indonesia has been given by Kartawinata, (1978). Some workers focus their study in distribution patterns were Odum, (1971); Saxena and Singh, (1982) and Kershaw, (1973).

Material and methods

Study area

In the present field survey of the Mandakni, Kakdagad and Lastar valley with existing five sites viz; Badhani, Bhanaj, Kakda, Khaliyan and Kot of district Rudraprayag Grahwal Himalaya.

Distribution pattern

The abundance to frequency ratio was used to represent the distribution pattern (Whitford, 1949) of any species in the community. The ratio 0.025 indicates regular, between 0.025 to0.05 indicates random and 0.05 indicates contagious distribution of species (Curtis and Cottom, 1956). If we divide the abundance by frequency, this provides a measure of the relative randomness of the species of the interest.

$$R = \frac{A}{F}$$

Results

After the survey and tabulation of occurrence, associating plants species of *ringal* (hill-bamboo) and distribution patterns the following results were recorded.

Distribution pattern

Table- 1.1 explains the distribution pattern of the *Ringal* (hill bamboo) in winter, summer, and rainy season of all sites. In site 1 (Badhani), *Ringal* (Hill-bamboo) distributed also majorly in random (75.00%) and contagious (25.00%) form in winter, random (50.00%) and contagious (50.00%) in summer season and random (50.00%) and contagious (50.00%) form in rainy season respectively. In site 2 (Bhanaj) *Ringal* (Hill-bamboo) distributed also in random (75.00%) and contagious (25.00%) in winter season, regular (25.00%) random (25.00%) and contagious (50.00%) form in summer and random (50.00%) and contagious (50.00%) in rainy season. *Ringal* (Hill-bamboo) distributed also in regular (25.00 %) random (5.00%) and contagious (25.00%) form in winter, random (50.00%), and contagious (50.00%) form in summer and random (75.00%) and contagious (25.00%) form random (75.00%) and contagious (25.00%) form in site 3 (Kakda-kund). In site 4 (Khaliyan), *ringal* (Hill-bamboo) distributed also majorly in random (75.00%) and regular (25.00%) form in winter season, random (50.00%) and contagious (50.00%) form in summer season and random (75.00%) and contagious (25.00%) form in rainy season respectively. Whereas *ringal* (Hill-bamboo) distributed also in regular (50.00%) random (25.000%) and contagious (25.00%) form in winter season, random (75.00%) and contagious (25.00%) in summer season and random (75.00%) and contagious (25.00%) in rainy season from site 5 (Kot),

Conclusion

A/F ratio was used to assess the distribution pattern of the species. The conducted investigation shows that the distribution patterns of the *ringal* (hill-bamboo) strata.

The present investigation shows the distribution patterns of the *ringal* (Hill-bamboo) of all the study sites in winter season regularly ranges Kakda-kund and Khaliyan (25.00%) to Kot (50.00%), randomly ranges Kot (25.00%) to Badhani, Bhanaj and Khaliyan (75.00%) and contagious distribution from Badhani, Bhanaj, Kakda-kund and Kot (25.00%). In summer season, it regularly recorded from Bhanaj (25.00%), randomly ranges Bhanaj (25.00%) to Kot (75.00%) and contagiously ranges Kot (25.00%) to Badhani, Bhanaj, Kakda-kund and Khaliyan (50.00%). in rainy season, it randomly ranges Badhani and Bhanaj (50.00%) to Kakda-kund, Khaliyan and Kot (75.00%) and contagiously ranges Kakda-kund, Khaliyan and Kot (25.00%) to Badhani and Bhanaj (50.00%). Across the sites and seasons the distribution patterns regularly ranges Bhanaj, Kakda-kund and Khaliyan (25.00%) to Kot (50.00%), randomly ranges Bhanaj and Kot (25.00%) to Badhani, Bhanaj, kakda-kund, Khaliyan and Kot (75.00%) and contagiously ranges Badhani, Bhanaj, Kakda-kund, Khaliyan and Kot (25.00%) to Badhani, Bhanaj, Kakda-kund and Khaliyan (50.00%), respectively.

Distribution pattern indicated that most of the plant's species were distributed randomly followed by contagiously. Regular distribution of plants species was rare. The contagious pattern of distribution is general characteristic pattern of nature (Odum, 1971). Even in the present investigation, all study sites show this pattern for *ringal* (Hill-bamboo). Most of *ringal* (Hill-bamboo) species contagiously distributed, very fewer was distributed randomly and regularly at different sites. Contagious distribution in natural vegetation has been reported by Greig-Smith (1957), Kershaw, (1973) and Singh and Yadava, (1974). The present investigation show dissimilarity with the result of Raturi, (2002) recorded the high regular and random distribution pattern and less contagious distribution pattern from Rdudraprayag District, Garhwal Himalaya.

Table-1.1-Distribution pattern of different woody and herbaceous vegetation indifferent study sites

Sites	Season	Distribution pattern (%)		
		Regular	Random	Contagious
Badhani (Site 1)	Winter season	---	75.000	25.000
	Summer season	---	50.000	50.000
	Rainy season	---	50.000	50.000
Bhanaj (Site 2)	Winter season	---	75.000	25.000
	Summer season	25.000	25.000	50.000
	Rainy season	---	50.000	50.000
Kakda- kund (Site 3)	Winter season	25.000	50.000	25.000
	Summer season	---	50.000	50.000
	Rainy season	---	75.000	25.000
Khaliyan (Site 4)	Winter season	25.000	75.000	---
	Summer season	---	50.000	50.000
	Rainy season	---	75.000	25.000
Kot (Site 5)	Winter season	50.000	25.000	25.000
	Summer season	---	75.000	25.000
	Rainy season	---	75.000	25.000

Acknowledgment

Author is thankful to Dr. J. P. Mehta Assistant Professor Department of Botany, H. N. B. Garhwal University (A Central University), Srinagar (Garhwal), 246174, India for providing necessary facilities and conducting reconnaissance survey during the study.

Corresponding to

Dr. Subodh Kumar Kandari

Department of Botany

O.N.S. Government Degree College Devprayag, Tehri Garhwal -249301 Uttarakhand, India.

E-mail: *subodhkandari96@gmail.com*

Contact No. - 9084401913

References

Balapure, K.M. (1985) A botanical exploration to Madh Maheshwar (Garhwal Himalaya), Uttar Pradesh. *J. Econe. Tax. Bot.*, 7: 407-24.

Bhattacharya, U.C. and Malhotra, C.L. (1982) A botanical exploration enroute Roopkund Lake (North-East Garhwal). *In*: Paliwal, G. S. (Ed.), *the vegetational wealth of the Himalayas.* Puja Publication, New Delhi, pp. 161-174.

Bor, N.L. (1960) The Grasses of Burma, Ceylon, India, and Pakistan (excluding Bambuseae). Pergamon Press, Oxford.

Curtis, J.T. and Cottam, G. (1956) *Plant Ecology Work Book.* Laboratory field reference manual (Burgess Publication Co). Minnesota, 193.

Dey, A.C., Uniyal, M.R. and Shankar, V. (1969) Flora of Bhillangana Valley erstwhile Tehri Garhwal State. *J. Bombay. Net. His. Soc.,* 65: 284-407.

Duthie, J.F. (1906) *Catalogue of the plants of Kumaun and of the adjacent portions of Garhwal and Tibet based on the collections made by Strachey and Winterbottom during the years 1846-1849s.* London. Rep. Bishen Singh Mahendra Pal Singh, Dehradun, 1994.

Duthie, J.F. (1903-1929) *Flora of Upper Gangetic Plain and of the adjacent Siwalic and Sub-Himalayan tracts.* Culcutta.

Ghildiyal, B.N. (1957) a botanical trip to valley of flowers. *J. Bombay Nat. Hist. Soc,* 54: 365-386.

Greig-Smith, P. (1957) *Quantitative Plant Ecology.* 2nd Edition. Bulterworth, London.

Gupta, R.K. (1957) Botanical exploration in the Bhilangana valley of the erstwhile Tehri Garhwal State-II. *Journal of Bombay Natural History Society* 54 878-886.

Hajra, P.K. and Balodi, B. (1995) Plant wealth of Nanda Devi Biosphere Reserve. In *Flora of India,* 16 Series-4, Botanical Survey of India, Calcutta, pp. 1–441.

Joshi, D.N. (1982) Preliminary study of the alpine flora of Rudranath Bughyal of district Chamoli (North Garhwal). *The Vegetational Wealth of the Himalayas* (eds.: G. S. Paliwal). Puja Publishers Delhi.

Kartawinata, K. (1978) *A note on a Kerangas (health) Forest Sebulu, East Kalimatan, Reinwarditia, 9.*

Kersaw, K.A. (1973) Quantitative and dynamic plant ecology. Second edition. Edward Arnold Limited, London.

Mehrotra, B.N; and Aswal, B.S. (1985) The flora of Mandakani Valley (Garhwal Himalaya). *J. Econ. Tax. Bot.,* 6: 579- 627.

Mishra, O.P. and Joshi, P. (1974) Enumeration of grasses of Kumaun and Garhwal divisions and Dehradun dist. In U.P., *J. Indian Forester,* 100: 736-753.

Odum, E.P. (1963) Ecology, Modern Biology Series, Holt, Rinehart and Winston, New York.

Odum, E.P. (1971) Fundamentals of Ecology, (3rd eds.) W. B. Saunders Co. Philadelphia. 574.

Osmaston, A.C. (1927) A forest flora for Kumaun. International Book Distributors, Dehradun, pp 605.

Raturi, G.P. (2002) *Landscape Characterization of Rudraprayag District of Uttarancha and GIS Techniques.* Unpublished D.Phil. Thesis, HNB Garhwal University, Srinagar (Garhwal).

Rau, M.A. (1974) Vegetation and phyto-geography of the Himalaya. *In: Ecology and Biogeography in India.* (*ed.*) M.S. Mani, *The Hague.*

Rau, M.A. (1975) *High Altitude Flowering Plants of Western Himalaya.* B.S.I. India, Hawrah. Calcutta.

Saxena, A. K. and Singh, J. S. (1982) A phytosociological analysis of woody species in forest communities of a part of Kumaun Himalaya. *Vegetatio,* 50: 3-22.

Singh, J.S. and Yadava, P.S. (1974) Seasonal variation in composition, plant biomass and net primary productivity of a tropical grassland at Kurukshetra, *India. Ecol. Monogr.* 44: 351-376.

Smythe, F. S. (1932) *KametConquered* Gollanz, London.

Smythe, F. S. (1938) *The Valley of the Flowers* Hodder, London.

Whitford, P.B. (1949) Distribution of woodland plants in relation to Succession & clonal growth. *Ecology,* 30: 199-208.

CHAPTER 10

LILIUM POLYPHYLLUM D. DON EX. ROYLE: A CRITICALLY ENDANGERED MEDICINAL PLANT OF WESTERN HIMALAYA

Anand Singh Bisht

College of Forestry, VCSG Uttarakhand University of Horticulture and Forestry, Ranichauri, Tehri Garhwal Uttarakhand 249199

Abstract

Lilium polyphyllum D. Don ex Royle, a Critically Endangered medicinal plant of the Himalayan Region is exploited for its roots to meet the demand of pharmaceutical industries. It is described under *Astavarga* plants in Ayurvedic system of medicine. "*Astavarga*" a group of 8 medicinal plants which are used in the preparation of various Ayurvedic tonics and have rejuvenating effects. It is harvested for its bulbs which have soothing, astringent properties and are used in the preparation of various traditional and modern medicines as a refrigerant, aphrodisiac, diuretic tonics. Its bulbs also contain various phytochemicals such as linalool, saponins etc. which are also responsible for its medicinal value. This plant is used in the preparation of various Ayurvedic tonics such as Chyavanprash, Astvarga Churan, Jivaniya ghrita. Due to its vast medicinal values, it is being over exploited at very high rates that gives it a status of a critically threatened.

Keywords: *Lilium polyphyllum*, Astavarga, Kshirkakoli, Ayurveda, Medicinal values, Endangered plants

Introduction

Lilium polyphyllum D. Don ex Royle which is known as *Kshirkakoli* in Indian traditional system of medicine that belongs to family Liliaceae. It is a perennial bulbous herb, grows up to 90 cm high, leaves sessile, narrowly lanceolate, acuminate arranged in alternate manner. Flowering period is during June–July, flowers are pendulous, white speckled with pink, in terminal racemes with long stalked. Seed pod is 4–7 cm long with winged seeds (Gaur 1999). It is a critically threatened herb which is also mentioned in the IUCN's Red list of endangered plant (Ved et al., 2003). The genus *Lilium* reportedly contains approximately 110 species and is classified into 5-10 subgenera which is geographically scattered in northern hemisphere (Europe, Temperate Asia and North America) (Dhyani et al. 2013). It is found between North-Western Himalayas of Uttarakhand (a state in India) up to west of Afghanistan (Rana & Samant 2011).

In India it is found in western Himalayan states, Himachal Pradesh, Jammu and Kashmir and Uttarakhand (Balakrishnan et al. 2018). A monograph on genus Lilium was also published (Elwes et al., 1880). It is observed that this herb grows better in cooler, moist and porous acidic soil (pH around 6.5-6.8) on northward gentle slope (30-35 degree) underneath the shady sites of *Cedrus deodara* forest where sunlight received for only 2 to 4 hour a day (Sourabh *et al.*, 2015).

Lilium polyphyllum belong to Astavarga plants of Ayurvedic medicinal system. The "Astavarga" plants suffer a lot of confusion in their identification and authentication in accordance with Ayurvedic literature (Shau et al., 2016). According to various Ayurvedic literatures, the Astavarga drugs are used in the formation of various Ayurvedic formulations like *Satavari ghritam, Shiva ghritam, Ashok ghritam, Piplayangh ghritam, Balatailam* etc. (Dhyani *et al.*, 2010). *L. poylphyllum* (Kshirkakoli) along with another Astavarga plant *Roscea procera* (Kakoli) comes under "*Brmhaniya varga*" (drug responsible for formation of body mass i.e. Mansadhatu) (Vij et al., 2019). The medicinal value of these plants is due to their chemical constituent which produces a definite physiological action in human body (Shau et al., 2016). The bulb of species is used in more than thirty Ayurvedic medicinal formulations as an expectorant, astringent, aphrodisiac, antipyretic and in the treatment of general debility (Warrier et al. 1997). Ethno-botanical study reveals its uses in treatment of kidney problems, impotency, promote strength and disease resistance, treatment of fever, pain and redness of eyes (Dhyani et al. 2010). First report of The State of the World's Plant by Royal Botanic Gardens, Kew, informs, one in five of the world's plant species is at the risk of extinction (Royal Botanic Gardens Kew 2016). The Global Strategy for Plant Conservation (GSPC 2012) aims to halt the continuing loss of plant diversity. Target two of the GSPC advises, assessment of conservation status is important to guide conservation actions of plant species and the accepted standard for this is the categories and criteria of the IUCN Red List of Threatened Species (IUCN 2001). *L. polyphyllum* has been exploited for its medicinal purposes to meet the demand of pharmaceutical industries (Rana & Samant 2011). Thus, overexploitation from natural habitat has caused population depletion to great extent. Thus, the persent article about the distribution data, population size, habitat characterization, mature wild individuals, threats identification and update conservation status of *L. polyphyllum*.

Medicinal values

Lilum polyphllum D. Don Royle is known to have various medicinal properties. Sahu et al. (2016) conducted a series of experiments on the powdered part (rhizome) of *Kshirkakoli* to find out its immunomodulation activity. They prepared various extracts of this plant. These extracts were then evaluated and estimated quantitatively and qualitatively for their chemical constituents. Results indicated that the extract of *kshirkakoli* shows the presence of Saponins, Phytosterols, Phenolic compounds, Flavonoids, Alkaloids, Steroids and Carbohydrate compounds. These extracts were then used in the study for acute toxicity of *kshirkakoli* extracts; its effect on delayed type hypersensitivity reaction (DTH), on hematological parameter and on lymphoid organ weights. These tests were conducted on Swiss albino mice (25-30g) of either sex, and results indicated that EAEKS (Ethyl acetate extract of *Kshirkakoli*) significantly increases the total WBC's count, suggesting that extract can possibly stimulate hematopoietic system. The extract was also found to be effective for improving humoral and cell-mediated immunity. Moreover, it was also reported that weight of spleen and thymus was increased suggesting that the EAEKS extract might have stimulated the formation of immune cell.

Further HPTLC analysis of *Kshirkakoli* carried out by Shau et al. (2016) shows the presence of maximum constituents like sugars, steroids and saponins. They used the bulbs of *kshirkakoli* plant to carry out extraction. The rhizomes of *kshirkakoli* were dried and coarsely powdered, then extracted with ethyl acetate (50°-55°), petroleum ether (55°-60°C), acetone (55°-60°) chloroform (50°-55°C) and ethanol (60°-65°C) successively by soxhlation. After the ethanol extraction, marc was dried and hydroalcoholic extract was obtained by cold maceration. After this the extracts were then dried in oven(45°C). These extracts were then used in performing HPTLC analysis. The presence of sugar, saponins and steroids were reported in the ethyl acetate and chloroform of *kshirkakoli*. Presence of these active biological phytochemicals secondary metabolites are proved to be beneficial in maintaining human health and in protection from other diseases. The Anti- Diabetic and spectral analysis carried out by Mir *et al.* (2020) on various extracts of *Lilium polyphyllum* showed that it possesses a good inhibiting tendency for alpha amylase and alpha glycosidase, these are the two key enzymes which plays an important role in metabolism of glucose (mainly by breaking down oligosaccharides into glucose). The inhibition of these drugs decreases intestinal glucose absorption which naturally reduces the post prandial blood sugar levels. Further, when the ethanol extracts of this plant are subjected to GC-MS analysis, it showed the presence of*5-hydroxymethyl furfural, Methyl 2-furoate, methyl piperate,* *methyl palmitate, piperine and palmitic acid*. Their presence in the extracts of this plant justifies its anti-diabetic properties.

The presence of polyphenolic compounds such as *phloridzin, caffeicacid acid, 3-hydroxy cinnamic, chlorogenic acid, rutin, ferulic acid, gallic acid, vanillic acid, 3-hydroxy benzoic acid* in the extract of *Lilium polyphyllum* plant prove its role in DNA damage activity and anti-aging properties (Giri et al., 2017). The biological activity of some of these phenolic compounds is given in table (1). Further FRAP (Ferric reducing antioxidant power) essay shows the presence of maximum antioxidant activity in *Lilium polyphyllum* extract (Giri et al., 2017).

Table 1: Various phytochemicals and their biological effects.

Active Constituents	Biological effects	References
5-Hydroxymethyl furfural	Inhibit sickling of RBCs, relieve fatigue, antioxidative, anticarcinogenic action, antimicrobial, anti-inflammatory	Balkrishna et al. (2018)
Peiminine	Anti-cancer, anti-allergic and anti-inflammatory	Balkrishna et al. (2018)
Gallic acid	Antioxidant, antimutagenic, anti-Inflammatory	Mansouri et al. (2013); Verma et al. (2013)
Catechin	Anticancer, anti-inflammatory, antiobesity, antidiabetes and neuroprotective effects	Giri et al.(2017)
Chlorogenic acid	DNA protection and antimutagenic activity	Xu et al.(2012); Burgos- Moron et al. (2012)
Linalool	Anti-parasitic, anti-convulsant, anti-cancer, anti-bacterial, neuroprotective, anti-oxidant, anti-inflammatory, anti-Alzheimer, anxiolytic, hepatoprotective, anti-hyperalgesic, neuroprotective	Balkrishna et al. (2012), Dutra (2016) Kamatou & Viljoen (2018)
Alpha-terpineol	Cardiovascular and antihypertensive effects, Antioxidant activity, Anticancer activity, Antiulcer activity, Anticonvulsant and sedative activity, Anti- bronchitis activity, Insecticidal activity	Khaleel et al. (2018)
Kaempferol	Anti-apoptotic, pro-wound healing, anti-cancer, cardioprotective, anti-oxidant, pro-apoptotic, anti-allergic, anti-parasitic, antidiabetic, anti-adipogenic, anti-thrombotic, anti-inflammatory, antimetabolic syndrome, anti-bacterial, immunoregulatory, hepatoprotective, anti-atherosclerosis	Suchal et al (2016)

Active Constituents	Biological effects	References
Citronellal	Anti-fungal, insect repellant, hepatoprotective, anti-nociceptive, anti-inflammatory, anti-bacterial	Du et al 2015, Singh et al 2016
Caryophyllene	Anti-cancer, anti-mutagenic, anti-bacterial, oxygen deprivation protective, neuroprotective, hepatoprotective, anti-convulsant, antidiabetic, anti-microbial, anti-Alzheimer, pro-longevity, analgesic, nephroprotective.	Tian et al 2016
Humulene	Inseccidal, anti-cancer, anti-inflammatory	Rogerio et al 2009, Lan et al 2011, Govindrajan et al., 2016.
Neridiol	Anti-parasitic, antioxidant, neuroprotective, pro-wound healing, anti-microbial	Kaur et al 2016, Baldissera et al 2017

It is also reported that bulbs of this plant is used in the preparation of more than 30 Ayurvedic tonics (Dhyani et al., 2017). These bulbs are used as refringent, aphrodisiac galactogogue, antipyretic, expectorant and tonic in cough in many traditional and modern medicines. Bulbs are also utilized in the preparation of revitalizing night cream and *Chywanaprasha*(Rana & Samant 2011). Various ethnobotanical studies reveals that it can be used in the treatment of kidney problems, pain and redness of eye, impotency, enhance strength and disease resistance and in cooling down of fever (Dhyani et al., 2010).

Lilium polyphyllum along with other plants namely *Roscoea purpurea* Smith, *Crepidium acuminatum* (D. Wear) Szlach, *Malaxis muscifera* (Lindl)Kuntze, *Polygonatum verticillatum* (L.) Allioni, *Polygonatum cirrhifolium* (Wall.) Royle, *Habenaria intermedia* D. Don, *Habenaria edgeworthii* Hook.f. ex-Collett belongs to Astavarga group of plants in Ayurvedic medicinal system (Vij D. *et al.*, 2019). These Astavarga plants are *Jivaniya* (Vitality promoter) which are known to maintain a balance between three body *Dosa* i.e., *Vata, Pitta and Kapha*. They are *Bramhaniya* i.e., they promote body mass and is described under *Bemhaaniyavarga*. They are *Ayusya* i.e., promote longevity (Balkrishna A *et al.*, 2018).

The knowledge of therapeutic uses of *Lilium polyphyllum* and its mode of usage lies in the traditional medicinal knowledge which is available either in local dialects or is transferred from one generation to next generation orally. But due to the reluctance of next generations in accruing the knowledge of its therapeutic uses many of its potential medicinal uses are lost (Dhyani et al., 2010).

Table 2: Therapeutic uses and formulation of Lilium polyphyllum in accordance with traditional medicinal knowledge (Dhyani et al., 2010).

Part Used	Therapeutic Uses	Mode of Usage
Bulb	Pain	Blending It With *Asparagus Racemosus* To Make Oil, Which Can Be Applied After A Bath.
	Menstrual Pain	0.5G Of Powder Blended With 1G *Smilax Ovalifolia* In Hot Water (Boiled).
	Impotency	2G Of Itspowder Is Added To Astavarga Plant, That Can Consume After Night Meal With Milk.
	Eye Pain And Redness	Powdered With Leaves Of *Cinnamomum Tamala, Gallium Elegans* And *Glycyrrhiza Glabra,* Then Boil It With Goat Milk And Apply On Eyes.

In traditional medicine system *Lilium polyphyllum* had been described for its use in treatment in sexual disorders and intermittent fever (Warrier et al., 1997). In Gangatori region of India, it is also reported that the bulbs of *Lilium polyphyllum* is consumed as food which can be cooked in vegetable oil along with potato that could possibly enhance sexual potency and can be used during winters to keep body warm (Dhyani et al., 2010). Due to its increasing demand in various Ayurvedic tonics and its less accessibility, many of its substitutes are being used. Also, its declining population and lack of traditional medicinal knowledge about the therapeutic uses of this plant are the reasons for its substitution. These substitute plants have nearly same biological activity as of this herb. These substitutes may include Aswagandh*a (Withania somnifera* (Linn.) Dunal), Safeda *(Chloroohytum arundinaceum* Baker), *Fritillaria roylei* Hook (Balkarishna et al., 2012).

Conclusion and Future Perspective

This article describes the medicinal usage, the species is traded at local, regional and national level. *L. polyphyllum* had been harvested in indiscriminate way, to prepare vitality promoting product Chyawanprash (a herbal combination used in Indian medicine) since Charak Samhita (ancient Ayurvedic treatize) period fourth century BC. In addition, its dominant associate *Cedrus deodara* populations have also been reducing drastically due to indiscriminate harvesting to meet high industrial demands. This species has active constituents known to provide resistance in our body against various ailments and boost our immunity. However, detailed information about its medicinal properties is restricted as its traditional medicinal knowledge had conveyed orally from one generation to next generation so there is very a smaller number of written text available (Mathur, 2003; Shastri, 2002) It has a great medicinal potential and requires an immediate protection and conservation strategy. A detailed survey is required on its natural habitat to ensure its availability and serious research study is required on its medicinal importance. Further, overexploitation of this plant for commerce and trade, degradation of its habitat due to overgrazing (Rana & Samant 2010) restricts its scope of scientific studies.

References

Balakrishna A, Mishra RK, Sharma N, Sharma VK and Misra L. Phytochemical, Botanical and Biological Paradigm of Astavarga Plants- The Ayurvedic Rejuvenators. *Journal of Natural & Ayurvedic medicine*. 2018; 2(6): 000145.

Balakrishna A, Srivastava A, Mishra RK, Patel SP, Vashistha RK, Singh A, Jadon V. and Saxena P. Astavarga plants – threatened medicinal herbs of the North-West Himalaya. *International Journal of Medicinal and Aromatic Plants*. 2012; 2(4): 661-676.

Buras-Moron E, Calderon-Montano JM, Orta ML, Pastor N, Perez-Guerrero C, Austin C, Mateos S and Lopez-Lazaro M. The Coffee Constituent Chlorogenic Acid InducesCellular DNA Damage and Formation of Topoisomerase I- and II- DNA Complexes in Cells. *Journal of Agricultural and Food Chemistry*. 2012; 60(30): 7384-7391.

Dhyani A, Bahuguna YM, Semwal DP, Nautiyal BP and Nautiyal MC. Anatomical features of *Lilium polyphyllum* D. Don ex Royle (Liliaceae). *Journal of American Science*. 2009; 5(5): 85-90.

Dhyani A, Nautiyal BP and Nautiyal M. Importance of Astavarga plants in traditional systems of medicine in Gharwal, Indian Himalaya. *International Journal of Biodiversity Science, Ecosystem Services & Management*. 2010;6, Nos. 1-2: 13-19.

Dhyani A, Nautiyal BP and Nautiyal MC. Distribution, status and conservation of *Lilium polyphyllum* (Liliaceae), a critically endangered medicinal plant from India. *Plant Biosystems*. 2017; 152(4): 608-611.

Dhyani A, Phartyal SS, Nautiyal BP, Nautyal MC. Epicotyl morphophysiological dormancy in seeds of *Lilium polyphyllum* (Liliaceae). *Journal Bioscience*. 2013; 38(1): 13-19.

Elwes HJ, Fitch WH and Bourne S. A Monograph of the Genus *Lilium*, Taylor and Francis, London. 1880.

Giri L, Belwal T, Bahukhandi A, Suyal R, Bhatt ID, Rawal RS and Nandi SK. Oxidative DNA damage protective activity and antioxidant potential of Astavarga species growing in the Indian Himalayan Region. *Industrial Crops and Products*. 2017; 102: 173-179.

Li JW, Zhang XC, Wang MR, Bi WL, Faisal M, Jaime A. Teixeira da Silva, Volk GM and Wang QC. Development, progress and future prospects in Cryobiotechnology of *Lilium* spp. *Plant Methods*. 2019; 15:125, doi: 10.1186/s13007-019-0506-9.

Kamatou GPP and Vijoen AM. Linalool- A Review of Biologically Active Compound of Commercial Importance. *Natural Product Communications*. 2008;3(7): 1183-1192.

Khaleel C, Tabanca N and Buchbauer G. Alpha- terpineol, a natural monoterpene: Areview of its biological properties. *Open Chemistry*. 2018;16: 349- 361.

Mansouri MT, Naghizadeh B, Ghorbanzadeh B, Farbood Y, Sarkaki A and Bavarsad K. Gallic acid prevents memory deficits and oxidative stress induced by intracerebroventricular injection of streptozotocin in rats. *Pharmacology Biochemistry and Behavior*. 2013; 111: 90-96.

Mathur DR. Yogtarangini. Varanasi (India): Chaukhamba Vidhyabhawan. 2003.

Mir AM, Ashraf MW, Singh SP, Madwal SP and Mir BA. Investigation of the Antibiotic Activity and GC- MS Analysis of Extracts od *Lilium polyphyllum*. *Tropical Journal of Natural Product Research*. 2020; 4(12): 1190-1195.

Rana MS and Samant SS. Threat categorisation and conservation prioritisation of floristic diversity in the Indian Himalayan region: a state of art approach from Manali Wildlife Sanctuary. *Journal for Nature Conservation*. 2010; 18(3): 159-168.

Rana MS and Samant SS. Population biology of *Lilium polyphyllum* D.Don ex Royle- a critically endangered medicinal plant in a protected area of Northwestern Himalaya. *Journal for Nature Conservation*.2011; 19(3): 137-142.

Sahu P, Itankar P, Mishra R, Maliye A and Sonekar M. Phytochemical Screening and HPTLC Finger Printing Analysis of *Roscea procera* (Kakoli) and *Lilium polyphyllum* (Kshirkakoli). *International Journal of Pharmacy & Pharmaceutical Research*. 2016; 8(1):53-69.

Sahu RA, Itankar PR, Mishra RR and Maliye AN. Pharmacological Evaluation of *Roscea procera* (Kakoli) and *Lilium polyphyllum* (Kshirkakoli) extracts for Immunomodulatory activity. *International Journal of Pharmacognosy and Phytochemical Research*. 2016; 8(11): 1788-1794.

Shastri V. Bhavprakashnighantu. New Delhi (India), Motilaal Banarasidash. 2002.

Sourabh P, Thakur J, Uniyal PL and Pandey AK. Biology of *Lilium polyphyllum*- A threatned medicinal plant. *Medicinal Plants*. 2015; 7(2): 158-166.

Ved DK, Kinhal GA, Ravikumar K, Prabhakaran V, Ghate U, Vijayshankar K, Indresha JH. Conservation assessment and management prioritization for the medicinal plants of Jammu & Kashmir, Himachal Pradesh & Uttaranchal. Banglore, India: Foundation for Revitalization of Local Health Tradition. 2003.

Verma S, Singh A and Mishra A. Gallic acid: Molecular rival of cancer. *Environmental Toxicology and Pharmacology*.2013; 35(3): 473-485.

Vij D, Srivastava PK and Mishra RK. Astavarga (Group of eight medicinal plants): The wonder herbs of Ayurveda. *International Ayurvedic Medical Journal*. 2019; 7(9): 1564-1571.

Warrier PK, Nambiar VPK and Ramankutty C. Indian Medicinal Plants. A compendium of 500 medicinal plants. Hyderabad (India): Orient Longman, pp. 176. 1997.

Xu JG, Hu QP and Liu Y. Antioxidant and DNA- protective Activities of Chlorogenic Acid Isomers. *Journal of Agricultural and Food Chemistry*. 2012;60(46): 11625-11630.

Suchal, K.; Malik, S.; Gamad, N.; Malhotra, R.K.; Goyal, S.N.; Chaudhary, U.; Bhatia, J.; Ojha, S.; Arya, D.S. 2016. Kaempferol attenuates myocardial ischemic injury via inhibition of mapksignaling pathway in experimental model of myocardial ischemia-reperfusion injury. Oxid. Med. Cell. Longev.

Dutra, F.L.; Oliveira, M.M.; Santos, R.S.; Silva, W.S.; Alviano, D.S.; Vieira, D.P.; Lopes, A.H. 2016. Effects of linalool and eugenol on the survival of Leishmania (L.) infantumchagasi within macrophages. Acta Trop. 164,69–76

Singh, S.; Fatima, Z.; Hameed, S. Citronellal-induced disruption of membrane homeostasis in Candida albicans and attenuation of its virulence attributes. Rev. Soc. Bras. Med. Trop. 2016, 49, 465–472.

Du, E.J.; Ahn, T.J.; Choi, M.S.; Kwon, I.; Kim, H.W.; Kwon, J.Y.; Kang, K. The mosquito repellent citronellal directly potentiates drosophila trpa1, facilitating feeding suppression. Mol. Cells 2015, 38, 911.

ovindarajan, M.; Benelli, G. A-humulene and elemene from syzygiumzeylanicum (myrtaceae) essential oil: Highly effective and eco-friendly larvicides against anopheles subpictus, aedesalbopictus, and culex tritaeniorhynchus (diptera: Culicidae). Parasitol. Res. 2016, 115, 2771–2778.

Lan, Y.H.; Wu, Y.C.; Wu, K.W.; Chung, J.G.; Lu, C.C.; Chen, Y.L.; Wu, T.S.; Yang, J.S. Death receptor 5- mediated TNFR family signaling pathways modulate humulene-induced apoptosis in human colorectal cancer HT29 cells. Oncol. Rep. 2011, 25, 419–424.

Rogerio, A.P.; Andrade, E.L.; Leite, D.F.; Figueiredo, C.P.; Calixto, J.B. Preventive and therapeutic antiinflammatory properties of the sesquiterpene humulene in experimental airways allergic inflammation. Br. J. Pharmacol. 2009, 158, 1074–1087.

Baldissera, M.D.; Souza, C.F.; Grando, T.H.; Moreira, K.L.; Schafer, A.S.; Cossetin, L.F.; da Silva, A.P.; Da Veiga, M.L.; da Rocha, M.I.; Stefani, L.M.; et al. Nerolidol-loaded nanospheres prevent behavioralimpairment viameliorating Na+, K+-ATPase and AChE activities as well as reducing oxidative stress in the brain of Trypanosomaevansi-infected mice. Naunyn Schmiedebergs Arch. Pharmacol. 2017, 390, 139–148.

Kaur, D.; Pahwa, P.; Goel, R.K. Protective effect of nerolidol against pentylenetetrazol-induced kindling, oxidative stress and associated behavioral comorbidities in mice. Neurochem. Res. 2016, 41, 2859–2867.

CHAPTER 11

SYNERGIZING THE NEXUS: UNPACKING THE INTERPLAY BETWEEN ENVIRONMENTAL DEGRADATION AND SUSTAINABLE DEVELOPMENT

Manjeet Singh

Assistant Professor, Department of Economics
Government Degree College, Khirsu, Pauri Garhwal, Uttarakhand

Introduction

Environmental Degradation:

Environmental Degradation has a long association with human Economic Development. Unplanned Economic Development let the Environment into the grip of misery, industrial revolution praises as boon for humanity but conversely it has been the prime cause of sustainable environmental for all organisms. Industrial Revolution (1800s-1900s) pave the way of rapid industrialization which led to increased pollution, deforestation, and habitat destruction and introduction of fossil fuels and industrial agriculture accelerated environmental degradation. After post-WWII (1945-1980s), Economic growth and consumerism fuelled environmental degradation and Pesticides, plastics, and other synthetic materials became widespread as well as air and water pollution increased, and climate change began to emerge as a concern. Rachel Carson's "Silent Spring" (1962) raised awareness about pesticide impacts. The awareness raised the voice to dedicate a day, First Earth Day (1970) marked growing environmental consciousness. And International agreements like the Stockholm Convention (1972) and Montreal Protocol (1987) addressed specific environmental issues. But now Climate change became a prominent concern, with the Kyoto Protocol (1997) and Paris Agreement (2015). However, Biodiversity loss, deforestation, and ocean pollution continued to worsen, but now Sustainable development and environmental protection became integral to global policy agendas. At a global level numerous conventions, conferences and agreements took place such as 1972: United Nations Conference on the Human Environment (Stockholm), 1987: Montreal Protocol (ozone layer protection), 1992: Rio Earth Summit (Agenda 21), 1997: Kyoto Protocol (climate change), 2015: Paris Agreement (climate change) and 2015: Sustainable Development Goals (SDGs).

Research Objectives

1. To examine the interplay between environmental degradation and sustainable development.
2. To investigate the impact of human activities on the environment, including deforestation, pollution, climate change, and biodiversity loss.
3. To analyze the effectiveness of international agreements, conferences, and initiatives in addressing environmental degradation and promoting sustainable development.
4. To identify areas where India needs to improve its environmental management practices and sustainable development efforts.
5. To recommend strategies for promoting sustainable development and mitigating environmental degradation in India.

Research Methodology:

1. Literature review: Analyze existing research papers, reports, and data on environmental degradation, sustainable development, and international agreements.
2. Data analysis: Examine data on environmental indicators, such as air and water pollution, deforestation, and biodiversity loss.
3. Historical studies: Conduct historical studies on successful sustainable development initiatives and environmental management practices in India and globally.
4. Policy analysis: Analyze international agreements, conferences, and

initiatives to assess their effectiveness in addressing environmental degradation and promoting sustainable development.

Current status of Environment Degradation or exploration

Biodiversity loss: 1 million species of plants and animals face extinction due to human activities (IPBES, 2019). 60% of primate species, 30% of coniferous trees, and 25% of mammals are threatened with extinction (IUCN, 2020) and 75% of insect species have declined in population over the past 50 years (WWF, 2020).

Water degradation: Only 2.5% of the Earth's water is freshwater, and the current top three uses of freshwater account for 95% of its consumption. 71% of freshwater is used for agriculture, 16% for industry, and 13% for domestic use (FAO, 2020) and 2 billion people live in water-stressed areas, and this number is expected to increase to 5 billion by 2050 (UN, 2020).

Climate change: Global temperatures have risen by 1°C since the late 1800s (NASA, 2020). 2020 was the hottest year on record, with an average global temperature of 1.29°C above pre- industrial levels (NOAA, 2021) and Sea levels have risen by 15-20 cm since 1900, and are projected to rise by another 26-82 cm by 2050 (IPCC, 2019).

Population growth: The global population is projected to reach 9.7 billion by 2050 and 11.2 billion by 2100 (UN, 2020). The population is expected to increase by 50% in Africa, 25% in Asia, and 15% in Latin America and the Caribbean by 2050 (UN, 2020).

Agriculture: Agriculture accounts for 24% of greenhouse gas emissions (FAO, 2020) and 30% of global land is used for agriculture, and this is expected to increase to 40% by 2050 (FAO, 2020).

Water management: 40% of the global population lives in water-scarce areas (UN, 2020). 25% of global water withdrawals are non-renewable (FAO, 2020) and 80% of wastewater is released into the environment without treatment (UN, 2020).

Various Agreements, Conferences and their Targeted Area

S.N.	Agreements, Conferences and Initiatives	Year	Major Target Area
I.	Stockholm Conference	1972	First major international environmental conference
II.	United Nations Environment Programme (UNEP)	1972	Coordinates UN environmental activities
III.	Convention on International Trade in Endangered Species of Wild Fauna and Flora (CITES)	1973	Regulates trade in endangered species
IV.	Montreal Protocol	1987	Phased out ozone-depleting substances
V.	Intergovernmental Panel on Climate Change (IPCC)	1988	Provides scientific advice on climate change
VI.	Basel Convention	1989	Regulates hazardous waste trade

S.N.	Agreements, Conferences and Initiatives	Year	Major Target Area
VII.	United Nations Framework Convention on Climate Change (UNFCCC)	1992	Sets a framework for global climate cooperation
III.	Rio Earth Summit	1992	Set stage for modern environmental diplomacy
IX.	Agenda 21	1992	Plan for sustainable development
X.	Convention on Biological Diversity (CBD)	1993	Aims to conserve and sustainably use biodiversity
XI.	United Nations Convention to Combat Desertification (UNCCD)	1994	Addresses desertification and land degradation
XII.	Kyoto Protocol	1997	Set binding greenhouse gas emission targets
III.	Rotterdam Convention	1998	Regulates trade in hazardous chemicals
XIV.	Johannesburg Summit	2002	Focused on sustainable development and implementation
XV.	Nagoya Protocol	2010	Regulates access and benefit-sharing of genetic resources
XVI.	Rio+20	2012	Emphasized green economy and sustainable development
VII.	Sustainable Development Goals (SDGs)	2015	17 goals to end poverty, protect the planet, and ensure peace and prosperity
VIII.	Paris Agreement	2015	Aims to limit global warming to well below 2°C and pursue efforts to limit it to 1.5°C above pre- industrial levels
XIX.	COP21	2015	Adopted the Paris Agreement

Effects of Environmental Degradation Environmental Effects:

1. Loss of Biodiversity: Environmental degradation leads to habitat destruction, extinction of species, and disruption of ecosystems. 1 million species face extinction due to human activities (IPBES, 2019). 60% of primate species, 30% of coniferous trees, and 25% of mammals are threatened with extinction (IUCN, 2020)
2. Climate Change: Greenhouse gas emissions contribute to global warming, altering weather patterns, and causing extreme weather events.
3. Water Pollution: Contamination of water sources harms aquatic life, human health, and the environment.
4. Soil Degradation: Soil erosion, nutrient depletion, and salinization reduce fertility, affecting agricultural productivity.

5. Air Pollution: Release of pollutants harms human health, crops, and the environment.

Human Health Effects:

1. Respiratory Problems: Air pollution causes respiratory issues, such as asthma and lung cancer. Air pollution causes 25% of lung cancer deaths and 25% of chronic obstructive pulmonary disease (COPD) deaths (WHO, 2020).
2. Waterborne Diseases: Contaminated water sources lead to waterborne diseases, like cholera and typhoid fever. 844 million people lack access to clean water, leading to 829,000 deaths annually (WHO, 2020).
3. Cancer: Exposure to pollutants increases cancer risk. Exposure to pollutants increases cancer risk by 20% (IARC, 2020).
4. Mental Health: Environmental degradation affects mental well-being, causing stress and anxiety. Environmental degradation affects mental well-being, causing 15% of mental health disorders (WHO, 2020).
5. Nutritional Deficiencies: Soil degradation and water pollution impact food quality, leading to nutritional deficiencies. Soil degradation affects food quality, leading to 20% of nutritional deficiencies (FAO, 2020).

Economic Effects:

1. Loss of Productivity: Environmental degradation reduces agricultural productivity, impacting food security and economies. Environmental degradation reduced agricultural productivity by 20% (FAO, 2020).
2. Healthcare Costs: Environmental pollution increases healthcare costs, burdening individuals and societies. Environmental pollution increases healthcare costs by 10% (WHO, 2020).
3. Infrastructure Damage: Extreme weather events damage infrastructure, disrupting economic activities. Extreme weather events damage infrastructure, costing $150 billion annually (UN, 2020)
4. Tourism Impacts: Environmental degradation affects tourism, impacting local economies. Environmental degradation affects tourism, impacting local economies by 10% (UN, 2020).
5. Resource Depletion: Overexploitation of resources leads to scarcity, increasing costs and affecting economic growth. Overexploitation of resources leads to scarcity, increasing costs by 20% (World Bank, 2020).

Social Effects:

1. Displacement: Environmental degradation forces communities to migrate, leading to social and cultural disruption. Environmental degradation forces 25 million people to migrate annually (UN, 2020)
2. Conflict: Competition for resources and environmental degradation can

lead to conflict. Competition for resources and environmental degradation lead to 50% of conflicts (UN, 2020)

3. Cultural Heritage: Environmental degradation threatens cultural heritage sites and traditional practices. Environmental degradation threatens 20% of cultural heritage sites (UNESCO, 2020)
4. Human Rights: Environmental degradation violates human rights, particularly the right to a healthy environment. Environmental degradation violates human rights, particularly the right to a healthy environment (UN, 2020)
5. Social Inequality: Environmental degradation disproportionately affects vulnerable populations, exacerbating social inequality. Environmental degradation disproportionately affects vulnerable populations, exacerbating social inequality by 15% (WHO, 2020).

Table-1 Impact of Environmental Degradation on Various Fields

Year	Loss of Biodiversity	Contaminated Water (Globally)	Soil Degrada-tion	Respiratory Problems	Tourism Impacts	Cancer	Mental Health	Nutritional Deficiencies	Conflict	Resource Depletion
1990	20%	50%	20%	10%	5%	10%	5%	10%	20%	10%
2000	25%	60%	25%	15%	10%	15%	10%	15%	30%	20%
2010	30%	70%	30%	20%	15%	20%	15%	20%	40%	30%
2020	60%	80%	33%	25%	10%	25%	22% (Approx)	20%	50%	20%
Sources	IUCN	WHO	UN	WHO	UN	IAR C	WHO	FAO	UN	Worl d Bank

Sources: - INCN, WHO, UN, WHO, UN, IARC, FAO and World Bank

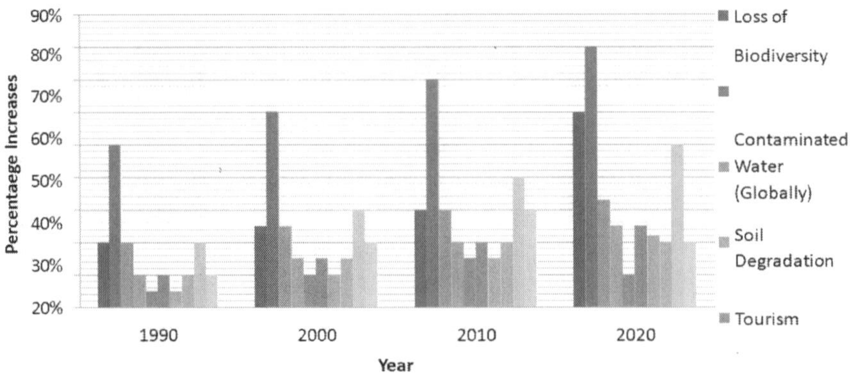

Graph-1 Impact of Environmental Degradation on Various Fields

Infrastructure Damage Healthcare Costs

Table-2 Infrastructure Damage and Healthcare Costs In Billion

Year	Infrastructure Damage	Healthcare Costs
1990	$50	$100
2000	$100	$200
2010	$150	$300
2020	$174	$310
Sources	UN	WHO

Sources- UN and WHO

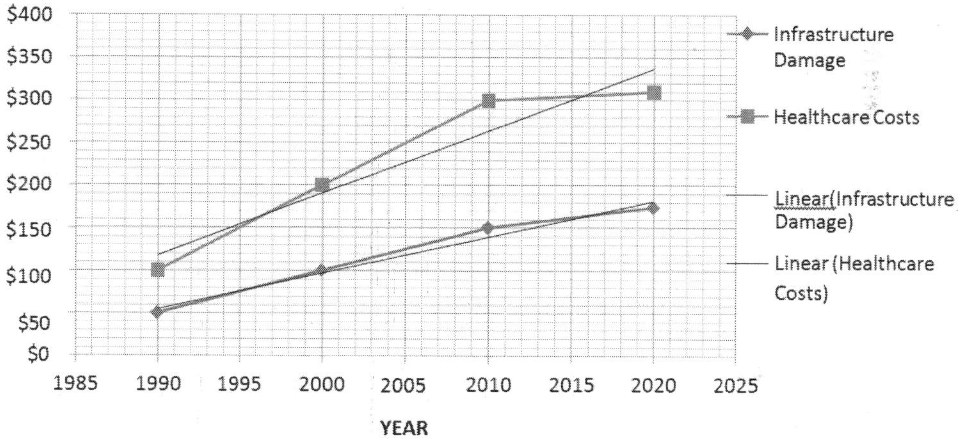

Graph -2 Infrastructure Damage and Healthcare Costs In Billion

India's Status on Environmental Degradation and Efforts Made:

1. National Clean Air Programme (NCAP): Launched in 2019 to reduce particulate matter (PM) levels by 20-30% by 2024.
2. Namami Gange Programme: Aims to clean the Ganga River by 2025 through infrastructure development and public awareness.
3. Swachh Bharat Abhiyan: Focuses on waste management, sanitation, and cleanliness, with a target of 100% waste collection and processing by 2025.
4. National Mission for Green India: Seeks to increase forest cover, improve biodiversity, and enhance ecosystem services.
5. Renewable Energy Targets: India aims to achieve 40% of installed power capacity from non- fossil fuels by 2030.

6. Electric Vehicle (EV) Policy: Encourages adoption of EVs through incentives and infrastructure development.
7. Climate Change Initiatives: India has set up the National Adaptation Fund to support climate resilience and adaptation efforts.

Challenges and Opportunities:

1. Implementation: Effective implementation of policies and programs remains a challenge.
2. Public Awareness: Raising awareness about environmental issues and promoting individual actions is crucial.
3. Technological Innovation: Encouraging innovation in clean technologies and sustainable practices can drive progress.
4. International Cooperation: Collaborating with global partners can facilitate knowledge sharing and resource mobilization.

In conclusion, while India faces significant environmental challenges, the country has made notable efforts to address these issues. Continued commitment, effective implementation, and innovative solutions are essential to mitigate environmental degradation and ensure a sustainable future for India.

Table-3 Percentage Increase in Environmental Degradation of World and India

Componants	Year	World	India
Air Pollution level (CPCB)	1990	20 %	50%
	2000	30%	70%
	2010	40%	90%
	2020	50%	120%
Forest Cove India (FSI) World (FAO)	1990	30%	20%
	2000	28%	18%
	2010	25%	20%
	2020	23%	22%
Biodiversity India (MoEFCC) World (IUCN)	1990	100 %	15%
	2000	95%	12%
	2010	90%	10%
	2020	85%	8%
Sustainable Agriculture India (FAO) World (FAO)	1990	20%	10%
	2000	25%	15%
	2010	30%	20%
	2020	35%	25%

Sources: - CPCB, FSI, FAO, MoEFCC and IUCN

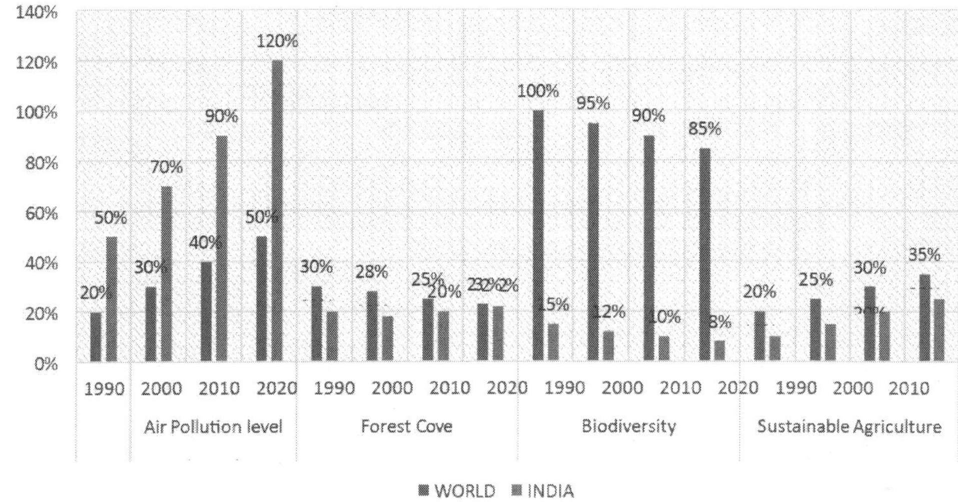

Table- 4 Environmental Degradation by World and by India
In billion

Componants	Year	World	India
Carbon Emissions (billion tons of CO2) (billion tons of CO2)	1990	20	1.2
	2000	30	2.5
	2010	40	5
	2020	50	7.5
Waste Management (billion tons of waste)	1990	0.1	0.01
	2000	0.2	0.02
	2010	0.4	0.03
	2020	0.4	0.04
Energy Consumption (billion tons of oil equivalent, GTOE)	1990	100	1
	2000	200	5
	2010	300	10
	2020	400	15
Greenhouse Gas Emissions (billion tons of CO2 equivalent, GtCO2e)	1990	20	1.2
	2000	30	2.5
	2010	40	5
	2020	50	7.5
Forest Restoration (billion hectares)	1990	0.01	0.001
	2000	0.02	0.002
	2010	0.03	0.003
	2020	0.04	0.004

Causes of this huge Environmental Degradation

1. Population growth and consumption patterns: Increasing population and consumption patterns lead to increased demand for resources, resulting in deforestation, water pollution, and waste generation.
2. Industrialization and urbanization: Industrial processes and urbanization lead to air and water pollution, solid waste generation, and loss of biodiversity.
3. Agricultural practices: Intensive agricultural practices lead to soil degradation, water pollution, and loss of biodiversity.
4. Climate change: Climate change is leading to rising temperatures, changing precipitation patterns, and increased frequency of extreme weather events, resulting in loss of biodiversity, water scarcity, and decreased agricultural productivity.
5. Lack of regulations and enforcement: Inadequate regulations and enforcement mechanisms lead to unchecked environmental degradation.
6. Poverty and lack of awareness: Poverty and lack of awareness among communities lead to unsustainable practices and environmental degradation.
7. Globalization and trade: Globalization and trade lead to increased resource extraction, transportation, and consumption, resulting in environmental degradation.
8. Technological advancements: While technological advancements can bring benefits, they also lead to increased resource consumption, e-waste generation, and environmental degradation.
9. Corruption and governance: Corruption and weak governance lead to environmental degradation and hinder efforts to address environmental issues.
10. Natural disasters: Natural disasters like floods, droughts, and landslides can lead to environmental degradation and loss of biodiversity.

Role of Economic Development to this Environment Degradation

1. Resource extraction: Economic development often leads to increased resource extraction, such as deforestation, mining, and drilling, which can harm the environment.
2. Industrialization: Industrialization processes can lead to air and water pollution, solid waste generation, and increased greenhouse gas emissions.
3. Consumerism: Economic development can lead to increased consumption patterns, resulting in more waste generation, pollution, and resource depletion.
4. Infrastructure development: Infrastructure development, such as road

construction and urbanization, can lead to habitat destruction and fragmentation.

5. Agricultural intensification: Economic development can lead to agricultural intensification, resulting in soil degradation, water pollution, and loss of biodiversity.
6. Global trade: Global trade can lead to increased transportation emissions, resource extraction, and waste generation.
7. Economic growth prioritization: Prioritizing economic growth over environmental concerns can lead to environmental degradation and losses.

However, economic development can also have positive effects on the environment, such as: Increased investment in environmental protection

- Development of sustainable technologies
- Improved resource efficiency
- Increased environmental awareness
- Economic incentives for conservation

To mitigate the negative environmental impacts of economic development, it's essential to adopt sustainable development practices, such as:

- Sustainable resource management
- Environmental impact assessments
- Green infrastructure development
- Eco-friendly technologies
- Circular economy practices
- Environmental education and awareness
- Policy frameworks supporting sustainable development

Effectiveness Measurement of Conferences, Agreements and Initiatives from out of 10.

We use a combination of quantitative and qualitative metrics to evaluate the effectiveness of international environmental agreements. Here's a breakdown of the methodology:

1. Goal Achievement (30%):

- Have the agreement's goals been met or are they on track to be met?
- Are the goals specific, measurable, achievable, relevant, and time-bound (SMART)?

2. Implementation and Enforcement (25%):

- Has the agreement been ratified and implemented by a sufficient number of countries?

- Are there robust enforcement mechanisms in place, such as penalties for non- compliance?

3. Impact and Progress (20%):

- Has the agreement led to tangible environmental benefits, such as reduced emissions or conservation of species?
- Are there measurable indicators of progress, such as increased renewable energy capacity or improved air quality?

4. Participation and Cooperation (15%):

- Is there broad participation from countries, including developed and developing nations?
- Are there mechanisms for cooperation, technology transfer, and capacity building?

5. Review and Revision (10%):

- Is there a regular review process to assess progress and identify areas for improvement?
- Can the agreement be revised or updated to address emerging challenges or new scientific information?

Effectiveness of Various Agreements and Conferences

Particulars	World Effectiveness (Scale out of 10)	India Effectiveness
Paris Agreement (2015)	6	7
Sustainable Development Goals (SDGs) (2015)	7	6.5
Montreal Protocol (1987)	9	8
Convention on International Trade in Endangered Species of Wild Fauna and Flora (CITES) (1973)	8	7
Basel Convention (1989)	5	6
United Nations Framework Convention on Climate Change (UNFCCC) (1992)	4	5.5
Convention on International Trade in Endangered Species of Wild Fauna and Flora (CITES) (1973)	8	7
Convention on Biological Diversity (CBD) (1993)	6	6
Global Environment Facility (GEF) (1991)	8.5	6.5
United Nations Environment Programme (UNEP) (1972)	8	6
Green Climate Fund (GCF) (2010)	8	6

Particulars	World Effectiveness (Scale out of 10)	India Effectiveness
Ramsey Convention (1971)	5.5	7.5
Sustainable Development Goals (SDGs) (2015)	7	6.5
United Nations Development Programme (UNDP) (1965)	6.5	6
Convention on the Law of the Sea (UNCLOS) (1982)	6	8
Green Economy	4.5	5

Sources: - Author Calculation

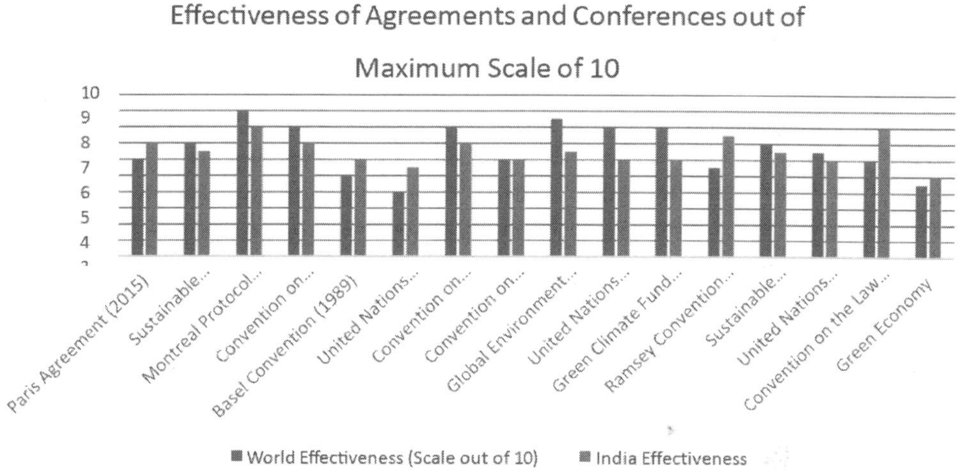

Achievements of Various Conventions and Conferences Paris Agreement (2015)

Goal: Limit global warming to well below 2°C and pursue efforts to limit it to 1.5°C above pre-industrial levels.

Achievements:

- 2020: 189 countries ratified the agreement, covering 95% of global greenhouse gas emissions.
- 2020: Countries submitted Nationally Determined Contributions (NDCs) outlining their emission reduction targets.
- India has set ambitious targets to reduce greenhouse gas emissions and increase renewable energy capacity.

Sustainable Development Goals (SDGs) (2015)

Goal: Achieve 17 goals, including ending poverty, protecting the planet, and ensuring peace and prosperity.

Achievements:

- 2020: 22% of SDG targets are on track to be met.
- 2019: Global poverty rate declined to 8.6%, from 35% in 1990.
- India has made progress in areas like poverty reduction, education, and healthcare.

Montreal Protocol (1987)

- **Goal:** Phase out ozone-depleting substances and protect the ozone layer.

Achievements:

- 2020: Ozone layer depletion slowed, with recovery expected by 2060. 2019: 98% of ozone-depleting substances phased out.

Convention on International Trade in Endangered Species of Wild Fauna and Flora (CITES) (1973)

Goal: Regulate international trade in endangered species.

Achievements:

- 2020: CITES protected over 35,000 species.
- 2019: Global wildlife crime arrests increased by 20% since 2015.
- India has implemented measures to regulate trade in endangered species.

Basel Convention (1989)

Goal: Regulate international hazardous waste trade.

Achievements:

- 2020: Basel Convention ratified by 187 countries.
- 2019: Global hazardous waste generation increased by 10% since 2015.

United Nations Framework Convention on Climate Change (UNFCCC) (1992)

Goal: Address climate change through international cooperation.

Achievements:

- 2020: UNFCCC has 197 member countries.
- 2019: Global carbon emissions increased by 1% since 2015.
- India has implemented some measures to reduce greenhouse gas emissions.

United Nations Environment Programme (UNEP) (1972)

Goal: Coordinate UN environmental activities and promote sustainable development.

Achievements:

- 2020: UNEP has 193 member countries.
- 2019: UNEP launched the Clean Air Initiative.
- India has participated in UNEP initiatives and implemented some recommendations.

Convention on the Law of the Sea (UNCLOS) (1982)

Goal: Regulate the use of the world's oceans.

Achievements:

- 2020: UNCLOS has 168 member countries.
- 2019: UNCLOS implemented marine protected areas.

Rio+20 (2012)

Achievements:

- Emphasized green economy and sustainable development.
- Led to the creation of the Sustainable Development Goals (SDGs). Established the High-Level Political Forum on Sustainable Development.

COP21 (2015)

Achievements:

Adopted the Paris Agreement.

Established a framework for climate finance, technology transfer, and capacity building. Facilitated international cooperation on climate change.

Areas on which India needs to Improvement

1. Water Management: India's water management practices need improvement to ensure efficient use and conservation of water resources.
2. Waste Management: India needs to enhance its waste management practices, including recycling, composting, and proper disposal of waste.
3. Disaster Risk Reduction: India needs to strengthen its disaster risk reduction measures to minimize the impact of natural disasters.
4. Sustainable Consumption and Production: India needs to promote sustainable consumption and production patterns to reduce waste and pollution.
5. Responsible Consumption and Production: India needs to encourage responsible consumption and production practices to reduce environmental degradation.
6. Climate Change Mitigation: India needs to enhance its efforts to reduce greenhouse gas emissions and transition to renewable energy sources.

7. Green Economy: India needs to promote green economy initiatives to create sustainable jobs and reduce environmental impact.
8. Education for Sustainable Development: India needs to integrate education for sustainable development into its education system to raise awareness and promote sustainable practices.
9. Sustainable Cities and Communities: India needs to promote sustainable urban planning and development to create livable and sustainable cities.
10. Ocean Conservation: India needs to strengthen its efforts to protect and conserve its ocean resources and marine biodiversity.

Conclusion

The data analysis reveals significant trends and patterns in India's and the world's environmental performance across various parameters. Carbon emissions, waste management, energy consumption, fossil fuel subsidies, greenhouse gas emissions, and forest restoration show varying degrees of growth and improvement. India's performance is notable, but the country still lags behind global averages. Addressing these environmental challenges requires concerted efforts, policy changes, and technological innovations.

Suggestions

1. Enhance renewable energy sources to reduce dependence on fossil fuels.
2. Implement effective waste management strategies, including recycling and waste-to-energy conversion.
3. Promote sustainable agriculture practices and reforestation efforts.
4. Encourage eco-friendly technologies and green infrastructure development.
5. Strengthen international cooperation to address global environmental issues.

References

1. United Nations Framework Convention on Climate Change (UNFCCC). (2022). Greenhouse Gas Data Interface.
2. International Energy Agency (IEA). (2022). Energy Consumption and CO2 Emissions.
3. World Health Organization (WHO). (2022). Waste Management.
4. United Nations Environment Programme (UNEP). (2022). Forest Restoration.
5. International Renewable Energy Agency (IRENA). (2022). Renewable Energy Market Analysis.

CHAPTER 12

SPECTRO-THERMAL CHARACTERIZATION AND MOLECULAR MODELING OF BIOACTIVE METAL COMPLEXES OF 4-DIETHYLAMINOSALI CYLDEOXIME

Jagdish Chandra[1], Pinkee[2]

1 PGT Chemistry, G.I.C, Kathpuriya, Almora, Uttarakhand.
2 PGT Chemistry, A.I.C Almora, Uttarakhand

Abstract

Newly synthesized ligand 4-diethylaminosalicyldeoxime (4-DEASDOX) and their Co(II), Ni(ii), Cu (II), and Zn(II) complexes have been prepared and characterized by different physicochemical techniques. The ligand 4-diethylaminosalicyldeoxime behaves as a bidentate ligand forming neutral metal chelates through the phenolic oxygen and the oxime nitrogen. The M–O stretching frequencies for the transition metals show good agreement with the Irving– William's stability order. Similar trend is seen for the M–N stretching frequencies in vibrational spectra and the shift in transitions from electronic spectral data for the metal complexes. Mass spectrum explains the successive degradation of the molecular species in solution and justifies ML_2 complexes. Electronic spectra and magnetic susceptibility measurements suggest square planer geometry of the metal complexes. Thermodynamic activation parameters were computed from the thermal data using Coats and Redfern method, which confirm first order kinetics. The geometry of the complexes has been optimized with the help of molecular modeling. The bio-efficacy of free ligand and its metal complexes have been examined against the growth of bacteria and pathogenic fungi *in vitro* to evaluate their anti- microbial potential.

Keywords: Bioactivity, Metal complexes, Molecular modeling, Salicyldeoxime, Spectra, Thermal studies

Introduction

There is currently a renewed interest in the coordination chemistry of structurally modified bio-ligands. Oximes play an important role in the development of transition metal coordination chemistry due to their versatile bonding mode. Oximes and their derivative are important intermediates in organic synthesis[1]. Many drugs have an oxime group in their structure, which is then frequently used for molecular modifications to develop new drugs and for synthesis of many functional groups as a starting point [1]. Oximes have both Nitrogen and Oxygen donor atoms; therefore, they link metals forming chelate complexes. On the other hand, some of the oximes and their metal complexes exhibit liquid crystalline properties and having photo or electro-driven emitting properties[2].

Anti-tumour copper salicylaldoxime complex which induces Topoisomerase I (topo II) to form single-strand nicks in DNA and poisons its activity, which could be one of the possible mechanisms for the anti-cancer activity of this complex. Copper salicylaldoxime complex inhibits the enzyme activity through induction of enzyme linked single-strand breaks in the DNA and may inhibit topo II dimerization, which may lead to the formation of single- strand breaks[3]. Topoisomerase II is a nuclear enzyme which is crucial for resolving DNA knots in the chromosomes during replication, transcription and cell division during chromosome segregation[3]. In acid medium, salicylaldoxime functions as a molecular ligand (H_2Sal), which brings about a decrease in electron density on donor atoms when compared with the HSal anion. Thus the ability of the used ligand to stabilize the central atom in unusually high oxidation states is impaired. The reaction of $Ni(HSal)_2$ with bromine involves a preferential attack on the aromatic ring of ligand. The possibility of oxidizing with peroxodisulfate is negatively affected by separation of nickel(II) salicylaldoximate from the water- organic solvent (ethanol, acetone) mixture[4]. However, with increasing use of oxime as drugs and pesticides, the intake on these chemical followed by enzymatic oxidation may result in the formation of a variety of reactive intermediates, which may lead to cell and tissue damage [5].

Spectral and thermal characterization serves as important tools for the interpretation of the structures of bio-molecules. Therefore, we report here in the synthesis and characterization of the complexes of 4-diethylaminosalicyldeoxime (DEASDOX) bidentate ligand with transition metal (II) ions. The geometry of the complexes has been optimized by molecular modeling studies. The bioactivity of the ligand and its metal complexes possess antimicrobial properties.

Experimental

All the chemicals used were of analytical grade and used as procured. Solvents used were of analytical grade and were purified by standard procedures. The stoichiometric analyses (C, H and N) of the complexes were performed using Elementar vario EL III (Germany) model. Metal contents were estimated on an AA-640-13 Shimadzu flame atomic absorption spectrophotometer in solution prepared by decomposing the respective complex in hot concentrated HNO_3. Their IR spectra were recorded on Perkin–Elmer FT- IR spectrophotometer in KBr and polyethylene pellets in the range 4000-400 cm^{-1} and 400-100 cm^{-1}, respectively.^{1}H-NMR spectra were recorded in DMSO-d_6 solvent on a Bruker Advance 400 instrument. Proton chemical shifts are reported in parts per million(ppm) relative to an internal standard of Me_4Si.The UV-Visible spectra were recorded in DMSO on Beckman DU-64 spectrophotometer with quartz cells of 1 cm path length from 200-900nm and mass spectra (TOF-MS) were recorded on Waters (USA) KC-455 model with ES^+ mode in DMSO. Magnetic susceptibility measurements were carried out at room temperature in powder form on a vibrating sample magnetometer PAR 155 with 5000G-field strength, using $Co[Hg(SCN)_4]$ as the calibrant (magnetic susceptibility $\approx 1.644 \times 10^{-5} cm^3 g^{-1}$). Rigaku model 8150 thermoanalyser (Thermaflex) was used for simultaneous recording of TG-DTA curves at a heating rate of 10°min^{-1}. For TG, the instrument was calibrated using calcium oxalate while for DTA, calibration was done using indium metal, both of which were supplied along with the instrument. A flat bed type aluminium crucible was used with - alumina (99% pure) as the reference material for DTA. The number of decomposition steps was identified using TG. Molecular modeling of the proposed structure of the complexes was performed using HyperChem professional version 7.51 program package[6]. *In vitro* antibacterial activity of the synthesized compounds against *Streptococcus, Staph, Staphylococcus aureus* and *Escherchia coli* bacteria were carried out using Muller Hinton Agar media(Hi media).

The activity was carried out using paper disc method[7]. Base plates were prepared by pouring 10ml of autoclaved Muller Hinton agar into sterilized Petri dishes (9 mm diameter) are allowing them to settle. Molten autoclaved Muller Hinton had been kept at 48°C was incubated with a broth culture of the *Streptococcus, Staph, Staphylococcus aureus* and *Escherchia coli* bacteria and then poured over the base plate. The discs were air dried and placed on the top of agar layer. The plates were incubated for 30 h and the inhibition zones (mm) were measured around each disc. The Antifungal activity of the complex was checked by dry weight method for the *Alternarie alternate, Aspergillus flavus, Aspergillus nidulans* and *Aspergillus niger* fungi. The complexes were directly added to the growth medium in varying concentration. The actively growing mycelia (of the test fungi) were placed on the medium with the help of inoculums needle and incubated at 27±2 ^{0}C for 7 days. The medium with the test solution served as treated while without them as control or check. The resulting mycelia mats in each set were carefully removed, washed, dried and then weighed separately. The fungal growth was calculated from the following relation:

Fungal growth inhibition (%) = $\underline{Cg - Tg}$ × 100

Cg where *Cg* is the average growth in the 'control' or 'check' set

Tg is the average growth in the treated set.

Synthesis of ligand and complexes

Synthesis of ligand

The ligand(Fig.1) has been synthesized by mixing 4-diethylaminosalicyldehyde (0.01M) with hydroxylamine hydrochloride(0.01M) in ethanol(25ml) at room temperature and also added triethylamine drop wise with constant stirring by keeping the pH 8-9 for 48*h*. The white precipitate was filtered, washed with ethanol and dried under vacuum over silica gel.

Synthesis of complexes

To 50 ml 0.2M aqueous solution of the salts of respective metal chlorides, 50ml. of 0.4M solution of the ligand 4-diethylaminosalicyldehydeoxime in 50% ethanol was added. The precipitated complexes were digested, filtered, washed first with hot water and then with 20% ethanol and finally dried at 110–120 ^{0}C in an air oven. The Co(II) complex formed brown, Ni(II) complex formed Light green, Cu(II) complex formed Light blue and Zn(II) complex formed white precipitate. The solubility of these complexes is sparingly soluble in water and completely soluble in DMSO, Acetone etc.

Results and discussion

On the basis of elemental analysis data(Table1), the complexes have the general composition ML_2, where M = Co(II),Ni(II),Cu(II) and Zn(II); Ligand = 4-diethylaminosalicyldeoxime.The complexes were obtained in powder form. These were found to be sufficiently soluble in Acetone and DMSO for spectral measurements. Various attempts to obtain the single crystals of the complexes have so far been unsuccessful. However, the analytical and spectroscopic data enables us to predict possible structures. The molecular modeling helps to determine the bond angles, bond length and supports the possible geometry of the complexes.

Spectroscopic and magnetic characterizations

The ^{1}H NMR spectra of a DMSO-d_6 solution of the oxime ligand(DEASDOX) show well resolved signals as expected(Figure 2: ^{1}H NMR of the ligand). The spectrum of the oxime ligand shows a doublet at 1.2ppm, (6H), doublet at 3.39ppm(4H), singlet peak at 6.2ppm, 7.3ppm and 8.18ppm of benzylidenimin group, and singlet at 2.0ppm(alcoholic H) and 5.0 ppm(aromatic C-OH). The ^{1}H -NMR spectra of complex II and complex IV show the resonance with expected integrated intensities. In both the complexes, no signal of the phenolic hydrogen indicates deprotonation of the ortho-hydroxyl group on complexation. Furthermore, the aliphatic protons are not greatly affected upon complexation.

The vibrational spectrum of the free ligand showed a broad band between 3100- 3500 cm^{-1}, which can be attributed to phenolic OH group. This band disappears in all complexes, which can be attributed to the involvement of phenolic OH in coordination. The involvement of deprotonated phenolic moiety in complexes is confirmed by the shift of ν(C-O) stretching band observed at1205 cm^{-1} in the free ligand to a lower frequency to the extent of 10–20 cm^{-1}[8]. The shift of ν(C-O) band at 1205 cm^{-1} to a lower frequency suggests the weakening of ν(C-O) and formation of stronger M - O bond. The broad and low intensity bands due to ν(O-H) modes of N- OH groups in the 3300–3000 cm^{-1} frequency range shifted to the lower frequency region (2900– 3000 cm^{-1}) which suggests the weakening of N-OH bond and formation of M- N bond[9] . The medium bands observed in the 1650–1610 cm^{-1} frequency ranges in complexes were assigned to ν(C= N) mode (figure 3a-e: IR spectrum of the ligand and complexes). The shift of ν(C= N) vibration in all the complexes to a lower frequency suggests that the nitrogen atom of the ring contributes to the complexation. The lower ν(C= N) frequency also indicates stronger M -N bonding[10,11]. The IR spectrum of the ligand showed the ν(N-O) band appears at 860 cm^{-1}(Table 2: IR sprecral data of ligand and complexes). So, on M-N interaction, ν(N-O) appeared at higher frequencies, indicating that N-O bond length decreases. The positive ν(N-O) shift indicates strengthening of the M- N bond. In the IR spectra of the complexes, a band is observed between 285 and 295 cm^{-1} that is attributed to the ν(M- N) stretching vibrations. Another band appeared between 640 and 655 cm^{-1}, which is assigned to the interaction of phenolic oxygen to the metal atom, i.e.,the stretching vibrations ν(M-O)[8].

The electronic spectra of the complexes were recorded in DMSO. The bands in the 200-380 nm regions coincide with the bands observed in the free ligand. Two relatively less intense bands (n →π* and π → π*) occur in the 200-380 nm region, these can be assigned to phenolic oxygen to metal and C=N to the metal charge transfer transitions, respectively. The electronic spectral data the Co(II) complexes show absorption bands at 530 nm and 422 nm assignable to $^{2}B_{2g}\rightarrow{}^{4}E_{g}$(P) and $^{2}B_{1g}\rightarrow{}^{4}A_{2g}$ (P) transitions, respectively, suggest square planar geometry of the complexes [12,13], which is also corroborated by magnetic moment value given in Table 1. The reason for the departure from the spin only value lies partly in the existence of the second order Zeeman effect between the ground and the higher ligand field terms [14]. However, it lies mainly in the fact, that in the presence of spin-orbit coupling, the quenching effect of the ligand field cannot be complete. Spectra of the nickel(II) complex shows an absorption band at 655nm, assignable to a $^1A_{1g} \rightarrow {}^1A_{2g}$ transition and a shoulder in the region 550–560 nm corresponding to a $^1A_{1g} \rightarrow {}^1B_{1g}$ transition which are consistent with square planar stereochemistry about the nickel(II) ion. It can be explained with the planar ligand set causes one of the d-orbitals ($dx^2–y^2$) to be uniquely high in energy and the eight electrons can occupy the other four d-orbitals but leave this strongly antibonding one vacant. The diamagnetic nature of the Ni(II) complexes further supports its square planar nature. The electronic spectrum of the Cu(II) complex shows very intense absorption bands in the UV region, attributed to ligand transitions($n \rightarrow \pi^*$ and $\pi \rightarrow \pi^*$), and a characteristic broad bands in visible region at 650 nm and 515 nm assignable to $^2B_{1g} \rightarrow {}^2E_g$ and $^2B_{1g} \rightarrow {}^2A_{1g}$ transitions respectively, indicates the possibility of square planar geometry of metal complex. The Cu(II) complex has magnetic moment value 1.74 BM reveals square planar geometry around the metalion[12]. Zn(II) complex is diamagnetic in nature and their electronic absorption spectrum does not have d-d transition bands. These complexes assigned square planer geometry.

Mass spectrometry has been successfully used to investigate molecular species in solution [15]. The pattern of the mass spectrum gives an impression of the successive degradation of the target compound with the series of peaks corresponding to the various fragments. The intensity gives an idea of stability of fragments. The mass spectrum of the ligand having a molecular ion peak at 208{85% m/z} confirms purity of the ligand. The recorded molecular ion peaks of the metal complexes have been used to confirm the proposed formula. All the synthesized complexes containing metal ion were confirmed by good agreement between the observed and calculated molecular formula (Figure 4a-d: TOF-Mass spectrum of the complexes). Elemental analysis values are in close agreement with the values calculated from molecular formulas assigned to these complexes, which further supported by the TOF-mass studies. In the TOF–mass spectra of metal complexes initial fragmentation pattern is similar, a binuclear nature for these complexes $[M(L2)]^+$ can be deduced. The mass spectrum of the complexes having a molecular ion peak at 471/472{39% m/z in complex I}, 470/471{40% m/z in complex II}, 476/477{30% m/z in complex III) and 476/477{25% in complex IV} respectively, further confirming the purity of the complexes. The prominent peak in the middle of the degradation pathway appears in nearly all the complexes at positions (m/z values) 325-332(nearly 28% m/z), and 248-252(nearly 39% m/z) are degradation of $[C_8H_{22}N_2]^+$ molecule which are results of demetallation and subsequently a partial intramolecular hydrogen bonding. Likewise peaks around 266-274(m/z) and 208(m/z) are attributable to unstable monomeric species as $[M(L)]^+$ and $[H_2L]^+$ respectively, usually present in the mass spectra of these systems [16].

Kinetics of thermal decomposition

Thermogravimetric (TG) and differential thermogravimetric analyses (DTA) were carried out for metal complexes in ambient conditions. The correlations between the different decomposition steps of the complexes with the corresponding weight losses are discussed in term of the proposed formula of the complexes. In the complexes, loss of (C_4H_8+ NH_3) molecules takes place in the temperature range 380 – 450K followed by the decomposition of the organic moiety with the formation of metal oxide as residue in the temperature range 500- 680K[17]. This has been again confirmed by comparing the observed/estimated and the calculated mass of the pyrolysis product. On the basis of thermal decomposition, the thermodynamic activation parameters(Table 3) such as activation energy (E*), enthalpy of activation (ΔH*), entropy of activation (ΔS*), free energy change of decomposition (ΔG*) were evaluated by employing the Coats – Redfern relation [18]. The Coats and Redfern linearization plots (figure 5a-d), confirms the first order kinetics for the decomposition process. According to the kinetic data obtained from the TG curves, all the complexes have negative entropy, which indicate that the complexes are formed spontaneously. The negative value of entropy indicates a more ordered activated state that may be possible through the chemisorptions of oxygen and other decomposition products. The negative values of the entropies of activation are compensated by the values of the enthalpies of activation, leading to almost the same values for the free energy of activation [17].

Molecular modeling studies

Molecular mechanics attempts to reproduce molecular geometries, energies and other features by adjusting bond length, bond angles and torsion angles to equilibrium values that are dependent on the hybridization of an atom and its bonding scheme(Table 4: Bond length and bond angles of complex II and complex III). In order to obtain estimates of structural details of these complexes, we have optimized the molecular structure of complexes. Energy minimization was repeated several times to find the global minimum. The energy minimization value for square planar and without restricting the structure for the Ni(II) and Cu(II) complexes are almost same i.e, 6341kcal/mol with gradient 0.930 and 6210kcal/mol with gradient 0.987 respectively. This supports square planar geometry of the metal complexes [29]. The optimized molecular structure of the Ni(II) and Cu(II) complexes are represented in Figure 6a-b. The energy minimization values for copper complex has minimum than nickel complex than cobalt complex than zinc complex which has maximum energy. It indicates copper complex has maximum stability than other metal complexes.

Biological studies

The free ligand and its respective metal complexes were screened against *A. alternate*, *A. flavus*, *A. nidulence* and *A. niger* fungi and *Streproproteus*, *Staph*, *Staphylococcus* and *E. coli* bacteria to assess their potential antimicrobial agents. The results are quite promising. It is clear from the antifungal screening data (Fig. 7a), that the metal complexes are more fungitoxic than the chelating agent itself. The complexes showed maximum activity against *A. niger* fungi. The bacterial screening results (Fig. 7b) reveal that the free ligand showed the maximum activity against *Staphylococcus* bacteria. Its metal complexes also showed the maximum activity against *Staphylococcus bacteria* except zinc complex showed the maximum activity against *Streproproteus* bacteria. In general, the activity was the least against *Staph* bacteria. The presence of nitrogen and oxygen donor groups in the synthesized compounds inhibited enzyme production because enzymes require free hydroxyl group for their activity appear to be especially susceptible to deactivation by the metal ion [19]. Complexation reduces the polarity of the metal ion from partial sharing of its positive charge with the donor groups; π- electron delocalization in this chelating ring also increases the lipophilic nature of the metal atom, favoring permeation through the lipid layer of the membrane. The enhanced activity of the metal complexes may be ascribed to the increased lipophilic nature of these complexes arising due to chelation [19]. It was also noted that the toxicity of the metal chelates increases on increasing the metal ion concentration. It is probably due to faster diffusion of the chelates as a whole through the cell membrane or due to the chelation theory. The bounded metal may block enzymatic activity of the cell or else it may catalyze toxic reactions among cellular constituents.

Conclusion

With the help of various physico-chemical techniques, geometries of the newly synthesized compounds have been proposed (Figure 8). The ligand acts as a bidentate ligand involved coordination with metal ions through the phenolic oxygen and the oxime nitrogen forming square planar complexes. Kinetics of thermal decomposition was computed from the thermal data using Coats and Redfern method, which confirm first order kinetics. The Cu(II) complex was found to be most active against the growth of bacteria and pathogenic fungi.

Acknowledgements

The author (B K Singh) thankfully acknowledge to University Grants Commission(UGC), New Delhi, India for financial support.

References

1. (a) Durmus M. Dincer F. Ahsen V. Dyes and Pigments 2008,77, 402-407.(b) Abele E. Lukevics E. Org. Prep. Proceed Int. 2000,32,235-264.(c) Chakravorty A. Coord. Chem. Rev. 1974,13,1-46.(d) Ramesh V. Umasundari P. Das K. K. Spectrochimica Acta Part A 1998,54, 285–297.
2. (a) Martinez J. Aiello I. Bellusci A.Crispini A. Ghedini M.Inorganic Chim Acta 2008, 361, 2677-2682.(b)Gumus G. Ahsen V.*Mol*. Cryst Liq Cryst. 2000, 348, 167-178.(c)Ohta K, Hasebe H. Moriya M. Fujimoto T.Yamamoto I. Mol Cryst Liq Cryst 1991, 208, 43-54.
3. Jayaraju D.Kondapi A. K. Current Science 2001,81(7), 787-792.
4. B ezina F. Tillmannová H. And Šindelá Z.Chem. Papers ,1986, 40(6), 727—733.

5. Rajabi L. Courreges C.Montoya J. Aguilera R. J.Primm T. P. Lett. Appl. Microbiol. 2005, 40, 212- 215.
6. Hyperchem Release 7.51 Professional Version for Windows, Molecular Modeling System, Hyperchem Inc. Canada, 2005.
7. Saxena C.Sharma D. K. Singh R. V. Phosphorous Sulfur Silicon 1993, 85(1-4), 9-1
8. Omar M. M. Mohamed G. G. Spectrochim. Acta A,2005, 61, 929.
9. Georgieva I.Tredafilova N. Bauer G. Spectrochim. Acta A 2006, 63, 403
10. Nakamoto K. Infrared and Raman Spectra of Inorganic and Coordination Compounds, Wiley, New York, 1978.
11. Silverstein R. M. Bassler O. G.Morrill T. C. Spectrophotometic Identification of Organic Compounds, Wiley, New York, 1974.
12. Nakamoto K, McCarthy S.J, Spectroscopy & Structure of Metal Chelate Compounds, John Wiley & Sons, USA, 1968.
13. Lever A.B P, Inorganic Electronic Spectroscopy, Elsevier, Amsterdam, 1968.
14. Figgis B N, Introduction to Ligand Fields, 1st ed., 1966, p. 263.
15. (a) Sanmartin J. Novio F. Garcia- Deibe A. M.Fondo M.Ocampo N. Bermejo M. R. Polyhedron ,2006, 25, 1714-1722. (b) Singh B. K. Prakash A. Adhikari D. Spectrochim. Acta A 2009, 74, 657-664.(c) Singh B. K. Rajour H. K. Prakash A. Spectrochim. Acta A,2012, 94, 143- 151.(d) Beloso I. Castro J. Garcia-Vazquez J. Perez-Lourido A. P. Romero J. Sousa A, Polyhedron,2003, 22, 1099-1111.
16. (a) Yoshida N. Ichikawa K. Shiro M. J. Chem. Soc.Parkin *Trans*. 2000, 2, 17-26. (b) Yoshida N. Oshio H.Ito T.J. Chem. Soc. Perkin Trans.1999, 2, 975-984.(c) Singh B.K. Rajour H.K. Chauhan R. Goyal S. Chemical Sci. Transections 2015, 4(3), 806-818.
17. (a) Singh B K , Sharma R.K, Garg B S, J. Thermal Anal. Cal.2006, 84, 593. (b) Prakash A, Singh B. K., Bhojak N, Adhikari D, Spectrochim Acta A 2010, 76, 356-362.
18. Coats W, Redfern J, P, Nature, 1964,201 68-69.
19. (a) Ahamad T. Nishat N. Parveen S. J. Coord. Chem. 2008, 61, 1963-1972 (b) Singh B. K,.Rajour H. K, Chandra J. J. Applied Chemical Sci. International, 2016, 6(1), 45-57.(c) Singh B.K, Mishra P, .Prakash A,,Bhojak N, Arabian J . Chem.2017, 10,S472-S483.

Table 1: Analytical and physical data of ligand and their metal complexes

S No	Compound/ complex (Empirical formula)	Color	Decomposition temperature (°C)	Yield (%)	Analysis found (Calculated(%) C	H	N	M	μeff (BM)	Electronic spectra(nm)
1	Ligand ($C_{11}H_{16}O_2N$)	White	159	85	63.8 (63.78)	7.30 (7,32)	13.50 (13.48)	-	-	285,370
2	Complex I [$Co(C_{11}H_{14}O_2N_2)_2$]	Light Brown	246	80	28.03 (28.22)	2.97 (2.99)	5.94 (5.99)	12.50 (12.61)	2.34	290,372,422, 530
3	Complex II [$Ni(C_{11}H_{14}O_2N_2)_2$]	Light green	298	86	28.04 (28.23)	2.97 (2.98)	5.94 (5.98)	12.45 (12.51)	diamagnetic	283,375,655, 557
4	Complex III [$Cu(C_{11}H_{14}O_2N_2)_2$]	Light blue	287	82	27.76 (27.81)	2.94 (2.98)	5.88 (5.90)	13.35 (13.57)	1.74	280,369,515, 650
5	Complex IV [$Zn(C_{11}H_{14}O_2N_2)_2$]	White	315	80	27.65 (27.77)	2.93 (2.99)	5.86 (5.91)	13.78 (13.88)	diamagnetic	282,378

Table 2. IR spectral data (cm^{-1}) of ligand and its metal complexes

Frequency	Ligand	Complex I	Complex II	Complex III	Complex IV
ν $(OH)_{NOH}$	3280(s)	2988(w)	2994(w)	2996(w)	2971 (w)
ν (C=N)	1660(s)	1640(s)	1647(s)	1645(s)	1643(s)
δ (NOH)	1458	1452(s)	1455 (w)	1450(s)	1453(s)
ν (CH)	2852(s)	3022 (s)	2989(s)	2922(s)	3023(s)
ν (N-O)	860	868	890	872	878
δ (O-M-O)	-	185	188	192	179
ν (M-O)	-	326(m)	321(m)	388(m)	332(m)
ν (M-N)	-	287(m)	279(m)	276(m)	293(m)
ν (CH_3)	1376	1348	1351	1349	1358
ν (C-O)	1205	1195	1187	1192	1188
δ (O-M-N)		231	235	215	239

s = strong;m = medium, w = weak

Table 3: Thermodynamic activation parameter of metal complexes

Complexes	Order(n)	Steps	E*($Jmol^{-1}$)	A($\times 10^5 s^{-1}$)	ΔS^*($JK^{-1}mol^{-1}$)	ΔH^*($J\ mol^{-1}$)	ΔG^*($k\ J\ mol^{-1}$)
Complex I	1	I II	44.11 45.39	0.56 0.43	-213.44 -218.35	17.95 118.62	84.96 120.21
Complex II	1	I II	20.66 31.53	0.16 0.29	-243.05 -241.00	65.26 223.75	102.39 137.11
Complex III	1	I II	19.11 21.84	0.35 0.34	-258.94 -259.56	78.71 181.56	116.25 161.89
Complex IV	1	I II	20.91 22.05	0.15 0.12	-244.85 -248.75	112.33 442.57	116.42 157.90

Table 4: Selected bond length(Å) and angle(°)for complexes II and III

Complexes	Bond length (Å)		Bond Angles(°)	
Complex II	16-10	1.8558	1-16-10	91.8809
	16-1	1.8059	17-16-19	88.6387
	16-19	1.8339	17-16-1	170.418
	16-17	1.8585	19-16-10	167.155
	17-18	1.3165	17-16-10	81.7281
	19-20	1.3832	19-16-1	98.6763
	20-21	1.411	16-19-31	102.686
	18-21	1.4628	20-19-31	119.248
	19-31	1.4165	16-17-18	97.3437
	2-9	1.8059	16-1-9	104.432
	1-2	1.3882	16-10-4	103.043
	10-4	1.3211	16-1-2	122.544
	4-3	1.4735	14-11-12	113.974
	18-22	1.4198	28-26-27	110.337
	22-23	1.3808	9-1-2	119.174
	21-25	1.4013		
	4-5	1.4163		
	3-8	1.42		
Complex III	16-10	1.9191	1-16-10	98.2308
	16-1	1.9645	17-16-19	95.8283
	16-19	1.9177	17-16-1	100.396
	16-17	1.8772	19-16-10	99.5898
	17-18	1.3124	17-16-10	149.096
	19-20	1.3544	19-16-1	112.393
	20-21	1.4121	16-19-31	127.672
	18-21	1.444	20-19-31	107.656
	19-31	1.5305	16-17-18	122.303
	2-9	1.4616	16-1-9	147.288
	1-2	1.3052	16-10-4	79.9504
	10-4	1.3351	16-1-2	96.9305
	4-3	1.4575	14-11-12	113.102
	18-22	1.4207	28-26-27	108.002
	22-23	1.3822	9-1-2	115.748
	21-25	1.4118		
	4-5	1.4414		
	3-8	1.3989		

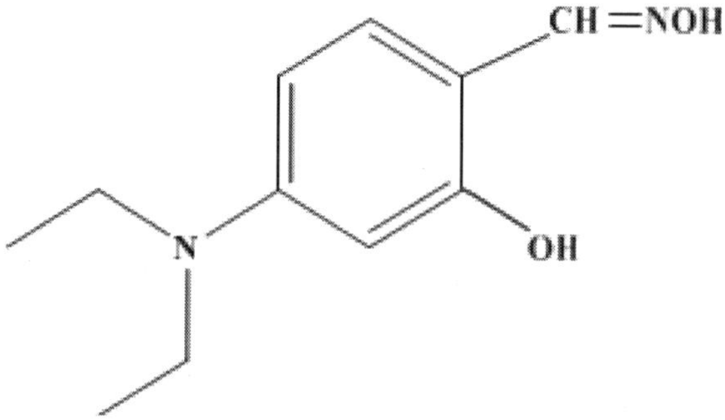

Fig.1 Structure of the ligand: 4-diethylaminosalicyldeoxime(4-DEASDOX)

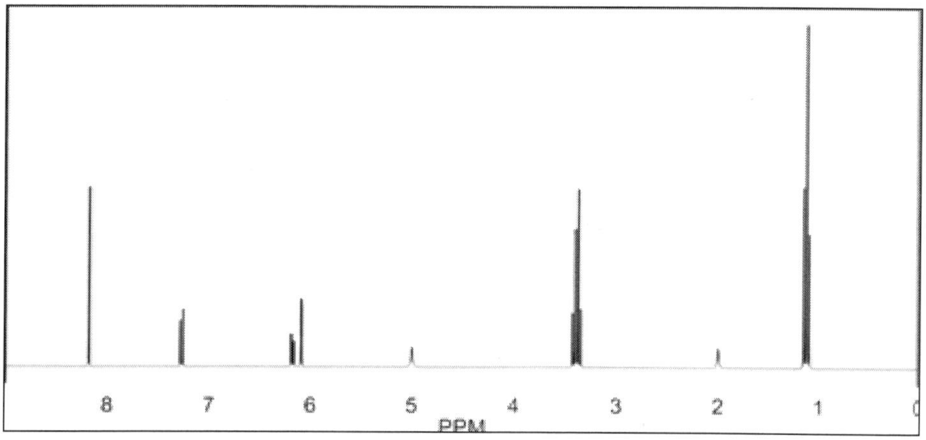

Fig.2 ^{1}H- NMR- Spectra of the ligand

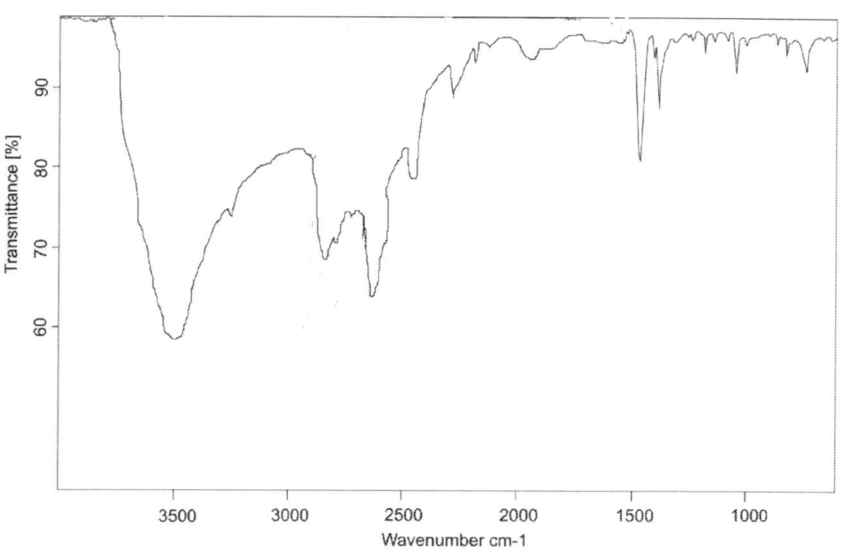

Fig.3a: IR spectrum of the ligand

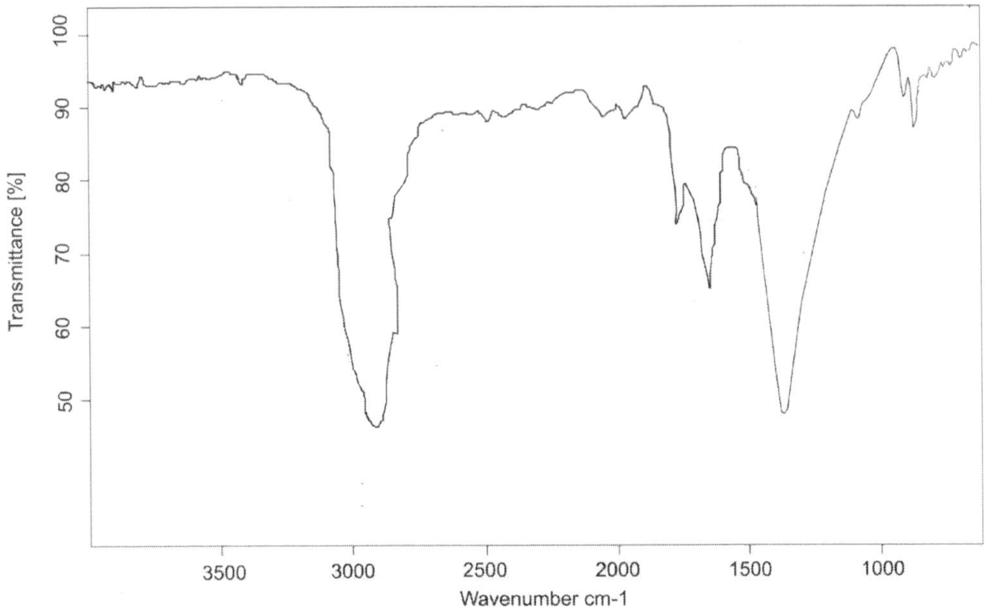

Fig.3b: IR Spectra of the complex I

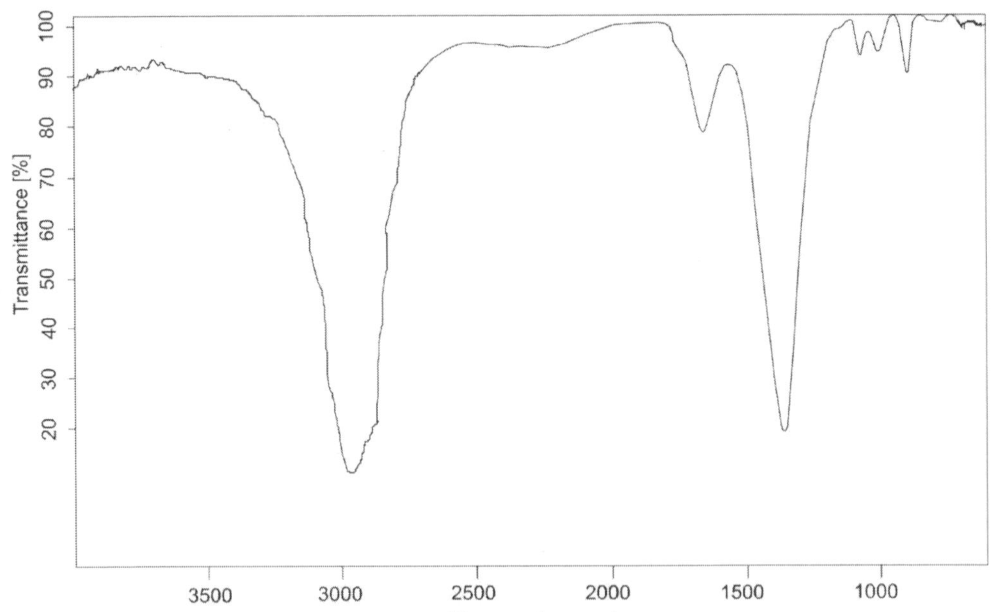

Fig.3c IR Spectra of the complex II

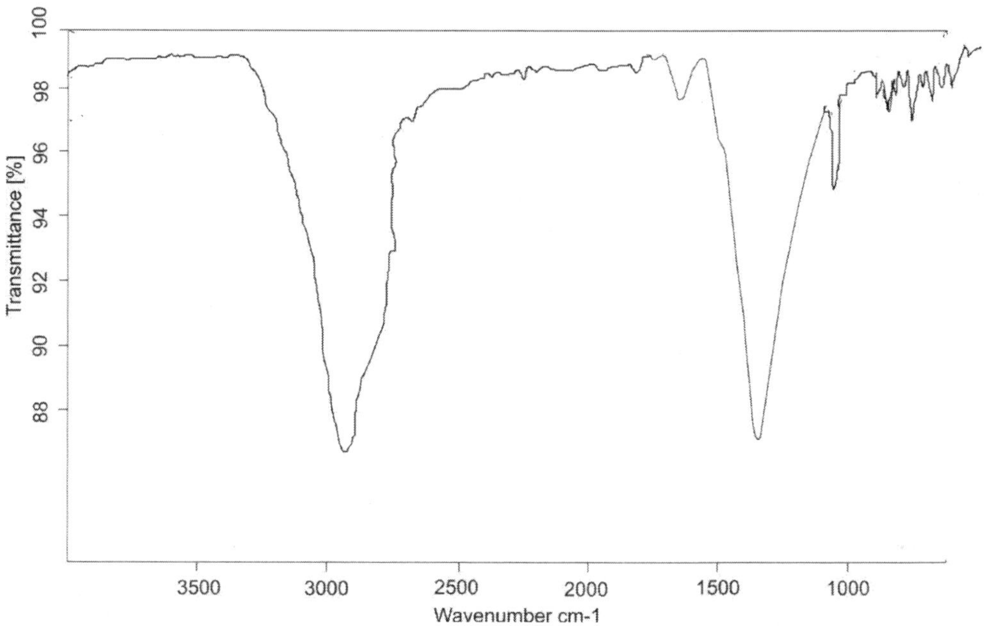

Fig.3d: IR Spectra of the complex III

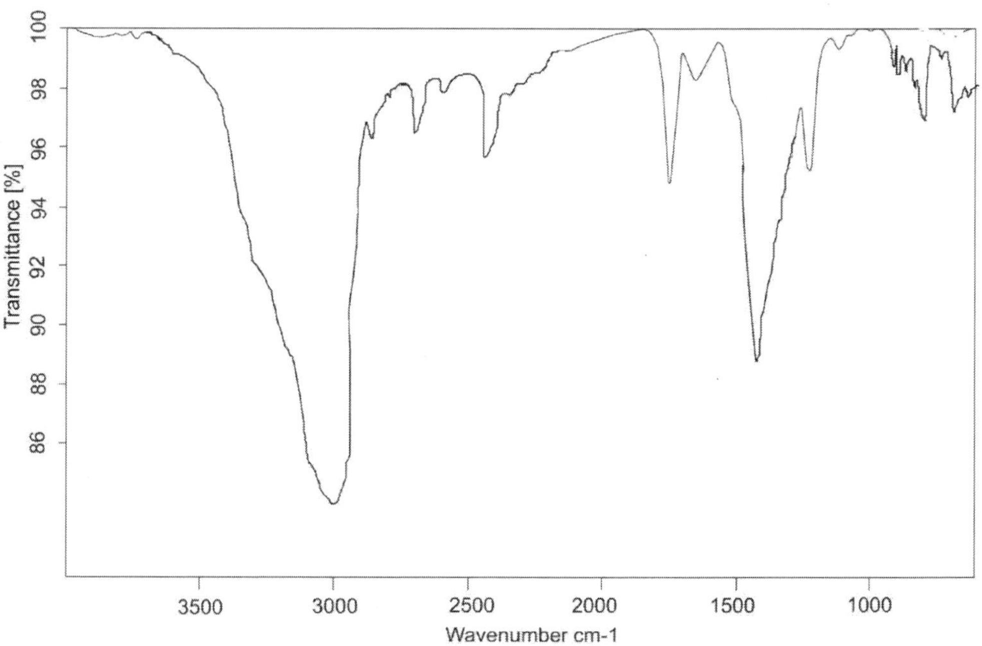

Fig.3e: IR Spectra of the complex IV

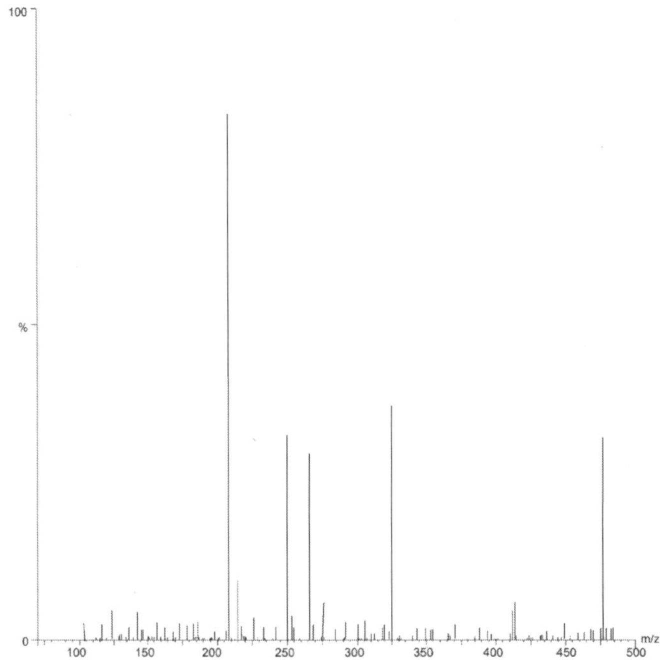

Fig.4 a: TOF- mass spectrum of the complex I

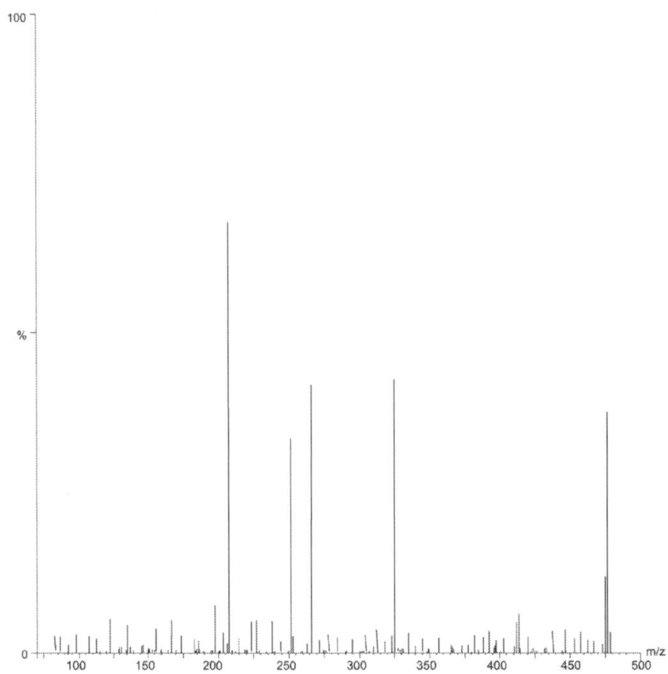

Fig.4b: TOF- mass spectrum of the complex II

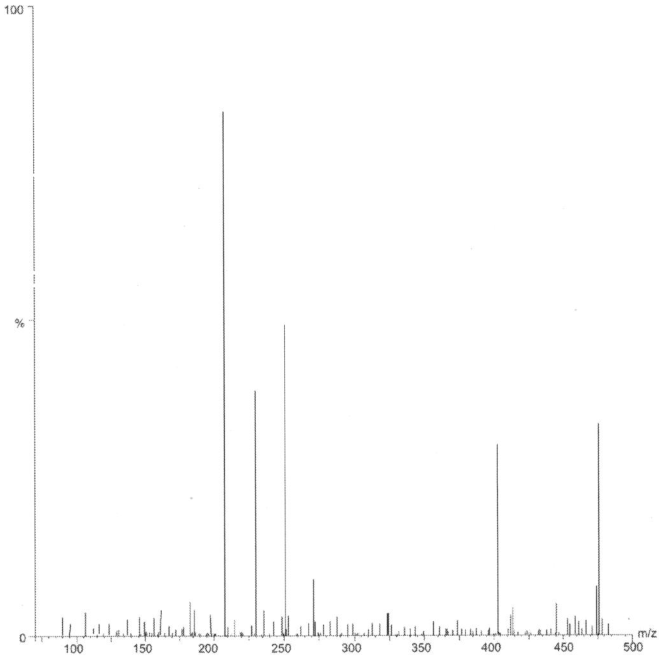

Fig.4c: TOF- mass spectrum of the complex III

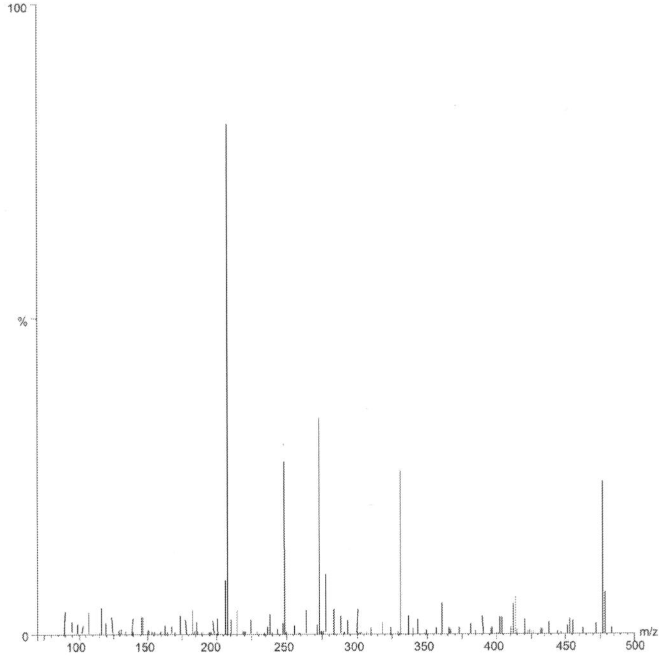

Fig.4d: TOF- mass spectrum of the complex IV

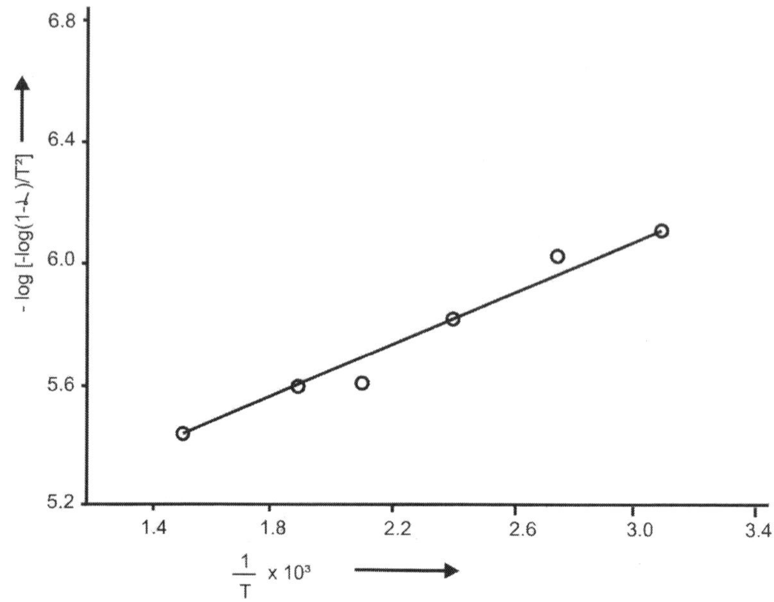

Fig. 5a: Linearization plot of Complex I

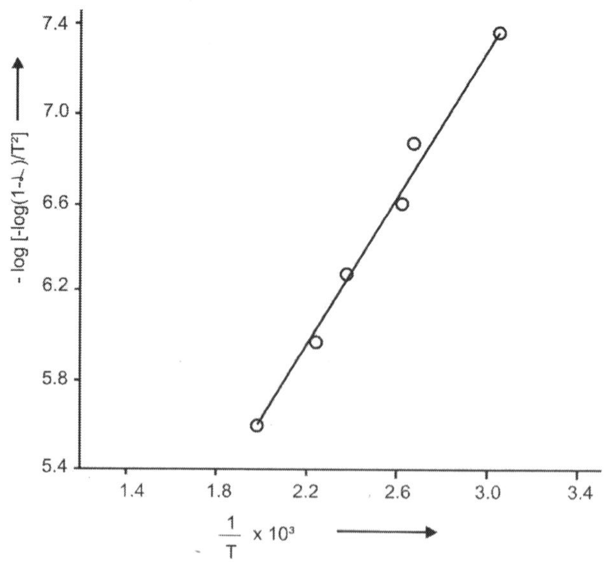

Fig. 5b: Linearization plot of Complex II

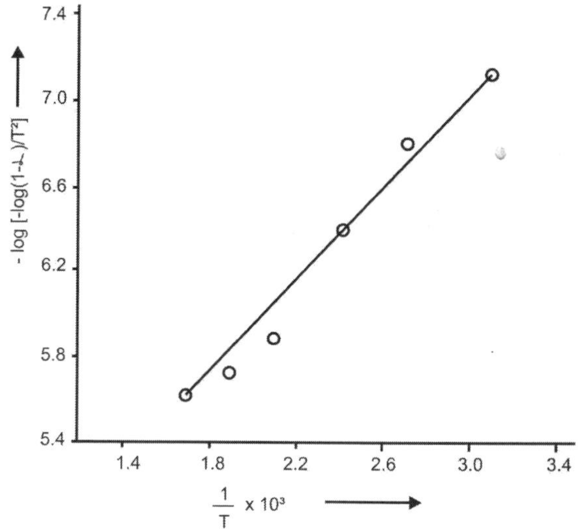

Fig 5c: Linearization plot of Complex III

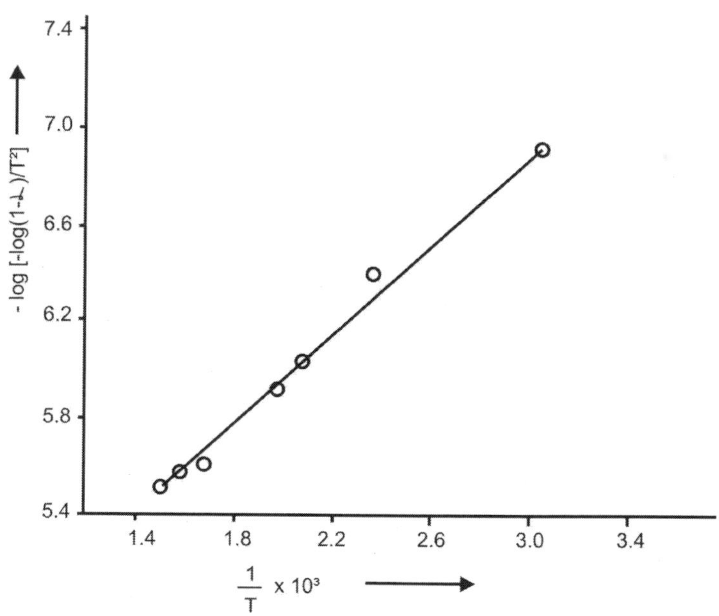

Fig 5 d: Linearization plot of Complex IV

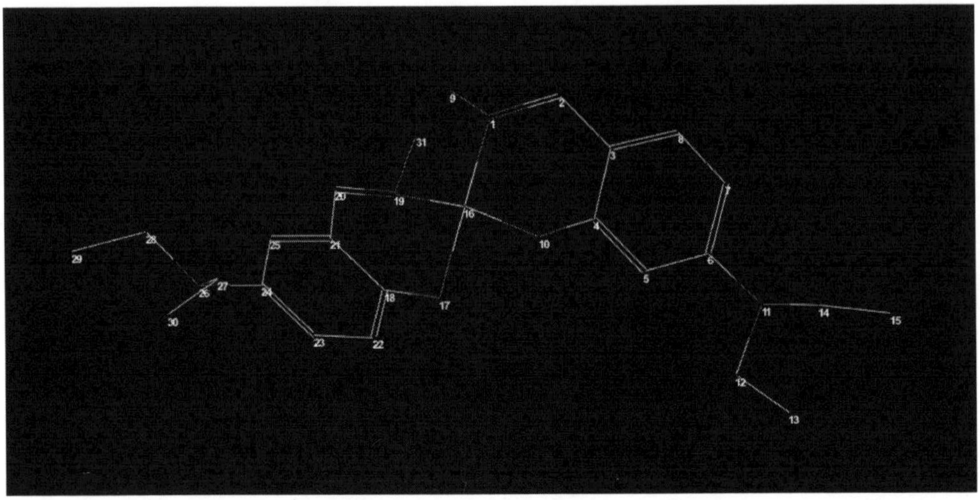

Fig. 6a: Molecular modeling of the complexes II

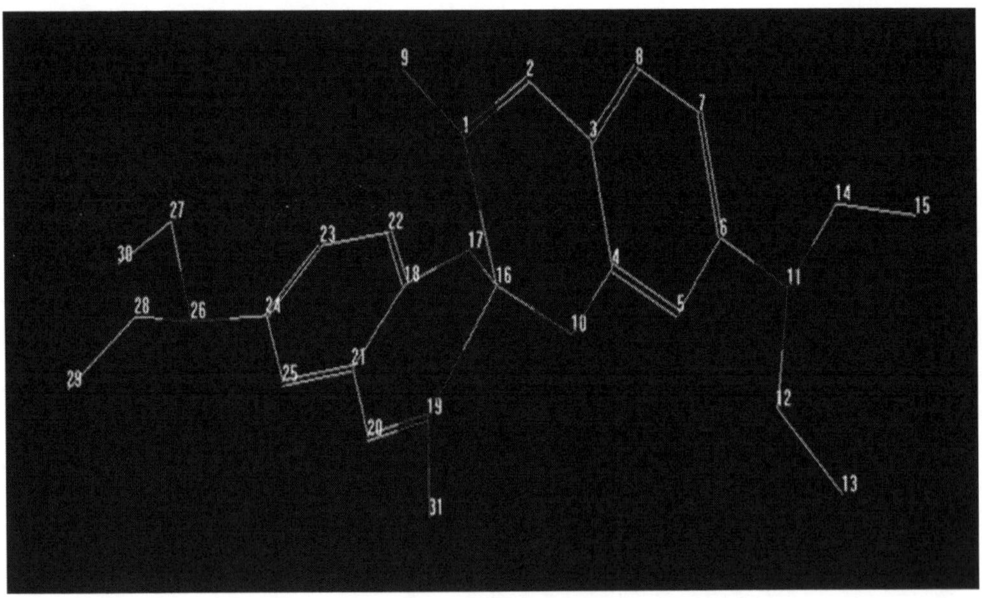

Fig 6b: Molecular modeling of complex III

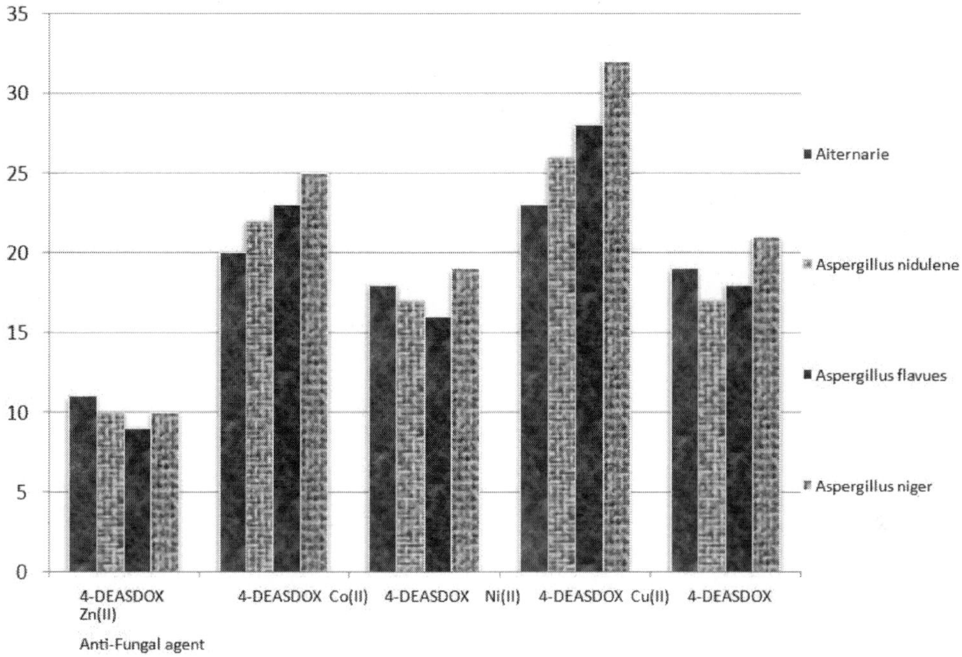

Fig.7a : Antifungal potential of ligand and its metal complexes at 0.6% concentration

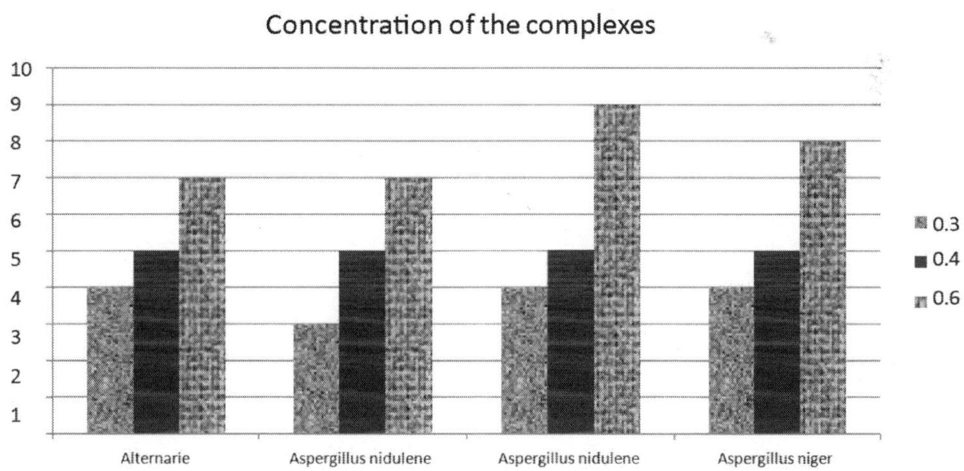

Fig.7b: Effect of concentration of complex III and its antifungal potential

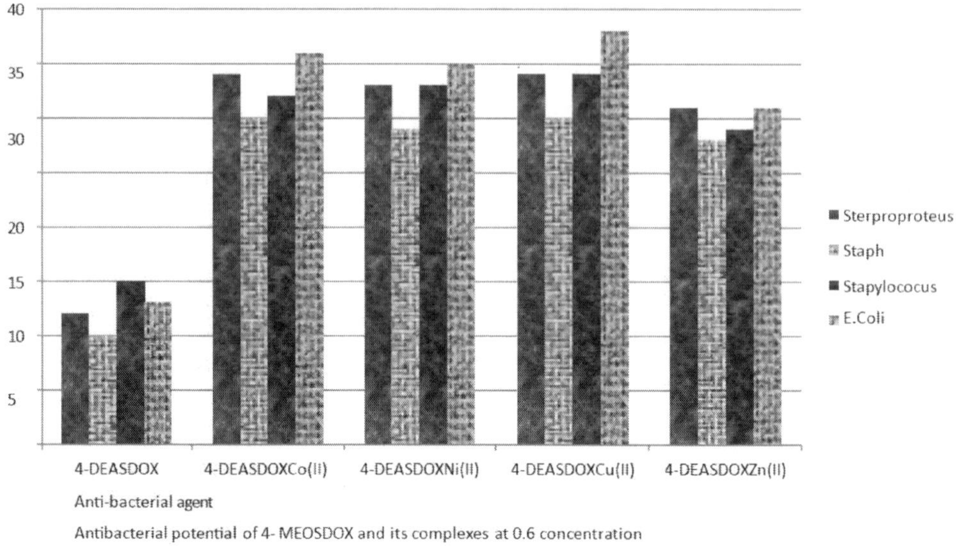

Fig.7c: Anti-bactrial potential ligand and its complexes at 0.6%

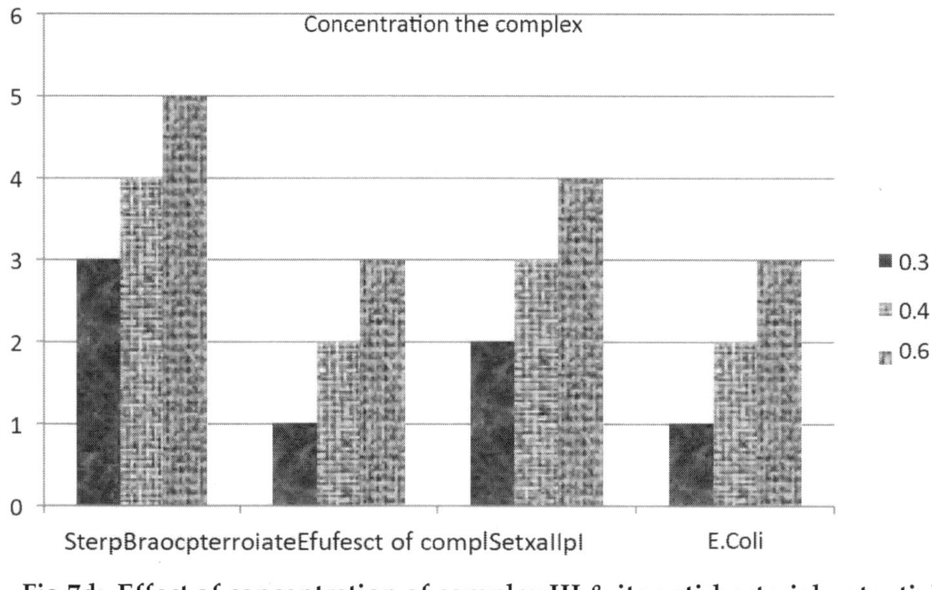

Fig.7d: Effect of concentration of complex III & its anti-bacterial potential

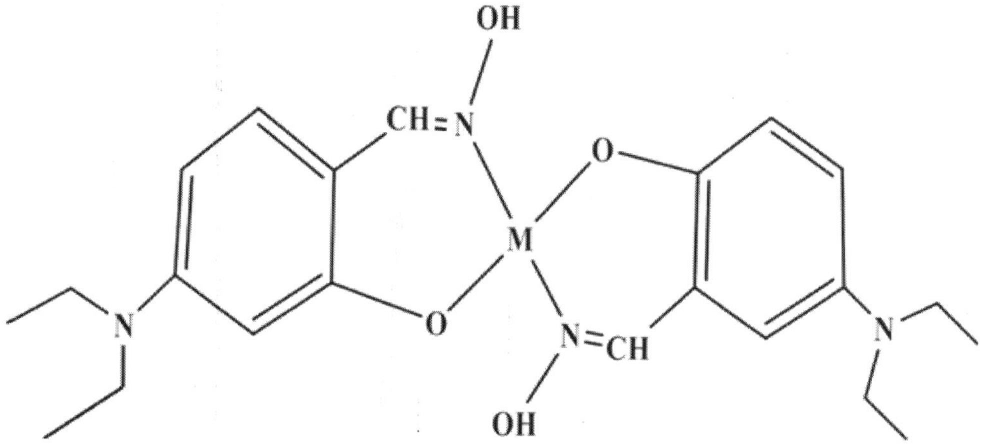

Fig.8: Proposed structure of metal- 4-DEASOX complexes, Where M= Cu(II),Co(II),Ni(II) and Zn(II) .

CHAPTER 13

SIGNIFICANT ROLE OF TRICHODERMA AS A BIOFERTILIZER ON THE GROWTH OF VIGNA MUNGO

Sandeep Sirohi[1*], Surya Prakash D.V[2] and Nitika Vats[2]

1 Department of Botany, Hariom Saraswati P.G. College, Dhanauri, Haridwar, Uttarakhand

2 Department of Biotechnology, Meerut Institute of Engineering and Technology, Meerut

---

Abstract

Microorganisms like bacteria and fungi are mostly thought of as invaders, but some fungi, like *Trichoderma viride*, can be used to help plants grow. The current research efforts are being undertaken to determine the effect of *Trichoderma viride* as a bio fertilizer on plant growth and development. Soil was collected and processed for experimental purposes, and triplicates were created. *Vigna mungo* was chosen for experimental purposes because it is easy to grow and maintain. Pre-treated *V. mungo* seeds were sown in treated soil, and seed germination percentage was calculated along with a number of parameters such as shoot height, root length, and vigour index after seed germination. All parameters except seed germination percentage were measured and calculated at regular intervals of days. Seed germination percentage, shoot height, root length, and vigour index all show significant increases. Based on these findings, it is possible to conclude that 10 gm *T.viride* in 1 kg soil is optimal for seed germination and *Vigna mungo* growth parameters.

Key Words: *Vigna mungo, Trichoderma viride*, Bio-fertilizer, Seed germination and Seed vigour index

Introduction

Trichoderma is a biocontrol agent used to manage soil-borne diseases. Trichoderma is a very important and effective biological mean for plant diseases, which restrict all the diseases, especially soil-borne illnesses. Trichoderma is a free-living fungus commonly found in soil and root systems. It is quite interacting in the root, soil, and leaf environments. It restricts the growth of pathogens, which disturb the plant growth and soil nutrition. (Gary E. Harman, Charles R. Howell, Ada Viterbo, Ilan Chet and Matteo Lorito, 2004). The biostimulant effects of Trichoderma spp. on horticultural crops are extremely varied. Thus, the practical application of Trichoderma sp. necessitates the development of workable formulations and appropriate substrates. *Trichoderma viride* work as a bio-fertilizer. Beneficial microorganisms are increasingly being used in agriculture, with some being commercialized as biopesticide and biofertilizer formulations (Ciccarese *et al.*, 1992). Trichoderma fungi are among the most studied because of their multiple effects on plants, acting as disease biocontrol agents (Harman, 2011) on plant growth and development, as well as their ability to induce defense responses against phytopathogens, insect damage, and abiotic stress. (Vitti *et al.*, 2015) Basically Trichoderma works as a fertilizers and that helps in the growth on plants. Plant species *Vigna mungo* was chosen for the experimental purpose because that plant species are easy to grow generally used by human beings for daily domestic purposes. The effect of Trichoderma on that growth notes in alternative days and checks them in daily bases and saw the effect of different-different amount. Trichoderma also promotes plant development, as seen by seed germination, roots, cutting sprouting, branch growth, increased leaf area, and organic matter buildup.

Taxonomical Status:

Kingdom	:	Fungi
Division	:	Ascomycota
Class	:	Sordariomycetes
Order	:	Hypocreales
Family	:	Hypocreaceae
Genus	:	Trichoderma
Species	:	*T. viride*

Benefits of Trichoderma

Disease control

Trichoderma is a biocontrol agent used to treat soil-borne illnesses. It is effective against pathogenic fungus in several genera, including Fusarium. Plant

Growth Promoter: Trichoderma also work as a fertilizer which enhance the plant growth. Plant has the ability to resist drought (Gawade *et al.*, 2000).

Bioremediation

Bioremediation is a process where biological organisms are used to remove harmful environmental pollutant by their metabolic activities. Trichoderma strains play a vital function in bioremediation of pesticide and herbicide-contaminated soils. Trichoderma can inhibit the growth of insecticides like organophosphates and carbonates (N. Rana Singh, A. Saurabh, and M. Nedunchezhiyan, 2006).

Bio fertilizer

Trichoderma species can easily improve the plant health even in the absence of harmful pathogens. Fungal can easily grow in soil and make such useful environment and work as a fertilizer. (Gawande *et al.*, 2003) Antibiotic production: Trichoderma produce the chemicals that are toxic to the fungi. Some compounds are volatile and travel through the air. Trichoderma is extremely beneficial and vital for the growth of all types of plants and vegetables, including cotton, tomato, potato, legumes, sunflower, ginger, turmeric, tea, pepper, red gram, black gram, etc. Trichoderma is also used for seed treatment.

Origin of Black Gram/ Vigna mungo

Black gram has been cultivated since ancient times all throughout the world, with the majority of it grown in India, Burma, Sri Lanka, Pakistan, China, Fiji, and Africa. India is the leading producer of this legume. our contribution is around 25%. In India almost all the state produce Black gram in large quantity. It is grown in huge amount. In India some states production rate is very high and those states are: Odisha, Maharashtra, Andhra Pradesh, Madhya Pradesh, Gujrat, Rajasthan and Bihar. Black gram is often referred to as mung bean or moong dal. Black gram is a richest source of protein (25%). Mung dal consumes in various ways as a Dal, as sprouts and also makes a good or healthy sweet dish.

Black gram is also utilized as a green manure crop, and it has the ability to fix atmospheric nitrogen while also helping to minimize soil erosion. Black gram is a short-lived crop that may be fed to cattle and has a number of health benefits. Mung dal is one of the richest sources of plant-based protein, containing 211 calories and 14.2 grams of protein. Without increasing the fat, Mung dal helps to increase the protein level in the body (Elad Y *et al.*, 1980). Mung dal also help to reduce the cholesterol level, reduce the symptoms of constipation and reduce the risk of developing heart diseases.

Pharmacological Activity

Antioxidants

Black gram is powerhouse of anti-oxides, and also very beneficial for our health and help to fight against all the diseases, rich source of protein, iron etc. Black gram also increases the immunity and help to protect from all diseases.

Blood pressure control

High blood pressure increases the risk of heart disease and stroke, both of which are primary causes of death. The abundance of potassium and magnesium regulates the blood.

Cognitive health

Black gram is high in iron and protein, which aids in the transport of oxygen and nutrients through the blood to all critical organs and tissues.

Weight Loss

Black gram is a rich source of potent nourishment that also aids in weight management in the body. Black gram is also use as a diet food, and for the healthy diet mung dal is also take as a sprouts and as a healthy salad in our diet.

Black gram is a full of nutrition values like vitamins, protein, magnesium, fibre etc. Adding mung bean is a simple and convenient approach to maintain overall wellness, and when consumed as part of a balanced diet, it may help to prevent obesity, lower cholesterol, regulate blood pressure and blood sugar levels, reduce inflammation, and protect against cancer. This adaptable dal pairs nicely with all of the components and can be made in a variety of ways; include it in your daily diet.

Figure1. *Vigna mungo*

Trichoderma also use as a bio pesticides which help to protect the various crop plants from the pest. Application of Trichoderma formulates with strain mixture perform better than the individual strain (Sanjeev kumar, *et al*, 2014). Now day rice cultivation faces many environmental problems caused by the chemical bio fertilizer. Trichoderma used in the rice field and observed that the increases the nutrition and soil fertility, which help in the plant growth. It has high biotechnological value, with several biological control agents. Trichoderma work as a biocontrol agents as well, they have other beneficial effects on plants and promote the plant growth (Waghunde *et al.*, 2016).

There have been fewer researches to discover root rot and anthracnose resistance sources, and they have been undertaken in both lab/glasshouse and field settings. The majority of the reported resistant materials were produced from cultivars/recombinant lines/breeding lines/land races, while some were from wild relatives (Chang, 1986) or mutant lines. Because screening trials for resistance to Alternaria leaf spot, anthracnose, and root rot infections are rare, more emphasis is needed on these. These resistant lines from various sources can be used as donors to produce resistant variations.

Material and Methodologies

Garden soil was collected from Herbal garden at backyard of Department of Botany and Microbiology, Hariom Saraswati P.G College, Dhanauri, Haridwar. Soil samples were collected in zipped poly bags in triplicate, and transferred to the laboratory for analysis. In laboratory soil samples were processed such as crushed, sieved, and properly mixed and allow to air dried until constant weight was achieved. Treatments (Tc, T5, T10, T15) were prepared as control (Tc) where 100 % pure herbal garden soil was taken, treatment (T5) made by adding 5gm Trichoderma fertilizer in 100% pure soil, and T10, T15 treatment contain 10gm and 15 gm Trichoderma fertilizer in 100% pure soil successively. After that control and treated soil was filled in plastic glasses in equal quantity and rest for some time after sprinkling water on them to make soil soft and favorable for seed germination.

Plant material

Seeds of selected plant species (*Vigna mungo*) and Trichoderma fertilizer were purchased online from Amazone.com

Seed germination experiment

Equal in size and healthy and size *Vigna mungo* was selected and dipped in water for 24 hours. Seeds were settle down in bottom of beaker were chosen for pot experiment. These selected seeds were aseptically sown in amended and unamended pots. Each experiment was performed in the variety of plants. Water spray when required in 24 hours interval starting from the 3 days of showing up to 7 days and germination percentage was obtained. Appearance of plume on the soil surface was considered as germination.

Ecology of Vigna mungo

Vigna mungo/ Mung bean or Black gram of family Fabaceae, is native to subcontinent. In India almost all the state produce black gram in large quantity. It may grow in a variety of well-drained soils, but it prefers fertile sandy loams. It is an upright annual legume with a height range of 15 cm to 1m; the mature plant has an average height of 0.9m. The junctions of the branches and stems are stipule. The first blossoms develop seven to eight weeks after planting, with the crop reaching maturity in 12 to 14 weeks. Pods are produced near the apex of the plant. The seeds are black and nearly spherical. Pods are covered with long, spreading, deciduous silky hairs from pulse food crop plants. Black gram is often referred to as mung bean or moong dal. Black gram is a richest source of protein (± 25%). Mung dal consumes in various ways as a Dal, as a sprouts and also make a good or healthy sweet dish.

Methodologies

PH of the Soil

Soil pH is determined using a 1:1 water to soil ratio. Each sample will take around five minutes to prepare and analyze. Plus fifteen minutes of shaking time for each individual sample. This analysis uses 10 g of soil. Each batch includes a single standard check sample.

Principle

Soil pH can be determined at a 1:1 water ratio. Soil ratios according to standard protocol, they are extremely similar. The pH of soil-water is 0.99 with a ratio of soil to water of 0.04. Can be tested in the field; however, pH measurement is intrinsically incorrect.

Procedure

Measure 10 grams of soil sample into the beaker. Add 20 ml of deionized water, mix often, or set on a shaker for 15 minutes, then calibrate the pH as directed. Insert washed pH and temperature probes into each sample. Keep the probes as still as possible, not touching the wall or bottom of the vessels. This may entail tipping the tank, using a tall vessel, or decanting the supernatant into a test tube. When the numbers have stabilized, record the results. When not measuring, ensure that the probes are washed and immersed in 3M KCL. Between measurements, thoroughly rinse each probe with DI water and gently wipe dry.

Percentage of Seed Germination

The percentage of germination is an estimate of seed viability computed as the ratio of germinated seeds to total seeds sowed represented as a percentage (Yaqub *et al.* 2009).

$$\text{Germination percentage (\%)} = \frac{\text{Total no. of germinated seeds}}{\text{Total no. of seeds}} \times 100$$

Seed Vigour Index (SVI)

This can be determined by adding up the germination percentage and seedling length from the same seed lot. Seed vigour index is computed by multiplying germination rate (%) by seedling length (shoot and root length).The seed lot with the highest seed vigour index is thought to be more vigorous (Shukla *et al.*, 2014).

Seed vigour index = germination percentage X seedling length (shoot height + root length)

Plant Attributes:

Plant shoot height and root length were measured with the help of measuring scale (cm).

Result and Discussion

Soil frequently becomes acidic or alkaline in nature, which can be determined by analyzing its pH level. A pH value is a measure of hydrogen ion concentration, which varies across a large range. When used of Trichoderma viride the nature of soil was acidic in nature. *Trichoderma viride* enhance the production of the soil and work also a pesticide. After using of Trichoderma soil fertility was also increased which help in crop production.

Seed germination is considered as first physiological activity where sensitivity of plant species can be observed clearly in respect of their growth rate. Seed germination percentage in all amendments was observed C (73%), T5 (99%), T10 (100%) and T15 with100 % for (Vigna mungo), whereas minimum seed germination percentage was observed in Vigna radiation Control (75%) and maximum seed germination percentage is 100 % in T10 and T15. Minimum shoot height in 3rdday of V. radiates (8.0cm) in Control and Maximum shoot height is observed in T10 (11cm), Minimum root length of Vigna mungo is observed in Control (2.9cm) and maximum observed in (3.46 cm) in T10, and Seed vigour index minimum in control (280) and maximum in T10 (670).

In 5th day observation minimum shoot height of control i.e. (9.0cm) and maximum shoot height is observe in T10 i.e. (10cm), and minimum root length observed in control i.e. (2.3cm) and maximum root length observed in T10 i.e. (3.2cm) and the seed vigour index in control observed minimum i.e. (890) and maximum in T10 i.e. (1400).

In 7th day observation minimum shoot height of control i.e.(10cm) and maximum observed shoot height in T10 i.e.(11.50cm), Minimum root length observed in control i.e.(2.4cm) and maximum root length observed in T10 i.e. (4cm), and the seed vigour index minimum observed in control i.e. (900) and maximum in T10 i.e. (1650).

Table 1: Seed germination percentage, seedling shoot height, seedling root length and seed vigour in different treatment at different interval in Vigna mungo. (Mean ±S.D)

Days	Parameter	Treatments			
		C	T5	T10	T15
3rd Day	Seed Germination	73.00±27.82	99±0.62	100.00±00.16	100.00±05.02
	Seedling shoot height (cm)	8.00±0.30	10±0.28	11±0.25	10±0.23
	Seedling Root length (cm)	2.9±0.23	3.22±0.25	3.46±0.27	3.6±0.24
	Seed Vigour index	280±3.60	600±3.50	670±3.77	500±2.51
	Parameter	**C**	**T5**	**T10**	**T15**
5th Day	Seedling shoot height(cm)	9.00±0.50	9.5±0.25	11.25±0.55	10.00±0.45
	Seedling Root length (cm)	2.3±0.24	3.1±0.40	3.2±0.5	3.00±0.5
	Seed Vigour index	890±0.25	1300±2.60	1450±3.21	1380±2.64

Days	Parameter	Treatments			
		C	T5	T10	T15
7th Day	Seedling shoot height(cm)	10.00±0.54	10.5±0.26	12.00±0.53	11.50±0.23
	Seedling root length (cm)	2.4±0.25	3.00±0.26	4.00±0.27	3.67±0.3
	Seed Vigour index	900±0.26	1320±2.64	1650±3.21	1400±2.64

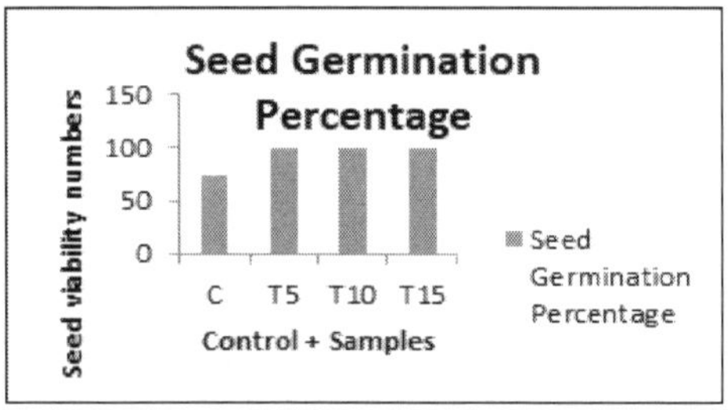

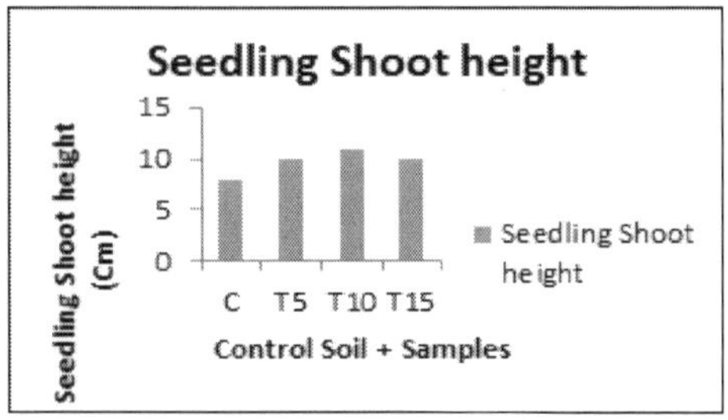

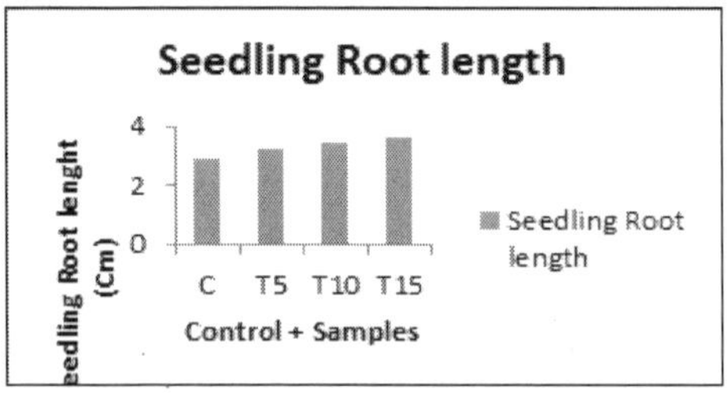

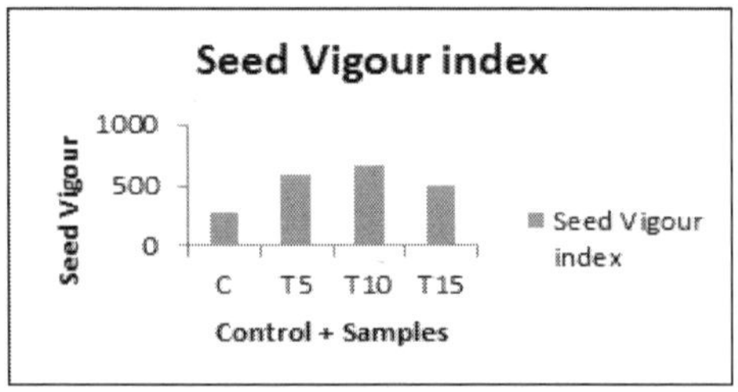

Graph:- 1 Representation of different parameters with control soil sample and deffirent concentration of treated soil on the growth of *Vigna mungo* after three days.

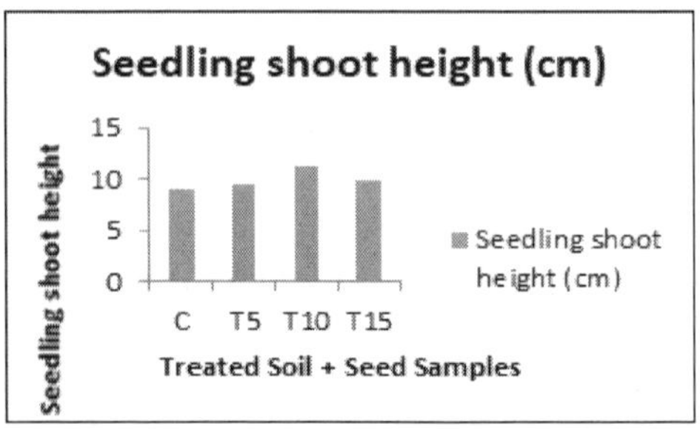

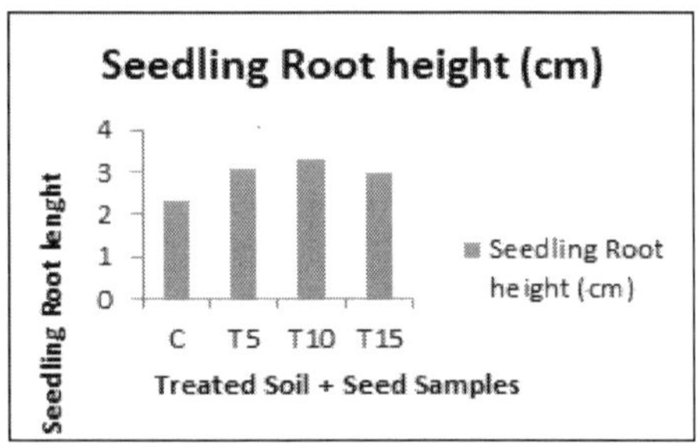

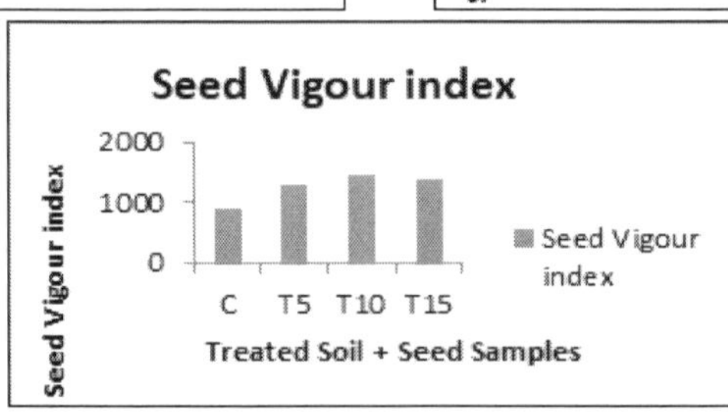

Graph:- 2 Representation of different parameters with control soil sample and

deffirent concentration of treated soil on the growth of *Vigna mungo* after Five days.

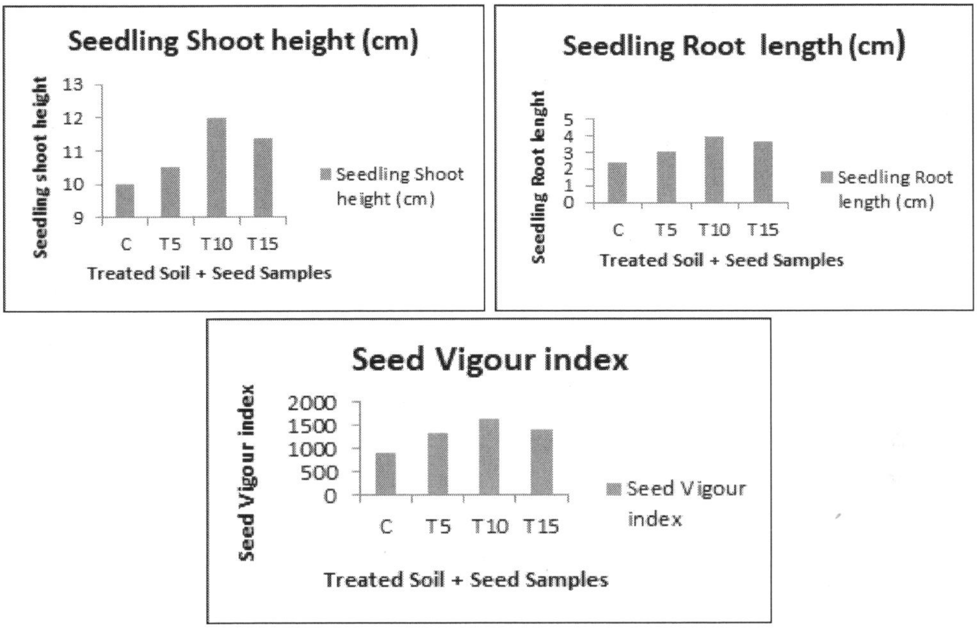

Graph:- 3 Representation of different parameters with control soil sample and different concentration of treated soil on the growth of *Vigna mungo* after Seven days.

Table 2: pH of the Soil treatments (Mean± S.D).

Parameters	Treatments			
	Control	T5	T10	T15
pH	6.9±0.12	6.7±0.15	6.7±0.15	6.6±0.15

Figure 2.pH measure of the herbal garden's soil.

Figure 3. Collected bulk fertile soil sample and processing of Soil from herbal garden of Botany Department.

Figure 4 Seed of Vigna mungo

5 gm. Treatment:

Figure 5 Vigna mungo Seed Sown and after 10 days of water treatment (spray of water to soil moist)

Figure 6 Depicted start of seed germination and consequently Seedling formation after 15 days

10 gm. Treatment:

Figure 7 Vigna mungo Seed Sown and after 10 days of water treatment (spray of water to soil moist)

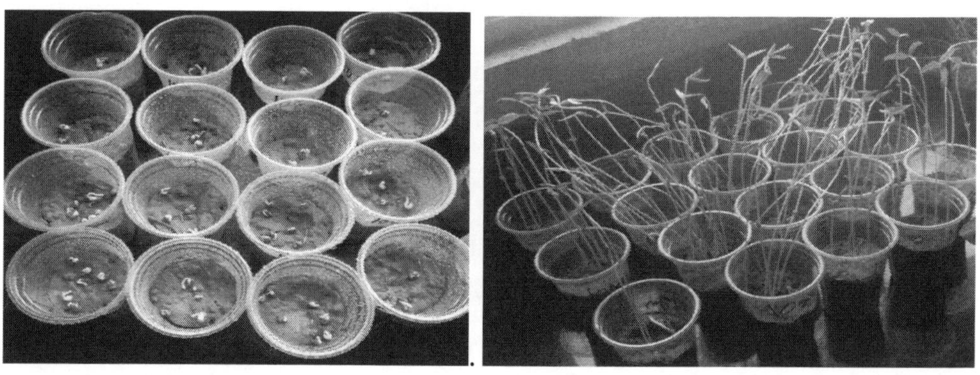

Figure 8 Depicted start of seed germination and consequently Seedling formation after 15 days

15 gm.Treatment:

Figure 9 Vigna mungo Seed Sown and after 10 days of water treatment (spray of water to soil moist)

Figure: 10 Depicted starts of seed germination and consequently Seedling formation after 15 days

Trichoderma work as a fertilizers and it helps to the plant in its growth, different amount of trichoderma works effectively. It helps plants to his proper growth within the month.

Some Trichoderma strains have been shown to boost plant growth, at least in part by enhancing nutrient absorption and nitrogen use efficiency (Sharma *et al.*, 2014; 2004). Our findings confirm the ability of the commercial biocontrol agent *T. viride* strain to increase canopy and stem growth in Moong, and show that *T.viride* is more effective than *T.viride* strain P1, a fungus primarily used in laboratory work and not commercially developed to a significant extent. Furthermore, our data show that the level of growth stimulation is substantially determined by the Moong genotype, indicating that the reaction to *Trichoderma viride* is genetically controlled. Shoot dry weight rose in most lines, with the exception of TA209 with P1 and the wild accession LA1777 with T22. Interestingly, P1 was able to shorten the stem length of LA1777 and had a similar, if less significant, effect on Corbarino, the second indeterminate line under investigation.

Conclusion

Trichoderma has an advantageous impact on the shoot and root lengths of Mung bean (*Vigna mungo*). Using 10 gm. Trichoderma in 1 kg soil yielded good results for seed germination, shoot height, root length, and seed vigour index, which means 10 gm. of Trichoderma in 1kg soil gave the beneficial result, which enhance the plant growth. Thus, it is concluded that Trichoderma is an useful fungus for agricultural and plant growth.

Hence it is indicated that *Trichoderma viride* can be a growth promoter and be used as a bio fertilizer, which enhance the plant growth and protect the plants to the pest. Many beneficial bacteria and fungi work as a pesticide and bio fertilizer, which enhance the crop production. One of them is Trichoderma, Trichoderma work as a pesticide, bio fertilizer and this is also help to make the fertilizer from the kitchen waste as well. *Vigna mungo* which is very beneficial for the health and which also provide the proteins to our body. In the study of Trichoderma effect on plant, one edible plant species *Vigana mungo* were chosen for the experimental purpose as these species are easily to grow and generally used by the human in their daily diet. This experiment's objective was Trichoderma effect on plant growth. The effects of Trichoderma in plants growth was much fined and also enhance the growth of plant, also amend the PH value of the soil. Amendment on chemical properties, Seed germination, Shoot height, Root length and Seed vigour index of the plant species was studied. Plant species were grown in the treatment (TC, T5, T10, T15), Seed germination percentage in all amendments where the maximum Seed germination percentage observed in the T10 and the minimum observed in the Tc. Maximum shoot height observed in the Tc and the Minimum short height observed in T10. Root length maximum observed in T10 and Minimum Root length observed in T10. Plant Vigour Index maximum observed in the T10 and the Minimum plant vigour index observed in the TC treatment. Trichoderma used for as plant promoter and a fertilizer, which enhance the production of crops and also work as pesticides.

References

Asaduzzaman M, Alam MJ, Islam MM (2010) Effect of Trichoderma on seed germination and seedling parameters of chilli. Journal of science foundation. 8(1-2), 141-150. https://doi.org/10.3329/jsf.v8i-2.14637

Ciccarese, F., Frisullo, S., Amenduni, M. and Cirulli, M. (1992): Biological control of Sclerotium rolfsii root rot of sugar beet withTrichoderma harzianum. (Eds.): E.C. Tjamos, G.C. Papavizas and R.J. Cook. pp. 243-248.

Chang YC, Chang YC and Baker (1986) Increased growth of plants in the presence of the Biological Control Agent Trichoderma harzianum. Plant disease.70 (2): 145-148. DOI: 10.1094/ PD-70-145

Elad, Y., Chet, I. and katan J. (1980): Trichoderma harzianum: A biocontrol agent effect against Sclerotium rolfsii and Rhizoctonia solani. Phytopathol, 70: 119-121.

Gawade SB, Zanjare SR, Suryawanshi AV and Shelar VR (2009). Efficacy of bioagents and botanicals on seed mycoflora and seed quality in mungbean. Agricultural science digest- A research journal 36(1), 30-34, 2016

Gawande, S. and Sharma, P. (2003): Changes in host enzyme activity due to induction of resistance against downy mildew in cauliflower. Ann. Agric. Res., 24(2):322-331.

Harman GE (2011). Overview of mechanisms and uses of Trichiderma spp. APS Publications 2006, 96(2), Doi: https://doi.org/10.1094/PHYTO-96-0190

Harman GE, Howell CR, Viterbo A *et al* (2004). Trichoderma species- Opportunities, Avirulent plant symbionts. Doi: 10.1038/nrmicro797

Johnson, L. F. and Curl, E. A. (1972): Methods for research and their ecology of soil borne plant pathogens.Burgess Pub. Com. Mim., pp. 247.

Kumar S, Thakur M and Rani A (2014). Trichoderma: Mass production, formulation, quality control, delivery and its scope in commercilization in india for the management of plant diseases. African journal of agriculture research. 9(53), 3838-3852 (2014). Doi: 10.5897/AJAR2014. 9061

Rana singh N, Saurabh A and Nedunchezhiyan M, 2006. Use of Trichoderma in Disease Management, 0rissa review 2006

Shukla K.K, Newsom RK, Naja M, Singh N *et al.* (2014) Estimation of the mixing layer height over a high altitude site in central Himalyan region bt ussing Doppler Lidar Journal of Atmospheric and Terrestrial physics, 109:48-53 doi:10.1016/j.jastp.2014.01.006

Sharma M, Ghosh R, Tarafdar A et al. (2019) Exploring the Genetic cipher of chickpea through identification and multi-environment validation of resistant sources against fusarium wilt. Front. Sustain. Food syst. (3:78), doi: 10.3389/fsufs.2019.00078

Sharma P, Sharma M, Manokaran R and Shanmugam V (2014), Status of Trichoderma research in india: A review.Indian phytopath. 67(1): 1-19 (2014).

Vitti A, Sofo A, Scopa A and Nuzzaci M (2015) Sustainable Agriculture practices in disease defence of traditional crops in southern italy: The case study of tomoto cheery protected by Trichoderma harzianum T-22 against cucumber mosaic virus (CMV) The sustainability of agro-food and natural resource system in the mediterranen basin (9); doi: 10.1007/978-3-319-16357-4_9

Waghunde R.R., Shelake R.M and Sabalpara A.N (2016) Trichoderma: A significant fungus for agriculture and environment. African journal of Agricultural Research.11 (22):1952-1965. DOI: 10.5897/AJAR2015.10584

Yaqub F and Shahzad S (2009) Effect os solar heating by polyethylene mulching on sclerotial viability and pathogenicity of sclerotium rolfsii on mungbean and sunflower. Pak. J. Bot., 41(6):3199-3205.

CHAPTER- 14

DIVERSITY OF AQUATIC HYPHOMYCETES ALONG WITH PHYSICO-CHEMICAL PARAMETERS IN FRESH WATER STREAMS OF KUMAUN HIMALAYA, UTTARAKHAND, INDIA

Ruchi Jalal and Saraswati Bisht*

Department of Botany, Indira Priyadarshini Govt. Girls PG College of Commerce, Haldwani- 263139, India

---

Abstract

Aquatic hyphomycetes or water borne conidial fungi are important group of micro-organisms playing key role as bio-monitors in any fresh water body. They are present preferentially in clean and well aerated aquatic habitats (Barlocher, 2005). The diversity of aquatic hyphomycetes in any freshwater body is directly corelated with its physico-chemical parameters. In the present study different physico-chemical parameters of some water bodies of Kumaun Himalaya were assessed viz: Temperature, pH, TDS, DO and BOD. The water analysis of these streams revealed that physio-chemical parameters can play an important role in the diversity of aquatic hyphomycetes. A total of 30 species belonging to 19 genera were isolated from different freshwater streams of Kumaun Himalaya during the period of February 2019- February 2023. It is observed that there is a common trend in these fungi to sporulate more during winters when pH, temperature, TDS and BOD of water were low and DO was high. Although low tempearture, pH, TDS, BOD and high DO supported maximum growth of these fungi still temperature was proved to be the most important governing factor for the sporulation of these fungi.

Keywords: Aquatic Hyphomycetes; Physico-chemical parameters; Freshwater streams

Introduction

The aquatic ecosystem has its own mycoflora in which Aquatic Hyphomycetes, a phylogenetically heterogenous group of fungi occupies an important place. These fungi are mainly the inhabitants of well-aerated cold water-bodies present dominantly on submerged, dead and decomposed plant material (Gulis et al., 2005). They have particular significance in decomposition of allochthonous dead plant litter which are the main source of carbon and energy in aquatic bodies (Gulis et al., 2019; Gossiaux et al., 2019). They are usually negatively correlated with the amount of nutrients in water, but positively correlated with dissolved oxygen concentration (Sole et al, 2008).

Temperature and pH are considered to be the most influential physico-chemical parameters affecting the diversity and species composition of aquatic hyphomycetes (Madigan et al., 2009). These fungi are very much sensitive to high temperatures and their sporulation peaks at lower temperatures (15-25 °C) (Dang et al., 2009). These fungi thrive in slightly acidic to neutral pH conditions, typically around 5 to 7. The enzymes produced by these fungi may become less effective at very low or high pH levels, reducing their ability to break down organic material. Changes in TDS concentrations in natural waters often result from industrial effluent thereby increasing water salinity, changing ion composition and ion toxicity (Weber-Scannell and Duffy, 2007). It has been reported earlier that increasing water salinity supresses the growth of aquatic hyphomycetes, potentially leading to shifts in their community composition (Derry et al., 2003).

Aquatic hyphomycetes require a fresh oxygenated environment to break down organic matter and carry out their decomposition processes (Sati and Arya, 2009). Being inhabitants of well-oxygenated water bodies, these fungi are sensitive to changes in DO levels, thus in areas with oxygen-depleted conditions, their diversity get minimized. Similarly, Biological Oxygen Demand, which is a measurement of oxygen consumed by microorganisms to decompose organic matter in water, is also an important physico-chemical parameter which influences the growth of aquatic hyphomycetes. High BOD levels can cause oxygen depletion in water, leading to hypoxia (absence of oxygen). In such oxygen-deprived conditions, the diversity of certain aquatic hyphomycetous species that are sensitive to low oxygen, may decline.

The changing physico-chemical parameters of any water body greatly affect the species composition of microbial community and may also result in the decline of some sensitive species. The presence or absence of aquatic hyphomycetes in any water body indicates, that physico-chemical parameters of that water body is changing. Though, a lot of information is available on the physico-chemical parameters of water but very few reports are available on the response pattern of aquatic hyphomycetes on physico-chemical parameters (Sati and Bisht 2005; Sati and Bisht, 2006; Sati and Arya, 2009; Arshad and Barreen, 2009). Thus, the present study aims to assess water quality of the selected water bodies so as to co-relate the diversity of aquatic hyphomycetes with physico-chemical parameters of water.

Material and Methods

Sampling: Samples were collected seasonally from some freshwater bodies of Kumaun Himalaya.

For effect of seasonal variation on diversity of aquatic hyphomycetes relative contribution was calculated as follows:

Relative Contribution of Different Seasons in Diversity of Aquatic hyphomycetes = $\frac{\text{Number of species reported from a season}}{\text{Total number of species}} \times 100$

Water analysis and measurement of water parameters:

Temperature:

Temperature was measured on site by dipping a degree centigrade (°C) thermometer at a depth of 10 cm in water for 2-3 minutes.

pH and TDS:

pH and TDS were measured on site using portable pH and TDS meters from Hanna. The meters were dipped in water at a depth of 5 cm for 2-3 minutes.

DO:

Dissolved Oxygen (DO) or Initial Dissolved Oxygen (DO_i) was measured on site by using Winkler's test based Dissolved Oxygen test kit from Aquasol.

BOD

For the measurement of BOD, water samples were collected in 500 ml polylab polypropylene bottles and were carefully brought to laboratory so as to measure BOD following Winkler's test.

BOD was calculated using the formula given below:

BOD= $DO_i - DO_f \times$ Dilution factor

DO_i = initial dissolved oxygen measured on-site

DO_f = final dissolved oxygen measured in laboratory.

Readings of all the parameters (Temperature, pH, TDS, DO and BOD) were taken in three replicates and their mean values were calculated to record the final readings.

Results and Discussions

In the present study it is observed that lower temperature, pH, TDS and higher Dissolved Oxygen during winter season helps these fungi to thrive best in winters. The results are summarised in Table 1 and 2 and Graphically presented in Figure 1, 2 and 3.

Water temperature of all the study sites during summer season was recorded to be 22-25 °C while during winter it was 10-10.6 °C (Table 2.; Fig. 1.). As the result shows minimum occurrence of these fungi during summers (Table 1.) therefore, it can be concluded that high temperature could be a main reason for the less occurrence of these fungi in summer season. Previous studies have also reported, temperature between 10-20 °C to be optimum for the growth of these fungi (Barlocher et al., 2008; Jalal et al., 2020). Several other workers have reported the influence of water temperature on species composition of aquatic hyphomycetes and demonstrated that cooler temperature is most preferable for the growth of these fungi (Seena et al., 2019; Khalil et al., 2023; Barreto et al., 2023).

The present results stipulate that water pH is relatively high in streams during summer season. This could be a possible reason that presence of aquatic hyphomycetes is less during summers. In contrast to summers, water pH was slightly neutral in winter season, supporting the sporulation of aquatic hyphomycetes in winters (Table 2.; Fig. 2.). Earlier studies have also reported that diversity of aquatic hyphomycetous species is generally high within a pH range of 5-7 and declines above it (Chuaseeharonnachai et al., 2013).

Water TDS ranged between 95-190 ppm during summer season and between 96-184 ppm during winter season (Table 2.; Fig. 3.). A greater number of aquatic hyphomycetous species were isolated when TDS values were between 100-184 ppm, suggesting that this could be the suitable TDS range for these fungi. This is in accordance with previous studies (Meride and Ayenew, 2016), which stated that water with TDS value between 100-195 is ideal for drinking purpose and survival of aquatic organisms.

Dissolved oxygen concentration of water bodies greatly affects the diversity of aquatic hyphomycetes. In all the streams, maximum water DO (5.2-8.3 mg/l) was recorded during winter season supporting the greater diversity of aquatic hyphomycetes. Similarly, minimum water DO (3.1-4.4 mg/l) was recorded during summer season in all the water bodies which supported less diversity of aquatic hyphomycetes. Low to moderate water DO (4.2-6.8) support moderate diversity of aquatic hyphomycetes in rainy season (Table 2.; Fig. 4.). Thus, it can be concluded that DO level decreases during summer season when water temperature and pH are higher, ultimately leading to the lesser sporulation of aquatic hyphomycetes during summers. Similar results were also observed in previous studies that aquatic hyphomycetes require a fresh oxygenated environment for their occurrence and sporulation (Medeiros et al., 2009; Sridhar et al., 2013; Dubey, 2016, Bai et al., 2018; Breda et al., 2021).

In most of the streams lower BOD was recorded during winter season as compared to that of summer and rainy seasons, as the diversity of aquatic hyphomycetes was observed maximum during winter season, thus it can be concluded that lesser BOD value of water may be responsible for the increased growth of aquatic hyphomycetes during winters (Table 2.; Fig. 5.). In contrary, higher BOD was observed in streams of Ramgaarh during winter season due to high anthropogenic activities and a greater number of aquatic hyphomycetes were recorded in the winters, this indicates that some of these fungi can also persist in highly stressed conditions and become tolerant species (Chauvet et al., 2016).

Conclusion

The present study, concludes that physico-chemical parameters of water have significant effect on diversity of aquatic hyphomycetes. Although, low temperature, pH, TDS, BOD and high DO supports the maximum growth of these fungi still temperature was proved to be the most important governing factor for the sporulation of these fungi. The observed water quality parameters showed significant seasonal variations, as higher temperature, TDS, pH and BOD were recorded in summer season, while higher values of DO was recorded in winter season. The influe nce of water quality parameters on species-composition of aquatic hyphomycetes may also change the functional attributes of running aquatic bodies, since these organisms play a significant role in decomposing the submerged leaf litter present in streams (Breda et al., 2021).

Acknowledgements

The authors are thankful to CURIE Project of DST (Ref: DST/CURIE-PG/2022/90) and Jawaharlal Nehru Memorial Fund, New Delhi (Jawaharlal Nehru Scholarships for Doctoral Studies- Ref: SU-1/76/2021-22/24) for providing financial support.

References

Arshad, M. and Bareen, F. (2009). Ecological studies of aquatic hyphomycetes in a canal and its connecting irrigation channels. *Biological Diversity and Conservation.* 2 (3): 99-106. 1.

Bai, Y., Wang, Q., Liao, K., Jian, Z., Zhao, C. and Qu, J. (2018). Fungal community as a bioindicator to reflect anthropogenic activities in a river ecosystem. *Frontiers in Microbiology.* 9. 3152. https://doi:10.3389/fmicb.2018.03152.

Barlocher, F. (2005). Freshwater fungal communities. *The Fungal Community: Its organization and Role in the Ecosystem. 3rd edn (Dighton J, White J.F. and Oudemans*

P. eds.). CRC Press, Boca Raton, F.L. 39-59. ISBN: 9780429116407. https://doi.org/10.1201/9781420027891.

Barlocher, F., Seena, S., Wilson, K.P. and Williams, D.D. (2008). Raised water temperature lowers diversity of hyporheic aquatic hyphomycetes. *Freshwater Biology*. 53: 368-379. https://doi:10.1111/j.1365-2427.2007.01899.x

Barreto, G.G., Hepp, L.U., Rezende, R. de S., Goncalves Junior, J.F., Moretti, M. da S., Moretto, Y., Loureiro, R.C., Restello, R.M. and Medeiros, A.O. (2023). The cooler the better: Increased aquatic hyphomycete diversity in subtropical streams along a neotropical latitudinal gradient. *Fungal Ecology*. 62: 101223. https://doi.org/10.1016/j.funeco.2022.101223.

Breda, M., Binotto, A.C., Biasi, C. and Hepp, L.U. (2021). Influence of environmental predictors on hyphomycete assemblages in subtropical streams. *Acta Oecologica*. 113, 103778. https://doi.org/10.1016/j.actao.2021.103778.

Chauvet, E., Cornut, J., Sridhar, K.R., Selosse, M.A. and Barlocher, F. (2016). Beyond the water column: aquatic hyphomycetes outside their preferred habitat. *Fungal Ecology*. (19): 112-127. http://dx.doi.org/10.1016/j.funeco.2015.05.014.

Chuaseeharonnachai, C., Yamaguchi, K., Sri-Indrasutdhia, V., Somrithipol, S., Okanec, I., Nakagirid, A. and Boonyuen, N. (2013). Diversity of aero-aquatic hyphomycetes from six streams in Doi Inthanon and Khao Yai tropical forests, Thailand. *Cryptogamie Mycologie*. 34 (2): 183-197. https://doi.org/10.7872/crym.v34.iss2.2013.183.

Dang, C.K., Schindler, M., Chauvet, E. and Gessener, M.O. (2009). Temperature oscillations coupled with fungal communities can modulate warming effect on litter decomposition. *Ecology*. 9: 122-131.

Dubey, A. (2016). Aquatic hyphomycetes communities occur as potential bioindicators of water quality. *International Journal of Applied Research*. 2 (8): 778-780.

Gossiaux, A., Jabiol, J., Poupin, P., Chauvet, E. and Guerold, F. (2019). Seasonal variations overwhelm temperature effects on microbial processes in headwater streams: insights from a temperature thermal spring. *Aquatic Sciences- Research Across Boundaries*. 81 (2). https://doi.org/10.1007/s00027-019-0627-2.

Gulis, V.I., Marvanova, L. and Descals, E. (2005). An illustrated key to the common temperate species of aquatic hyphomycetes. MAS Graca, F., Barlocher and MO Gessner (eds.) *Methods to study litter decomposition: A practical guide. Springer*. 153-167. https://doi.org.10.1007/1-4020-3466-021.

Gulis, V.I., Su, R. and Kuehn, K. (2019). Fungal decomposers in freshwater environments. *In: Hurst C (Ed.). The structure and Function of Aquatic Microbial Communities. Springer International Publishing, Cham*. pp 34. ISBN: 9783030167752. https://doi.org/10.1007/978-3-030-16775-2_5.

Jalal, R., Bisht, S., Altaf, S. and Tiwari, A. (2020). Diversity of water-borne conidial fungi in some freshwater bodies of Kumaun Himalaya in district Nainital (Uttarakhand), India. *Journal of Applied and Natural Science*. 12 (4): 484-490.

Khallil, A.R.M., Ali, E.H., Ibrahim, S.S. and Hassan, E.A. (2023). Seasonal fluctuations and diversity of Ingoldian mycobiota in two water bodies receiving different effluents at Assiut Governorate (Upper Egypt). *BMC Microbiology*. 23: 163. https://doi.org/10.1186/s12866-023-02903-z.

Madigan, M.T., Martinko, J.M., Dunlap, P.V. and Clark, D.P. (2009). *Biology of Microorganisms. (3rd Custom Ed. University of Central Florida) Pearson/Benjamin Cummings, San Fransisco.* ISBN: 9780536871244.

Medeiros, A.O., Pascoal, C. and Graca, M.A.S. (2009). Diversity and activity of aquatic fungi under low oxygen conditions. *Freshwater Biology*. 54: 142-149. http://dx.doi.org/10.1111/j.1365-2427.2008.02101.x.

Meride, Y. and Ayenew, B. (2016). Drinking water quality assessment and its effects on residents health in Wondo genet campus, Ethiopia. *Environmental Systems Research*. 5: 1-7.

Sati, S.C. and Arya, P. (2009). Occurrence of water borne conidial fungi in relation to some physico-chemical parameters in a fresh water stream. *Nature and Science*. 7 (4): 20-28.

Sati, S.C. and Bisht, S. (2005). Growth responses of aquatic hyphomycetes to different nitrogen sources. *Sydowia*. 58 (1): 123-132.

Sati, S.C. and Bisht, S. (2006). Utilization of various carbon sources for the growth of waterborne conidial fungi. *Mycologia*. 98 (5): 678-681.

Seena, S., Barlocher, F., Sobral, O., Gessner, M., Dudgeon, D., Mckie, B., Chauvet, E., Boyero, L., Ferreira, V., Frainer, A., Bruder, A., Matthaei, C., Fenoglio, S., Sridhar, K., Albsrino, R., Douglas, M., Encalada, A., Garcia, E., Ghate, S., Giling, D., Goncalves, V., Iwata, T., Landeira- Dabarca, A., McMaster, D., Medeiros, A., Naggea, J., Pozo, J., Raposeiro, P., Swan, C., Tenkiano, N.D., Yule, C. and Graca, MS. (2019). Biodiversity of leaf litter fungi in streams along a latitudinal gradient. *Science of the Total Environment*. 661: 306-315. https://doi.org/10.1016/j.scitotenv.2019.01.122.

Sole, M., Fetzer, I., Wennrich, R., Sridhar, K.R., Harms, H. and Krauss, G. (2008). Aquatic hyphomycete communities as potential bioindicators for assessing anthropogenic stress. *Science of the Total Environment*. 389: 557-565. https://doi:10.1016/j.scitotenv.2007.09.010.

Sridhar, K.R., Karamchand, K.S. and Seena, S. (2013). Fungal assemblage and leaf litter decomposition in riparian tree holes and in a coastal stream of the south-west India. *Mycologia*. 4 (2): 118-124. https://dx.doi.org/10.1080/21501203.825657.

Weber-Scannell, P.K. and Duffy, L.K. (2007). Effects of total dissolved solids on aquatic organisms. A review of literature and recommendation for Salmonid species. *American Journal of Environmental Sciences*. 3 (1): 1-6. https://doi.org/10.3844/ajessp.2007.1.6.

Table 1. Species Composition in Relation to Seasonal variation

S. No.	Species	Summer	Rainy	Winter
1.	*Acaulopage dichotoma*	+	-	-
2.	*Alatospora acuminata*	+	+	+
3.	*Alatospora constricta*	-	+	-
4.	*Anguillospora crassa*	+	+	-
5.	*Anguillospora longissima*	+	-	-
6.	*Anguillospora pseudolongissima*	-	+	-
7.	*Anguillospora rostrate*	-	+	-
8.	*Beltrania rhombica*	+	+	-
9.	*Campylospora chaetocladia*	+	+	+
10.	*Clavariopsis aquatica*	+	-	+
11.	*Cylindrocarpon aquaticum*	+	+	+
12.	*Flagellospora penicilloides*	-	-	+
13.	*Flagellospora curvula*	-	-	+
14.	*Flabellospora acuminata*	-	-	+
15.	*Helicomyces roseus*	-	+	-
16.	*Heliscus lugdunensis*	-	-	+
17.	*Lemonniera cornuta*	-	-	+
18.	*Lemonniera pseudofloscula*	-	+	+
19.	*Lemonniera terrestris*	-	-	+
20.	*Lunulospora curvula*	-	+	+
21.	*Lunulospora cymbiformis*	-	+	+
22.	*Pestalotiopsis submerses*	-	-	+
23.	*Setosynnema isthmosporum*	-	+	+
24.	*Subulispora procurvata*	-	-	+
25.	*Tetrachaetum elegans*	-	+	+
26.	*Tetracladium apiense*	-	-	+
27.	*Tetracladium breve*	-	-	+
28.	*Tetracladium marchalianum*	+	+	+
29.	*Triscelophorus acuminatus*	-	+	+
30.	*Triscelophorus monosporus*	-	+	+
	Total	9	17	22
	Relative Contribution (%)	30	56.6	**73.3**

+ = Present; - = Absent

Table 2. Physico-Chemical Parameters of Different Water Bodies along with Different Seasons

Site	Summer					Rainy					Winter				
	Temp.	pH	TDS	DO	BOD	Temp.	pH	TDS	DO	BOD	Temp.	pH	TDS	DO	BOD
Ramgaarh	24.6±0.3	7.9±0.03	123.3±8.8	4.16±0.16	0.56±0.23	21.3±0.3	**7.8±0.03**	110±0.6	6.86±0.03	0.56±0.26	10.03±0.03	**7.7±0.08**	100±3.3	**8.3±0.6**	**1.16±0.16**
Devdwar	23.3±0.3	7.8±0.03	95±7.6	3.06±0.06	0.5±0.11	18.3±0.3	7.1±0.2	104±3.04	4±0.1	1.4±0.14	10.4±0.2	7.3±0.2	96±5.7	7.8±0.4	0.5±0.28
Naikana	22.3±0.3	7.7±0.1	190±5.7	**4.4±0.06**	0.96±0.08	18.3±0.3	7.5±0.1	106±3.3	**4.9±0.03**	0.8±0.2	10.6±0.6	7.5±0.05	**184±3**	5.2±0.1	0.23±0.08

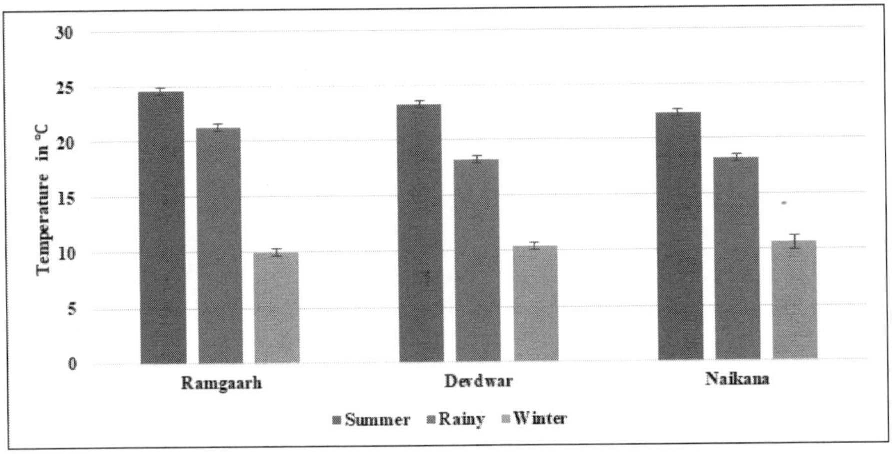

Fig 1. Water Temperature of Selected Streams in Different Seasons

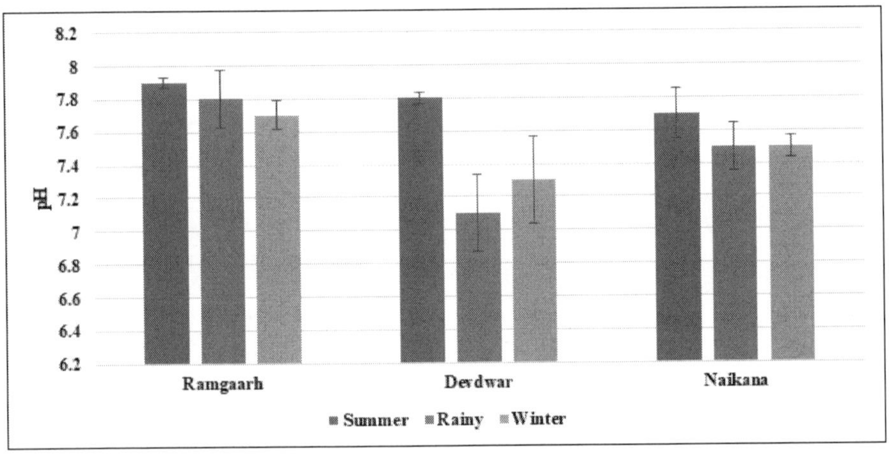

Fig 2. Water pH of Selected Streams in Different Seasons

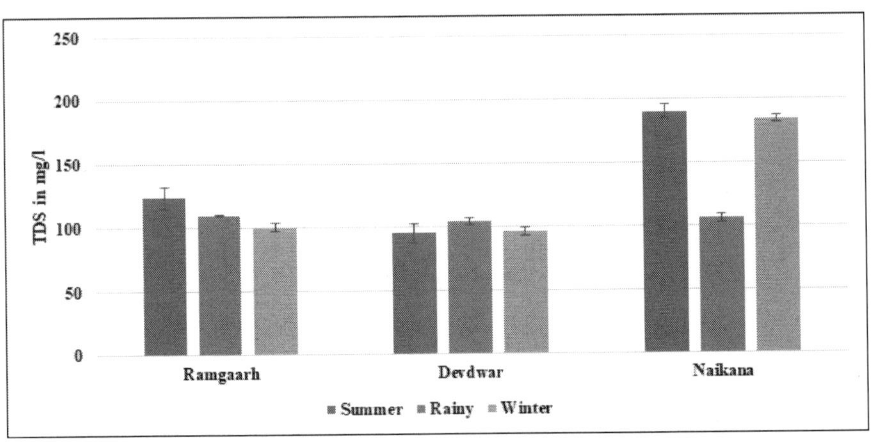

Fig 3. Water TDS of Selected Streams in Different Seasons

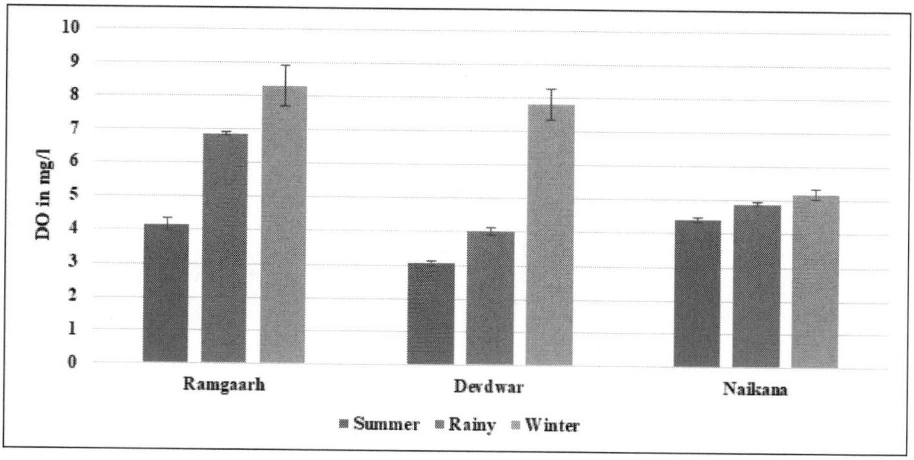

Fig 4. Water DO of Selected Streams in Different Seasons

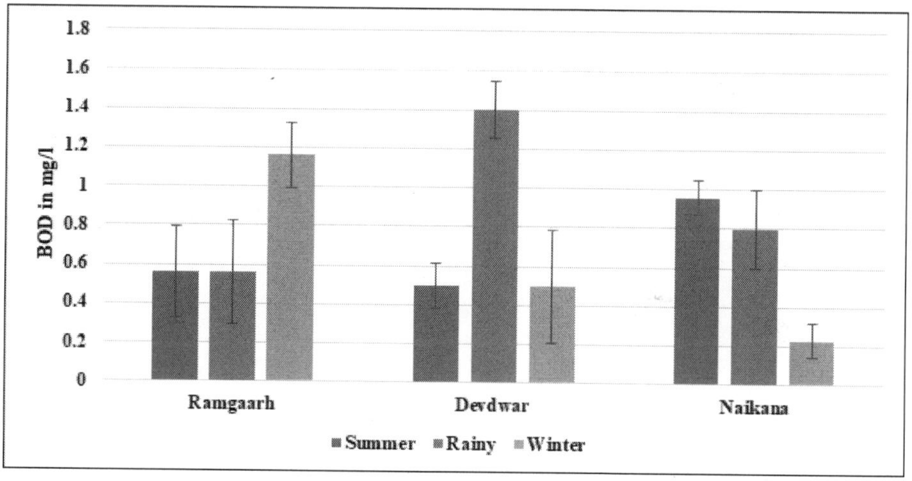

Fig.5. Water BOD of Selected Streams in Different Seasons

CHAPTER 15

A REVIEW ON THE HIDDEN POWER OF ENDOPHYTIC FUNGI: FROM TRANSMISSION TO SUSTAINABLE SOLUTIONS

Prabha Parki[1], Himani Verma[1], Kapil Khulbe[1]

Department of Botany, Kumaun University, Nainital, Uttarakhand, India

Abstract

Endophytic fungi, are prevalent within the healthy tissues of living flora and constitute vital constituents of plant micro-ecosystems. Throughout an extensive evolutionary timeline, certain cohabiting endophytes and their respective host plants have cultivated a unique symbiotic relationship, which can profoundly impact the biosynthesis of bioactive compounds within the plants. Endophytic fungi are inherently present within the tissues of plants and do not induce observable disease symptoms in their host. Nevertheless, it is posited that these fungi provide significant ecological and physiological advantages to the plant. The fungi impart considerable resistance to their host against pathogens, and pests by enhancing the growth rate of the plant and bolstering its adaptability and resilience to biotic and abiotic stresses. Moreover, the symbiotic interactions between fungal endophytes and their host plants facilitate the biosynthesis of various biologically significant secondary metabolites. In recent scholarly discussions, the attention on fungal endophytes has evolved to consider their ability to generate novel bioactive compounds. These properties involve antimicrobial activities, enzymatic, siderophore production, and antioxidant effects. The potential of these fungi to catalyze advancements in the sectors of medicine, agriculture, and industrialization is fundamentally rooted in the rich variety of biochemical compounds generated by endophytes. This review will emphasize the expanding comprehension of the interactions between endophytic fungi and host plants by conducting a comprehensive review of scholarly research findings accumulated over the past decade.

Keywords: Endophytic fungi, enzymatic activity, siderophore, agriculture

Introduction

An extensive variety of microorganisms engages in mutualistic interactions with numerous plant species, resulting in favourable effects for both organisms involved. Endosymbiotic microorganisms, predominantly consisting of bacteria and fungi, engage in symbiotic interactions that confer mutual benefits to their plant hosts, thereby enhancing host fitness. These entities are classified as non-pathogenic, as they are capable of completing a portion or even the entirely of their life-cycle within the host organism without inducing any pathological symptoms (Jha *et al.*, 2023; Nair and Padmavathy, 2014).

Endophytic fungi represent a critical component within plant micro-ecosystems, which are posited to exert considerable effects on the growth and developmental processes of their host plants; however, our comprehension of the precise interactions between endophytic fungi and their respective host plants remains markedly deficient. Comprehending and leveraging such interrelations will enable the optimal synthesis of enhanced pharmaceuticals through the modulation of the growth parameters of therapeutic flora, for instance, by introducing a specific assemblage of endophytic fungi to the plants to augment both the quality and quantity of the resultant medicinal compounds (Firáková *et al.*, 2007; Jia *et al.*, 2016).

Fungal endophytes form symbiotic relationships with host plants through a mechanism that entails the enzymatic breakdown of the cellular wall of the host plant. In order to successfully colonize the host plant tissues, endophytic fungi synthesize a variety of enzymes that degrade cell walls, such as cellulase, laccase, pectinase, and xylanase. These enzymes promote the infiltration, colonization, and expansion of endophytic fungi within the tissues of the host plant by modifying the physical and structural characteristics of the host's cell wall (Chow and Ting, 2019). The stability of endophytic fungal enzymes, alongside their extensive utilization across diverse industrial, pharmaceutical, and scientific sectors, represents a significant avenue for addressing our escalating demands in present era.

The emergence of pathogenic bacteria and fungi exhibiting resistance to commercially available pharmaceuticals constitute a significant challenge encountered by healthcare services, primarily attributed to the microorganisms developing novel mechanisms to evade antimicrobial agents (Elbasuney *et al.*, 2021; Costelloe *et al.*, 2021). Consequently, the identification of potent antimicrobial agents is imperative. Fungal endophytes possess the capability to inhabit plant tissues without manifesting any discernible symptoms of detrimental effects on their hosts (Sharaf *et al.*, 2022; Hashem *et al.*, 2023). Fungal endophytes are recognized as vita; reservoirs of bioactive compounds that exhibit a diverse array of biological activities (antimicrobial, antioxidant).

This review provides an overview of the therapeutic potentials of endophytes, to elucidating the emerging and prospective sectors of endophyte applications that could markedly improve healthcare and agriculture sustainability, consequently aiding in the attainment of a more innovative and sustainable future.

Host-endophyte interaction

Endophytic fungi closely interact with their host plants, primarily forming beneficial relationships that enhance plant health and resilience. These fungi reside within the plant tissues without causing harm. They assist the plants in absorbing more nutrients, managing environmental stress, and defending against diseases. This collaboration not only promotes the plants vitality but also helps maintain ecological balance and supports biodiversity.

These fungi colonize plant tissues in a way that does not produce visible symptoms, allowing them to coexist with the host plants. This symbiotic relationship offers several important benefits to the plants. Endophytes enhance the plant's ability to absorb nutrients, particularly under stress conditions (Sarkar *et al.*, 2021). They confer resistance to biotic stresses (e.g., pathogens, herbivores) and abiotic stresses (e.g., salinity, flooding) (Waqar *et al.*, 2023) (Kassem, 2023). Additionally, these fungi play a critical role in defending plants against various pathogens, including bacteria and viruses (Waqar *et al.*, 2023).

The mechanisms driving these beneficial interactions are intricate and involve multiple physiological processes within the plant. Fungi can trigger the plant's immune responses and stimulate the production of bioactive compounds that protect against disease. Overall, the relationship between fungi and plants is a complex interplay that significantly contributes to plant health, resilience, and productivity (Kassem, 2023; Galindo-Solis and Fernandez, 2022).

Rodriguez *et al* (2005), categorized three classes of fungal symbionts: mycorrhizae, class 1 endophytes, and class 2 endophytes. Mycorrhizal fungi are associated with plant roots, creating an extensive network of hyphae in the soil that facilitates nutrient sharing with their plant hosts. This relationship is mutualistic, indicating that both partners derive benefits from the interaction. Class1 endophytes are a group of clavicipataceous fungi that primarily infect grasses (Read,1999; Schardl *et al.*, 2004). In comparison, class 2 endophytes constitute the largest group of fungal symbionts and are thought to colonize a wide variety of plants in natural ecosystems (Petrini,1986; Rodiguez *et al.*, 2009; Rai and Agarkar 2014).

Endophytes can be categorized into two groups on their diversity, phylogeny, lifestyle, and mode of transmission: systemic (true endophytes) and non-systemic (transient endophytes). Systemic endophytes establish long-term mutualistic relationships with host plants, having coevolved alongside them over time. They remain asymptomatic throughout their lifecycle and typically employ a vertical mode of transmission. Conversely, non-systemic endophytes have a transient presence and exhibit short-term associations. They can shift from a symbiotic role to parasitism in response to changing environmental conditions (Higgins *et al.*, 2014; Wani *et al.*, 2015; Mishra *et al.*, 2021).

Plant-endophyte interactions can be complex and diverse, ranging from beneficial mutualistic symbiosis, where both the plant and endophyte derive advantages, to commensalism, in which one organism benefits while other remains unaffected. In certain circumstances, these interactions can even lead to parasitism, where the endophyte causes harm to the host plant. The type of relationship that develops between the plant and the endophyte is influenced by various factors, including the physiological state of the host plant and the specific environmental conditions it experiences. These dynamic interactions play a crucial role in the health and resilience of ecosystems, impacting plant growth, nutrient acquisition, and susceptibility to pathogens (Kogel *et al.*, 2006). The ecological dynamics and specific mechanisms of these interactions continue to be an area of ongoing research, emphasizing the complexity of plant-fungal relationships in natural ecosystems (Suryanarayanan, 2023).

Fungal endophytes and host physiology

Endophytes are fascinating microorganisms that reside within plant tissues without causing harm. They significantly enhance a plant's ability to absorb vital nutrients, particularly in challenging or stressful environments where resources may be limited (Sarkar *et al.*, 2021). These beneficial partners shield plants from a variety of threats, including biotic stresses, such as invasive pathogens and herbivorous pests-as well as abiotic stresses, including harsh conditions like salinity and flooding (Waqar *et al.*, 2023; Kassem, 2023). One of the remarkable ways endophytes bolster plant defences is by triggering a phenomenon known as induced systemic resistance (ISR). This process effectively activates the plant's innate defence mechanisms, empowering it to better withstand external challenges (Waqar *et al.*, 2023).

Moreover, endophytes play a crucial role in expanding the ecological tolerance of plants, allowing them to flourish across a wide array of habitats and conditions (Kassem, 2023). A key component of these interactions is the production of secondary metabolites, such as terpenoids, which are organic compounds that significantly influence plant health and resilience, enhancing the plant's ability to survive and thrive in its environment (Galindo-Solis and Fernandez, 2022).

Endophytes can cause plants to activate their immune systems or produce substances that might damage plant tissues. This can create challenges for the host plant (Petrini, 1991; Wani *et al.*, 2015). While these defences can help protect plants from serious harm, they may also hurt the plant's overall health. Additionally, the advantages and disadvantages of endophytes can change depending on the different living conditions that a host plant faces, especially for recently introduced alien plants (Nicolus., 2024).

Colonization

The process through which fungi colonize plants can occur in two primary ways: Horizontal and Vertical transmission. Horizontal transmission refers to the direct spread of fungi from one plant to another, as described by Wiewiora *et al* (2015). This method allows fungi to establish new infection quickly within plant populations, often influencing ecosystem dynamics and plant health. In contrast, vertical transmission involves the entry of fungi into the seeds of a parent plant, where they can reside and persist over several generations. When these seeds germinate, the fungi are transmitted to the offspring, continuing the cycle of colonization. This form of transmission highlights the long-term relationship between fungi and plants, as it allows for the propagation of fungal species directly through the genetic lineage of the plants. Various studies have documented this phenomenon, including research by Rodriguez *et al* (2009), Hodgson *et al* (2014), Gagic *et al* (2018), and Shahzad *et al* (2018).

Schulz and Boyle (2005) explain that fungi invade and colonize plants through one of the two main routes: [I] After attaching to the plant surface, the fungi, through their hyphae or spores, can penetrate the host using specialized structures. These include appressoria, penetration pegs, and intracellular haustoria (Schulz and Boyle, 2005). [II] Alternatively, fungi can directly penetrate the cell wall by utilizing secreted degradative enzymes and turgor pressure (Moy *et al.*, 2002; Suryanarayanan *et al.*, 2012; Favre-Godal *et al.*, 2020).

Fungi can infiltrate plant tissues through multiple pathways, which include tissue wounds, stomata, lenticels, root cracks, and germinating radicles. In order for endophytic fungi to effectively colonize plant tissues, they must first breach the plant's physical defences, primarily the rigid cell walls that protect plant cells. When a microbe comes into contact with the plant cell, several biochemical changes occur. The plant cell wall is modified, becoming less rigid and more permeable. As the plasma membrane invaginates, it facilitates the uptake of the endophytic fungi into plant cells.

Endophytic fungi play a crucial role in the decomposition of plant cell walls by producing specific enzymes known as cell wall degrading enzymes (CWDEs). These enzymes include cellulase, Laccase, pectinase, and xylanase. Through the secretion of these enzymes, endophytic fungi can effectively penetrate the rigid structures of plant tissues. This enzymatic activity not only alters the mechanical properties of the plant cell wall, making it easier for the fungi to colonize and proliferate, but it also facilitates the symbiotic relationship of fungi with the host plant, ultimately influencing the plant growth and health (Chow and Ting 2019).

The presence of this enzyme in various endophytes suggests a potential role in the diversity of sugar utilization, which may be beneficial for a functional endophyte (Chow *et al.*, 2018). Taghavi *et al* (2010) noted that the genome of *Enterobacter* sp. does not encode proteins involved in cellulose degradation. This finding aligns with its non-pathogenic behaviour in the interaction between the endophyte and the Poplar tree. An important pathway for endophyte colonization is the root system. Substances synthesized by plants are transported to the roots and released along with nutrients and water into the rhizosphere and soil (Busby *et al.*, 2017). These root exudates selectively attract specific fungi to the area around the roots (Chagas *et al.*, 2017; Igiehon and Babalola, 2017). The fungi grow chemotropically toward the root exudates, beginning to branch and invade upon contact with the roots (Ray *et al.*, 2018).

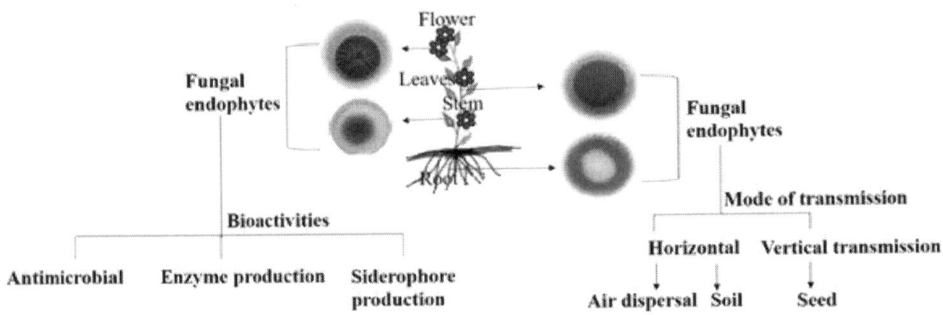

Bioactivities of endophytic fungi for sustainable applications

In recent years, endophytic fungi, which are microorganisms residing within plant tissues, have attracted considerable attention from microbiologists because of their remarkable and diverse capabilities. These fungi exist in a symbiotic relationship with plants, often enhancing plant growth and resilience. With advancements in biotechnology, researchers are exploring the various applications of these beneficial endophytic fungi across multiple sectors. In agriculture, they can improve crop yield, enhance nutrient absorption, and provide resistance against pathogens and environmental stressors. In industrial settings, endophytic fungi have potential uses in bioremediation and the production of biochemicals. Additionally, in the medicinal field, they are being investigated for their ability to produce bioactive compounds that could lead to new pharmaceuticals. Overall, the versatile roles of endophytic fungi present exciting opportunities for innovation and sustainability in various industries.

Endophytic fungi play important roles in agriculture, medicine, and industry, contributing to sustainable development. In the medical field, endophytic fungi and their products are utilized to combat diseases. The endophytic fungi could be applied as inoculants in the fields for crop production and protection.

Mycofungicides and fungal biofertilizers were used for agricultural uses to manage plant diseases as well as the environments defence its capability to improved crop production (Devi *et al.*, 2020; Kour *et al.*, 2020b; Suman *et al.*, 2016b).

The review provides a comprehensive analysis of the interactions between fungal endophytes and their host organisms. It emphasizes the significant contributions of these interactions to the field of medicine while also promoting advancements in agricultural practices and fostering environmental sustainability.

Antimicrobial activity

Bioprospecting for antimicrobial metabolites from endophytes begins with a systemic screening process to identify strains with antimicrobial potential. This involves cultivating endophytes in solid or liquid media to produce crude extracts. These extracts are tested for antimicrobial activity against selected test organisms using rapid screening methods such as disk diffusion and well diffusion assays. Extracts showing significant activity are further assessed to determine the minimum inhibitory concentration (MIC), which indicates the lowest concentration needed to inhibit microbial growth.

Antifungal activity

Gashgari et al. **(2016)** successfully isolated a diverse range of endophytic fungi, including *Alternaria, Aspergillus, Chaetomium, Drechslera, Emericella, Eupenicillium, Fusarium, Gibberella, Monilia, Mycelia sterilia, Penicillium, Scopulariopsis, Ulocladium, and Penicillium chrysogenum*, which exhibited stronger antifungal activity against the tested phytopathogens, specifically *Fusarium oxysporum, F. solani,* and *Alternaria*. The antifungal activity of these endophytic fungi was evaluated using a dual culture method, and the results indicated that these endophytic fungi exhibited stronger inhibition against plant pathogenic fungi. The results of their investigation revealed that the isolated endophytic fungi were not only effective in inhibiting the growth of the tested plant pathogenic fungi but markedly exhibited stronger inhibitory effects compared to control samples. This highlights the potential of these endophytic fungi as candidates for developing natural antifungal agents, offering promising avenues for agricultural applications and disease management in plants.

Uzma et al. **(2017)** In their study showed that, *Aspergillus sp.* and *Fusarium solani* isolated from *Tinospora cordifolia, Piper longum, P. nigrum,* and *Zingiber officinale*. The study revealed that these isolated fungi possess antifungal properties, effectively inhibiting the growth of *Candida albicans*. This finding highlights the potential of these plant-derived fungi as natural agents in combating fungal infections in plants, suggesting their use in developing alternative therapeutic strategies.

In a detailed investigation conducted by **Attia** et al. **in 2020**, the researchers explored the antifungal properties of ethyl acetate extracts of the fungal endophyte *Acremonium sp.* Their findings revealed that these extracts exhibited strong antifungal activity against two phytopathogens *Fusarium sp.* and *Candida albicans*. This study highlights the potential of *Acremonium sp.* as a source of bioactive compounds that could be developed into effective antifungal agents. *Trichoderma* cf. *harzianum* commonly used as biocontrol agents in the agriculture, exhibited a broad-spectrum antimicrobial activity and has potential application in biological control. The endophyte exhibited the broad spectrum antagonism against bacterial and fungal phytopathogen *Ralstonia solanacearum* and *Candida albicans* respectively (An et al., 2020).

4.3. Antibacterial activity

Dos Santos et al. **(2015)** conducted a study in which they isolated 65 endophytic fungi from *Indigofera suffruticosa* Miller. Of these isolates, 18 fungi exhibited antimicrobial activity against both Gram-negative and Gram-positive bacteria during preliminary screening. Notably, the most effective strains were *Nigrospora sphaerica* (URM-6060) and *Pestalotiopsis maculans* (URM-6061), which displayed significant inhibition against *Staphylococcus aureus*. The minimum inhibitory concentration (MIC) and minimum bactericidal concentration (MBC) for *Nigrospora sphaerica* were determined to be 1.56 mg/mL and 6.25 mg/mL, respectively.

Gashgari et al. **(2016)** demonstrated in their research the antibacterial activity of endophytic isolates from medicinal plants (*Alhagi graecorum, Cressa cretica, Citrullus colocynthis, Tamarix nilotica, Achillea fragrantissima, Artemisia sieberi, and Neurospora. Retusa*) against human pathogenic bacteria, suggesting a potential role of endophytes in the medicinal properties of these plants. Earlier, Jahanpour et al. (2015) showed that the ethanolic extract of *P. harmala* seeds and *P. granatum* peels exhibited significant activity against all Mycobacterium tuberculosis isolates, with mean inhibitory zones measuring 18.7 mm and 18.8 mm, respectively, at a concentration of 200 mg/mL.

In the comprehensive study conducted in 2017, Malhadas *et al.* explored the antimicrobial potential of three distinct fungal endophytes derived from the leaves of the olive tree, *Olea europaea* L. The researchers not only evaluated the endophytes' ability to inhibit microbial growth but also investigated how the extracts from the host plant influenced this antimicrobial activity. Their results revealed that *Penicillium commune* and *Penicillium canescens* were particularly effective, demonstrating significant inhibitory effects against both Gram-positive and Gram-negative bacteria—achieving up to 2.7 times more effectiveness compared to a standard concentration of 30 µg/mL chloramphenicol. In addition, *Alternaria alternata* stood out for its superior ability to inhibit yeast growth. Overall, the study highlighted that the antimicrobial activity of these endophytes was markedly enhanced when tested against the pathogens *Bacillus subtilis* and *Escherichia coli*, suggesting their potential as valuable sources of natural antimicrobial agents. Endophytic fungi belonging to the genus *Fusarium*, which were isolated from the plant *Sceletium tortuosum*, demonstrated significant antimicrobial properties against the bacterium *Pseudomonas aeruginosa*. This indicates a strong potential for these fungi as sources of natural antimicrobial agents (Manganyi *et al.*, 2019). Notably, these findings are consistent with previous research conducted by Zhang *et al.*, in 2016, which also highlighted the antimicrobial capabilities of endophytic fungi. Such insights underscore the importance of exploring plant-associated microbes for pharmaceutical applications.

A comprehensive study of **An** et al., **2020** demonstrated the antibacterial activity of *Trichoderma* cf. *harzianum* isolated from *Chloranthus japonica* Sieb. against several maleficent human pathogens (*Bacillus cereus, Escherichia coli, Bacillus subtilis, Staphylococcus aureus, Pseudomonas aeruginosa*). The findings highlight the potential of these antibacterial agents in combating these harmful pathogens.

In a comprehensive study by **Samira** et al., **2022** showed endophytic fungi, specifically *Phialocephala sp.*, isolated from the corms of *Crocus caspius* Fisch., demonstrated significant antibacterial activity against *E. coli* and *S. aureus*. Additionally, the methanolic extract of *Talaromyces purpureogenus* exhibited the highest antifungal activity, with a minimum inhibitory concentration (MIC) of 4 mg/ml against *C. albicans*. Following this, *Phialocephala sp., Talaromyces purpureogenus*, and *Fusarium oxysporum* also showed activity against *Aspergillus ochraceus*.

In **2024, Jigjiddori** et al., investigated endophytic fungal isolates from *Dioscorea bulbiflora* L. that demonstrated potential for biocontrol against bacterial pathogens. The endophytic fungi *Clonostachys pseudochroleucha* showed significant efficacy against Gram-positive bacteria, while *Parathyridaria percutanea* and *Curvularia lunata* proved effective against Gram-negative bacteria.

Table.1. Endophytic fungi and their antimicrobial activities activitiesactivityactivities

Endophyte	Host plant	Activity	Plant parts	Solvent	Reference
Penicillium commune, Penicillium canescens, Alternaria alternata	*Olea europaea* L.	Antimicrobial	Leaves	Ethyl acetate, aquous	Malhadas *et al.*, 2017
Alternaria, Aspergillus, Chaetomium, Drechslera, Emericella, Eupenicillium, Fusarium, Gibberella, Monilia, Mycelia sterilia, Penicillium, Scopulariopsis, Ulocladium, Penicillium chrysogenum	*Alhagi graecorum, Cressa cretica, Citrullus colocynthis, Tamarix nilotica, Achillea fragrantissima, Artemisia sieberi, and Neurospora retusa*	Antimicrobial	Stem and root	ethanolic extract	Gashgari *et al.*, 2016
Aureobasidium sp., Sodaria sp., Aspergillus sp., and Penicillium sp.	*Azadirachta indica*	Antimicrobial	Leaves	Ethyl acetate	Esimone *et al.*, 2020

Endophyte	Host plant	Activity	Plant parts	Solvent	Reference
Aspergillus, Diaporthe, Colletotrichum, Trichoderma, Aspergillus, Leptospora, Thanatephorus, Septoria, Nemania, Biscogniauxia, Hypoxylon, Clonostachys, Fusarium, Calonectria, Ilyonectria, Dactylonectria, Nectria, Neonectria, Trichoderma, Podospora, Diaporthe, Plectosphaerella, Cladosporium	*Chloranthus japonicus* Sieb	Antimicrobial	leaves	Ethyl acetate extract	An *et al.,* 2020
Fusarium, Alternaria, Cladosporium, Epicoccum, Colletotrichum,Aspergillus, Neopestalotiopsis chrysea, Xenoacremonium recifei	*Sonneratia apetala*	Antimicrobial	Leaves, root, bark, fruit	Ethyl acetate and methanolic extract	Nurunnabi *et al.,* 2020
Aspergillus, Alternaria, Chaetomium, Curvularia, Fusarium, Trichoderma harzianum, Penicillium, Lasiodiplodiatheobromae, Stemphylium vesicarium, Ulocladium sp.	*Solanum tuberosum L.*	Preliminary antimicrobial, Enzymatic	Leaves, Stem, Root	Amylase, Cellulase, Tyrosinase, Protease, Chitinase, Laccase	Yasser *et al.,* 2019
Schizophyllum commune, Talaromyces sp., *Aspergillus* sp., *Aspergillus terreus*	*Vernonia anthelmintica*	Antimicrobial, Antioxidant, Antidiabetic, Enzymatic	Root	Ethyl acetate extracts Tyrosinase	Rustamova *et al.,* 2020
Exserohilum turcicum, Colletotrichum capsici, Alternaria oleracea, Diaporthe citri, Ceratocystis paradoxa, Pestalotiopsis theae, Alternaria citri and Cochliobolus miyabeanus	*Stephania dielsiana*	Antimicrobial	Root	Ethyl acetate	Hongying *et al.,* 2020
Penicillium chrysogenum, Chaetomium sp., Aureobasidium sp., Nigrospora sphaerica, Acremonium falciforme, Periconia hispidula,. Allomyces arbuscula , .Nigrospora sphaerica, Acremonium falciforme, Allomyces arbuscula, Penicillium chrysogenum, Acrophialophora sp., Mycelia sterilia	*Litsea cubeba*	Antimicrobial	Leaves and Bark	Ethyl acetate	Eze *et al.,* 2019

Endophyte	Host plant	Activity	Plant parts	Solvent	Reference
Trichoderma harzianum, Phomopsis archeri,, Aspergillus niger, Penicillium fellutanum,, Nigrospora sphaerica,and Fusarium solani, Aspergillus flavus, Fusarium lateritium, Tritirachium oryzae, Cladosporium cladosporioides, and Marianea elegans	*Hancornia speciosa*	Antimicrobial	Stem, Bark	Methanolic extract	Chagas *et al.*, 2017
Fusarium sp.	*Lonicera japonica Thunb.)*	antimicrobial	Leaves	Methanol	Huiru *et al.*, 2016
Cladosporium sp., Phoma sp., Pestalotiopsis oryzae, Mucor fragilis, Botrytis cinerea, Phomopsis sp., Colletotrichum gloeosporioides, Trametes versicolor, Fusarium graminearum, Xylariaceae sp., Diaporthe phaseolorum, Didymella sp., Fusarium oxysporum, Alternaria alternata,	*Ageratina adenophora*	Antimicrobial	Root, Stem and Leaves	Ethyl acetate	Wen *et al.*, 2023
Aspergillus niger, Saccharomycopsis fibuligera, Talaromyces assiutensis, Aspergillus cejpii, Talaromyces trachyspermus, Fusarium solani	*Hedera helix L.*	Antimicrobial	Leaves, stem and root	Ethyl acetate	Osama *et al.*, 2024
Fusarium, Clonostachys, Penicillium, Alternaria, Aspergillus, and Madurella	*Cynomorium songaricum (Rupr.)*	Antimicrobial	Rhizome	Ethyl acetate	Jigjiddorj *et al.*, 2024
Aspergillus ochraceus, Aspergillus niger, Alternaria alstroemeriae	*Lagopsis supina*	Antimicrobial	Leaves and stem	Ethyl acetate	Zhang *et al.*, 2023
Fusarium solani, F. proliferatum, and Fusarium oxysporium	*Petiveria alliacea*	Antimicrobial	Leaves, stem and root	Ethyl acetate	Khadka *et al.*, 2023
Pestalotiopsi, Trichoderma yunnanense,, Schizopyllum commune	*Acanthus ilicifolius L*	Antimicrobial	Leaves, stem, bark and root	Ethyl acetate	Sandrawati *et al.*, 2023

Endophyte	Host plant	Activity	Plant parts	Solvent	Reference
Nigrospora sphaerica,Pestalotiopsis maculans, Phomopsis archeri, Curvularia pallescens,Colletotrichum dematium, Colletotrichum gloeosporioides	*Indigofera suffruticosa Miller*	Antimicrobial	Leaves	Methanol and Ethyl acetate	Dos Santos *et al.*, 2015
Thermothielavioides, Trichoderma, Cyphellophora, Cladosporium, Arcopilus, Plectosphaerella; Chaetomium, Sporothrix, Alboefibula, and Penicillium. Thermothielavioides spp.	*Peperomia argyreia (Miq.) É. Morren*	Antimicrobial	Leaves stem roots	Ethyl acetate	Barolo *et al.*, 2024
Alternaria sp.	*Eremophila longifolia*	Antimicrobial	Leaves	Ethyl acetate	Daniel *et al.*, 2023
Paecilomyces sp.	*Moringa oleifera*	Antimicrobial	Leaves	Ethyl acetate	Sumaiya *et al.*, 2023
Gliocladium catenulatum, Paecilomyces variotii , Penicillium bravicompatum, Aspergillus fumigatus, A. parasiticus , A. niger and Verticillium albo-atrum	*Salvadora persica*	Antimicrobial	Leaves stem root	Ethyl aetate, n-butanol	Fatma *et al.*, 2022
Xenoacremonium falcatus, Fusarium oxysporum,, Cadophora luteo-olivacea, Phialocephala sp., and Talaromyces purpureogenus.	*Crocus caspius Fisch.*	Antimicrobial	Corms	Methanol	Samira *et al.*, 2022
Alternaria, Fusarium, Chaetomium, Phoma, Diaporthe, Trichoderma, Gibberella, and Colletotrichum	*Artemisia argyi*	Antimicrobial	Leaves, stem and roots	Ethyl acetate	Gu *et al.*, 2022
Aspergillus, Penicillium rubens, Sordaria tomentoalba, Fusarium solani, Trichoderma asperellum, Lasiodiplodia theobromae	*Tinospora cordifolia (Willd.) Hook.f and Thomson (Menispermaceae), Piper nigrum L., Piper longum L. (Piperaceae) and Zingiber officinale Roscoe (Zingiberaceae)*	Antimicrobial and Antioxidant	Root, stem, leaves	Ethyl acetate	Uzma *et al.*, 2017

Endophyte	Host plant	Activity	Plant parts	Solvent	Reference
Fusarium proliferatum, Colletotrichu dematium, Aspergillus uvarum, Sistotrema, brinkmanni, Cylindrocarpon, Gibberella intermedia	*Ophiopogon japonicus*	Antimicrobial	Root, stem	Ethyl acetate	Liang *et al.*, 2012
Diporthe sp., Talaromyces funiculosus, Penicillium commune, Cochibolus sp., Phomopsis sp.,Clonostachys rosea,	*Cotyledon orbiculata L., Psychotria zombamontana (Kuntze) Petit, Tecomaria capensis (Thunb.) Lindl., Catha edulis (Vahl) Endl. and Melianthus comosus Vahl.*	Antimicrobial	Root, stem, leaves	Methanol	Abdalla *et al.*, 2020
Fusarium oxysporum, Acremonium sp., Aspergillus ochraceus	*Medicago sativa* L.	Antimicrobial	Root stem leaves	Ethyl acetate	Attia *et al.*, 2020
Acremonium sclerotigenum	*Terminalia bellerica*	Antimicrobial	Leaves	Ethyl acetate	Prathyusha *et al.*, 2015
Trichoderma spp.	*Glycine max* L.	Antimicrobial	Root	Ethyl acetate	Sallam *et al.*, 2021

5. Extracellular enzyme production

Endophytic fungi help the host plant to combat biotic and abiotic stress by synthesizing metabolites and extracellular enzymes, which can act as resistance mechanisms against pathogens and help to nourish the host plants (Jagannath *et al.*, 2021). Hence, endophytic fungi serve as a promising source for producing a wide range of extracellular enzymes (oxidoreductase and hydrolytic enzymes). The increasing use of dangerous artificial dye waste poses a significant risk to both environmental and human health. As an alternative, enzymes derived from fungi offer an environmentally friendly, suitable, and economical solution for use in industrial and agricultural sectors (Salem *et al.*, 2024). The biotechnological potential of fungal laccases is significant, particularly in bioremediation and dye transformation, owing to their wide-ranging substrate affinity and high redox potential. A notable application of these enzymes is in the treatment of water polluted with synthetic dyes (Sennaj *et al.*, 2023; Upadhyay *et al.*, 2016). Traditional methods for addressing this issue are costly, requiring substantial energy input and chemical reagents. In contrast, laccase enzymes offer a promising alternative for eliminating harmful or carcinogenic dye byproducts, including aromatic amines (Upadhyay *et al.*, 2016).

Sunitha et al., **2013** studied the degree of variability of enzyme production by the endophytic fungi isolated from medicinal plants. The research concludes that the enzyme production differs between fungi and often corresponds to the requirements of its habitat. This may be due to the many factors changing in the host as related to age, environmental factors such as, climatic conditions and geographical location may influence the biology of the fungi.

In their investigation, among the tested organisms, maximum extracellular protease activity was observed in *Aspergillus sp.* (Ci10), *Isaria sp.* (Ci12) followed by *Fusarium solani* (Ci24) of *Calophyllum inophyllum. Aspergillus sp.* (Ac32) and *Mycelia sterilia sp.* (Ac31) from isolates of *Alpinia calcarata, Cladosporium sp.* (Cr 3) from *Catharanthus roseus* also produced significant amount of protease. However, maximum pectinase activity was observed in *Taloromyces emersonii* (Ci1), followed by *Fusarium oxysporum* (Ci16) of *Calophyllum inophyllum*. Among *Alpinia calcarata* isolates, *Fusicoccum sp.* (Ac26), were the significant producers of pectinase activity. Microbial pectinases are important in the phytopathologic process, plant-microbe symbiosis and in the decomposition of dead plant material (Bezerra *et al.*, 2012). Degradation of host tissue by phytopathogens generally begins with the production of pectinolytic enzymes, which are the major enzymes involved in plant attack (Hoondal *et al.*, 2005, Brett *et al.*, 1990a). If an endophyte can degrade pectic substances, this implies that the fungus is likely to be a latent pathogen (Choi *et al.*, 2005). Additionally, the amylolytic potential of the endophytes may help them to degrade starch which is available when the plant senesces.

In a study conducted by **Taware** et al., **2017**, endophytic fungi were extracted from *Azadirachta indica A. Juss.* The isolated fungi included *Aspergillus flavus, Alternaria sp., Cladosporium sp., Fusarium sp., Colletotrichum truncatum,* and *Penicillium sp.* Among these, only *Alternaria sp.* showed positive results when tested for tyrosinase and laccase enzyme production.

In a study conducted by **Attia** et al., **2019**, researchers extracted six endophytic fungi from healthy roots, stems, and leaves of *Medicago sativa* L. The researchers then evaluated the isolated fungi's capacity to generate extracellular oxidative enzymes. Most of the fungi exhibited significant peroxidase activity, while *Aspergillus ochraceus* MSEF6 was the only one to demonstrate laccase enzyme production. These findings suggest that the endophytic fungus *A. ochraceus* MSEF6 could have applications in both pharmaceutical and industrial sectors.

In a recent study, **Salem** et al. **(2024)** discovered a novel fungal endophyte with laccase producing capabilities. This endophyte was initially isolated from the cladodes of *Opuntia ficus-indica* and subsequently identified as *T. harzianum* AUMC14897. The newly isolated endophytic strain demonstrates exceptional laccase activity, exhibiting distinctive properties that make it a promising candidate for various environmental and industrial applications.

Table.2. Endophytic fungi and their extracellular enzyme activities

Endophyte	Host plant	Activity	Plant parts	Enzyme	Reference
Doritis pulcherrima, Dendrobiuma phyllum, Dendrobium anosmum, Ascocentrum curvifolium, Aerides falcata.	Orchids	Enzyme	Leaves, root, stem	Amylase, cellulose, pectinase, protease and lipase	Sopalun and Iamtham 2020
Nectria,Guignardia, Nigospora, Ampelomyces, Xylaria, Corynescora, Chaetomium, Fusarium, Diaporthe, Colletotrichum	*Cananga odorata,* Terminalia catappa, *Terminalia mantaly*	Enzyme	Leaves, stem bark, root, flower	Amylase, cellulase lipase, laccase	Toghueo et al., 2017
Aspergillus flavus, Aspergillus fumigatus, Aspergilus niger, Cladosporium sp., Rhizopus sp., Mucor sp.	Ziziphus spina	Enzyme	Leaves	Amylase, lipase, protease, pectinase, cellulase	Hawar 2022
Cladosporium sp., Rhizoctonia sp., Aspergillus sp., Chaetomium sp., Biosporus sp., Fusarium sp., Curvularia sp., Cladosporium sp., Colletotrichum sp.	*Azadirachta indica, Citrus limon, Gossypium hirsutum, Magnolia champaca, Datura stramonium, Piper betle, Phyllanthus emblica*	Enzyme	Leaves, root, stem	Amylase, lipase, cellulase,	Patil et al., 2015
Phomopsis, Sordariomycetes, Diaporthe, Colletotrichum	*Sapindus saponaria*	Enzyme	Leaves	Amylase, cellulase, pectinase	dos SANTOS, et al. 2019
Talaromyces flavus, Mortierella hyaline, Penicillium sp., Paecilomyces variabilis	*Potentilla fulgens, Osbeckia stellata, Osbeckia chinensis, Camellia caduca, and Schima khasiana*	Enzyme	Root, leaves	Amylase, cellulase, protease, lipase, xylanase	Bhagobaty et al.,2012
Aspergillus oryzae, A. terreus, Curvularia sp., Fusarium oxysporum, Gliocladium solani, Penicillium citrinum, Trichoderma viride, Verticillium dahliae	*Litsea glutinosa*	Enzyme	Root, leaves, stem	Amylase, cellulase, protease, lipase, laccase	Madasi et al., 2021
Xylaria cubensis, Muyocopron laterale, Cyphellophora guyanensis, Acremonium hennebertii	*Calamus castaneus Griff.*	Enzyme	Spine	Amylase, cellulase, lipase, pectinase, protease	Azuddin et al., 2024

Endophyte	Host plant	Activity	Plant parts	Enzyme	Reference
Clonostachys pseudochroleucha, Parathyridaria percutanea, Fusarium proliferatum, Curvularia lunata, Fusariumequiseti	*Dioscorea bulbifera L.*	Enzyme	Root, stem, leaves	Amylase, cellulase, protease, lipase, chitinase	Sharma et al., 2023
Aspergillus flavus, A. fumigatus, A.niger, Cladosporium sp., Rhizopus sp., Mucor sp.	*Ziziphus spina*	Enzyme	Leaves	Amylase, cellulase, laccase, en-doglucanase, xylanase	Hawar 2022
Diaporthe phaseolorum, Schizophyllum commune, Phomopsis sp., Fomitopsis meliae, Nigrospora sphaerica, Lasiodiplodia theobromae,Fusarium brachygibbosum, Xylaris longipes, Pseudofussicoccum adansoniae, Daldinia eschscholtzii, Colletotrichum gleosporiodes,C. gigasporum, C. boninense, Xylaria longipes, Pseudofussicoccum adansoniae, Daldinia eschscholtzii.	*Dillenia indica* L.	Enzyme	Leaves, stem, fruits	Amylase, cellulase, lipase, asparaginase, protease, ligninolytic	Kumar and Prasher 2022
Aspergillus japonicus,A. sydowii, A. terreus, Curvularia lunata, Nigrospora spp., Colletotrichum gleosporidioides, Rhizoctonia spp., Xylaria spp., Trichoderma spp., Fusarium chlamydosporum, Penicillium citrinum, Helminthosporium spp., Curvulariaspp., Cladosporium sp., Alternaria alternata, Fusarium oxysporum, Penicillium chrysogenum, Colletotrichum truncatum, Bipolaris spp., Taloromyces rotundus, Penicillium purpurogenum, Cylindrocephalum spp.	*Cymbidium aloifolium*	Enzyme	Root leaves, flowers	Amylase, protease, pectinase, laccase, lipase, cellulase, phosphatase	Shubha and Srinivas 2017

Endophyte	Host plant	Activity	Plant parts	Enzyme	Reference
Diporthe sp., Talaromyces funiculosus, Penicillium commune, Cochibolus sp., Phomopsis sp.,Clonostachys rosea,	*Cotyledon orbiculata L., Psychotria zombamontana (Kuntze) Petit, Tecomaria capensis (Thunb.) Lindl., Catha edulis (Vahl) Endl. and Melianthus comosus Vahl.*	Enzymes	Root, stem, leaves	Xylanase, cellulase	Abdalla et al., 2020
Fusarium oxysporum, Acremonium sp., Aspergillus ochraceus	*Medicago sativa L.*	Enzymes	Root stem leaves	Laccase, peroxidases	Attia et al., 2020
Trichoderma spp.	*Glycine max* L.	Enzymes	Root	Amylase, cellulase, pectinase, protease,	Sallam et al., 2021
Aspergillus flavus, Alternaria sp., Cladosporium sp., Fusarium sp., Colletotrichum truncatum, Penicillium sp.	*Azadirachta indica A. Juss.*	Enzymes	Leaf, petiole, stem	Amylase, cellulase, laccase, tyrosinase, asparaginase, xylanase	Taware et al., 2017

6. Siderophore production

Iron (Fe) is the fourth most abundant element on earth and represents an essential nutrient for life. As a fundamental mineral element for cell growth and development, iron is available for uptake as ferric ions, which are usually oxidized into complex oxyhydroxide polymers, insoluble under aerobic conditions. So, to overcome the iron deficient condition, microorganisms have evolved the mechanism to obtain ferric irons from the environment by the production of small molecules known as siderophores. According to Gokarn and Pal 2018, The siderophores may act as facilitators for antibiotics across the cell membrane because of the increase in permeability caused by iron deficiency. The siderophore molecules combined with antibiotics are used as a 'Trojan horse' for targeted drug delivery.

Aramsirirujiwet et al., **2016** reported out of 12 endophytic fungi isolated from leaf, root, and stem of *Hottuunia cordata* Thunb.seven isolates (*Glomerella cingulata, Colletotrichum sp., Aeschynomene sp., Lasiodiplodia pseudotheobromae, Dothideomycetes sp., Fusarium sp., Xylaria sp.*) exhibited orange halo zone formed around the fungal colonies on Chrome azurol's agar.

Chowdappa et al., **2020** isolated endophytic fungi from *Cymbidium aloifolium* out of which the highest siderophore percentage was 57% in *Penicillium chrysogenum* (CAL1), 49% in *Aspergillus sydowii* (CAR12), 46% in *Aspergillus terreus* (CAR14). The study concludes that the solvent-extracted siderophores inhibited the virulent plant pathogens *Ralstonia solanacearum*, which causes bacterial wilt in groundnut and *Xanthomonas oryzae* pv. which causes bacterial blight disease in rice. Thus, bioactive siderophore-producing endophytic *P. chrysogenum* can be exploited in the form of formulations for development of resistance against other phytopathogens in crop plants.

A study conducted by **Turbat** et al., **2020** highlights high siderophore production by *Fusarium* sp. SZMC 26656 and *Alternaria* sp. SZMC 26651 producing hydroxamate and carboxylate type of siderophores.

Siderophore production by fungal endophytes is a neglected frontier that has not been explored extensively. However, there is a growing interest in this field, it remains a relatively niche research area.

Table.3. Endophytic fungi and their siderophore production potential

Endophyte	Host plant	Activity	Plant parts	Reference
Aspergillus japonicus, A. sydowii, Curvularia sp., Nigrospora sp., Trichoderma sp., Xylaria sp., Rhizoctonia sp., Penicillium citrinum, Helminthosporium, Cladosporium sp., Alternaria alternata, Fusarium oxysporum, Bipolaris sp., Cylindrocephalum sp.	*Cymbidium aloifolium*	Siderophore	Root, leaves, flower	Chowdappa et al., 2020
Acremonium sclerotigenum	*Terminalia bellerica*	Siderophore	Leaves	Prathyusha et al., 2015
Trichoderma spp.	*Glycine max* L.	Siderophore	Root	Sallam et al., 2021
Alternaria sp., Didymella sp., Fusarium sp., Xylogone sp.	*Sophora flavescens*	Siderophore	Leaves, branches, root, flower	Turbat et al., 2020

Conclusion

A comprehensive understanding of the complex interactions between endophytic fungi and their host plants is essential for leveraging their capabilities to confront global issues such as food insecurity and environmental degradation. The dynamics of endophyte colonization within plant ecosystems need further investigation. Endophytic fungi represent a valuable resource for sustainable agriculture and biotechnology, their potential to protect plants against pests and diseases, makes them a valuable asset in modern agriculture. The continuing research on endophytic fungi has the potential to revolutionize agriculture and contribute to a more sustainable future. This review elucidates that various biological activities exhibited by endophytic fungi serve as a significant reservoir of extracellular enzymes possessing medical and agricultural relevance. Furthermore, the extracts derived from these isolates were distinguished by their antimicrobial properties. Additionally, it is imperative to broaden the applications of fungal endophytes and enhance the methodologies for their formulation to ensure that such applications are safer, more user-friendly, economically viable, and of heightened efficacy.

Acknowledgements

The authors are thankful to the Head of Department of Botany, Thakur Dev Singh Bisht Campus, Kumaun University, Nainital for providing necessary lab facilities and instruments.

References

Abdalla, M.A., Aro, O.A., Gado, D., Passari, A.K., Mishra, V.K., Singh, B.P., McGaw, L.J. (2020). Isolation of endophytic fungi from South African plants, and screening for their antimicrobial and extracellular enzymatic activities and presence of type I polyketide synthases, *South African Journal of Botany*, Volume 134, 336-342, ISSN 0254-6299.

Adeleke, B.S., Ayilara, M.S., Akinola, S.A. *et al.* Biocontrol mechanisms of endophytic fungi. *Egypt J Biol Pest Control* **32**, 46 (2022). https://doi.org/10.1186/s41938-022-00547.

An, C., Ma, S., Shi, X., Xue, W., Liu, C., Ding, H. (2020). Diversity and Antimicrobial Activity of Endophytic Fungi Isolated from *Chloranthus japonicus* Sieb in Qinling Mountains, China. *International Journal of Molecular Sciences*. 21(17):5958. https://doi.org/10.3390/ijms2117595.

Attia, E.Z., Farouk, H.M., Abdelmohsen, U.R., El-Katatny, M.H. (2020). Antimicrobial and extracellular oxidative enzyme activities of endophytic fungi isolated from alfalfa (Medicago sativa) assisted by metabolic profiling, *South African Journal of Botany*, Volume 134, 156-162, ISSN0254-6299.

Azuddin, N.F., Masratul, H.M., Nik, F., Asyraf, M., Latiffah, Z. (2024). Extracellular enzymatic activity of endophytic fungi isolated from spines of rattan palm (Calamus castaneus Griff.). *Malaysian Journal of Microbiology*, doi: 10.21161/mjm.230209

Barolo, M.I., Castelli, M.V., López, S.N. (2024). Antimicrobial Activity of Fungal Endophytes Associated with *Peperomia argyreia* (Piperaceae). *Appl. Microbiol.*, 4(2):753-770.

Bezerra, J.D.P., Santos, M.G.S., Svedese, V.M., Lima, D.M.M., Fernandes, M.J.S., Paiva, L.M., Souza-Motta, C.M. (2012) *World J. Microbiol.*, 28: 1989-1995.

Bhagobaty, Ranjan, Joshi, SR. (2012). Enzymatic Activity of Fungi Endophytic on Five Medicinal Plant

Brett, C.T. 1990a. Cell wall structure and the skeletal functions of the wall. In: *Physiology and Biochemistry of Plant Cell Walls., Eds.*, C.T. Brett and Waldron, Unwin, Hyman, London, pp: 4-57.

Chagas, M. B. O., Prazeres Dos Santos, I., Nascimento da Silva, L. C., Correia, M. T. D. S., Magali de Araújo, J., Cavalcanti, M. D. S., & Lima, V. L. M. (2017). Antimicrobial Activity of Cultivable Endophytic Fungi Associated with *Hancornia Speciosa* Gomes Bark. *The open microbiology journal, 11*, 179–188. https://doi.org/10.2174/1874285801711010179

Choi, Y.W., I.J. Hodgkiss and K.D. Hyde, 2005. Enzyme production by endophytes of Brucea javanica. J. *Agricultural Technol.*, 1: 55-66.

Chow, Y. Y., and Ting, A. S. Y. (2019). Influence of fungal infection on plant tissues: FTIR detects compositional changes to plant cell walls. *Fungal Ecol.* 37, 38–47. doi: 10.1016/j.funeco.2018.10.004

Chow, Y.Y., Rahman, S., Ting, A.S.Y. (2018). Interaction dynamics between endophytic biocontrol agents and pathogen in the host plant studied via quantitative real-time polymerase chain reaction (qPCR) approach. *Biol. Control*, 125, 44–49.

Chow, Y.Y., Ting, A.S.Y. (2019). Influence of fungal infection on plant tissues: FTIR detects compositional changes to plant cell walls. *Fungal Ecol.*, 37, 38–47.

Chowdappa, S., Jagannath, S., Konappa, N., Udayashankar, A. C., & Jogaiah, S. (2020). Detection and Characterization of Antibacterial Siderophores Secreted by Endophytic Fungi from *Cymbidium aloifolium. Biomolecules, 10*(10), 1412. https://doi.org/10.3390/biom10101412

Costelloe, C., Metcalfe, C., Lovering, A., Mant, D., Hay, A.D. (2010). Effect of antibiotic prescribing in primary care on antimicrobial resistance in individual patients: systematic review and meta-analysis. *BMJ*. ;340: c2096.

Daniel, J., Caruso., Enzo, A., Palombo., Simon, E., Moulton., Peter, J., Duggan., Bita, Zaferanloo. (2023). Antibacterial and Antibiofilm Activity of Endophytic Alternaria sp. Isolated from Eremophila longifolia. *Antibiotics*, 12 doi: 10.3390/antibiotics12091459

Devi, R., Kaur, T., Kour, D., Rana, K.L., Yadav, A., Yadav, A.N. (2020b) Beneficial fungal communities from different habitats and their roles in plant growth promotion and soil health. *Microbial Biosystems* 5:21–47. https://doi.org/10.21608/mb.2020.32802.1016

Dos Santos, I. P., da Silva, L. C., da Silva, M. V., de Araújo, J. M., Cavalcanti, M.daS., & Lima, V. L. (2015). Antibacterial activity of endophytic fungi from leaves of Indigofera suffruticosa Miller (Fabaceae). *Frontiers in microbiology, 6*, 350. https://doi.org/10.3389/fmicb.2015.00350

Elbasuney, S., El-Sayyad, G.S, Tantawy, H., Hashem, A.H. (2021)P. romising antimicrobial and antibiofilm activities of reduced graphene oxide-metal oxide (RGO-NiO, RGO-AgO, and RGO-ZnO) nanocomposites. *RSC Adv.*, 11(42):25961–75. https://doi.org/10.1039/D1RA04542C.

Esimone, C., Okechukwu,T., Ujam, N., Abba, C. C., Maduabuchi, E. P., Nnamdi, O. A., Ebiye, E. C., Chigozie, U., Malachy, C., Okoye, F. B. (2020). The isolation identification and antimicrobial activities of endophytic fungi from *Azadirachta indica* .GSC *Biological and Pharmaceutical Sciences*, 11(03), 115-124

Eze, P.M., Nnanna, J.C., Okezie, U., Buzugbe, H.S., Abba, C.C., Chukwunwejim, C.R., Okoye, F.B.C. and Esimone, C.O. (2019). Screening of metabolites from endophytic fungi of some Nigerian medicinal plants for antimicrobial activities. The *EuroBiotech Journal, Sciendo*, vol. 3 no. 1, pp. 10-18. https://doi.org/10.2478/ebtj-2019-0002

Fatma, F., Abdel-Motaal., Fatma, Radi., Magdi, Tawfik, Abdelhamid, Elsayed. (2022). Isolation and Biological Activity of Endophytic Fungi isolated from Arak Medicinal Plant (*Salvadora persica*). *Aswan University Journal of Sciences and Technology*, 2(2):74-85. doi: 10.21608/aujst.2022.180186.1008

Gashgari, R., Gherbawy, Y., Ameen, F., Alsharari, S. (2016). Molecular Characterization and Analysis of Antimicrobial Activity of Endophytic Fungi From Medicinal Plants in Saudi Arabia. Jundishapur J *Microbiol.* 2;9(1): e26157. doi: 10.5812/jjm.26157. PMID: 27099679; PMCID: PMC4834135.

Gu, H., Zhang, S., Liu, L., Yang, Z., Zhao, F., & Tian, Y. (2022). Antimicrobial Potential of Endophytic Fungi From *Artemisia argyi* and Bioactive Metabolites From *Diaporthe* sp. AC1. *Frontiers in microbiology*, *13*, 908836. https://doi.org/10.3389/fmicb.2022.908836

Hashem, A.H., Attia, M.S., Kandil, E.K. *et al.* Bioactive compounds and biomedical applications of endophytic fungi: a recent review. *Microb Cell Fact* **22**, 107 (2023). https://doi.org/10.1186/s12934-023-02118-x

Hawar, S.N. (2022). Extracellular Enzyme of Endophytic Fungi Isolated from *Ziziphus spina* Leaves as Medicinal Plant. *International journal of biomaterials*, *2022*, 2135927. https://doi.org/10.1155/2022/2135927.

Hoondal, G.S., Tiwari, R.P., Tewari, R., Dahiya, N., Beg, Q.K. (2002). Microbial alkaline pectinases and their industrial applications: a review. *Appl. Microbiol. Biotechnol.*, 59: 409-18.

Jagannath, S., Konappa, N., Lokesh, A. (2021.) Bioactive compounds guided diversity of endophytic fungi from *Baliospermum montanum* and their potential extracellular enzymes. *Anal Biochem* 614:114024. https://doi.org/10.1016/j.ab.2020.114024

Jahanpour, S., Ghazisaidi, K., Davoodi, H., Mazandarani, M., Samet, M., Jahanpour, N. (2015). Antimicrobial Effects of Folk Medicinal Plants From the North of Iran Against Mycobacterium tuberculosis. *Arch Pediatr Infect Dis.*, 3(1 TB):e26157

Jha, P., Kaur, T., Ishita, C., Avirup, P., Sushreeta, P., Vijay, K., Tabarak, M.(2023). Endophytic fungi: hidden treasure chest of antimicrobial metabolites interrelationship of endophytes and metabolites. *Frontiers in Microbiology* VOL.4 DOI=10.3389/fmicb.2023.1227830

Ji, X., Xia, Y., Zhang, H., Cui, J.L. (2022). The microscopic mechanism between endophytic fungi and host plants: From recognition to building stable mutually beneficial relationships, *Microbiological Research*, Volume 261, , 127056, ISSN 0944-5013, https://doi.org/10.1016/j.micres.2022.127056.

Jia, M., Chen, L., Xin, H.L., Zheng, C.J., Rahman, K., Han, T., Qin, L.P. (2016) A Friendly Relationship between Endophytic Fungi and Medicinal Plants: A Systematic Review. *Front. Microbiol.* 7:906. doi: 10.3389/fmicb.2016.00906

Jigjiddorj, E.A., Maidarjav, A., Byambasuren, B., Nyamgerel, D. (2024). Identification, Antimicrobial and Plant Growth Promoting Activities of Endophytic Fungi Associated with *Cynomorium songaricum* Rupr., a Traditional Medicinal Plant in Mongolia. *Diversity, 16,* 122. https://doi.org/10.3390/d16020122

Karthika, M., Rasmi, A.R. (2022). Diversity and Extracellular Enzyme Production of Fungal Endophytes from the Genus Ocimum L. *Biosci Biotech Res Asia*, 19(4).

Kassam, R., Jaiswal, N., Hada, A., Phani, V., Yadav, J., Budhwar, R., et al., 2023. Evaluation of Paecilomyces tenuis producing Huperzine A for the management of root-knot nematode *Meloidogyne incognita* (Nematoda: meloidogynidae). *J. Pest. Sci.* 96 (2), 723–743.

Khadka, G., Annamalai, T., Shetty, K.G., Tse-Dinh, Y.C., Jayachandran, K. (2023). Isolation and Characterization of Fungal Endophytes from *Petiveria alliacea* and Their Antimicrobial Activities in South Florida. *Microbiol. Res., 14,* 1470-1482. https://doi.org/10.3390/microbiolres14030100

Kour, D., Rana, K.L., Yadav, A.N., Yadav. N., Kumar. M., Kumar, V. (2020b). Microbial biofertilizers: Bioresources and eco-friendly technologies for agricultural and environmental sustainability. *Biocatal Agric Biotechnol* 23:101487

Kumar, V., Prasher, I.B. (2022). Seasonal variation and tissues specificity of endophytic fungi of *Dillenia indica* l. and their extracellular enzymatic activity. *Arch Microbiol* **204**, 341 (2022). https://doi.org/10.1007/s00203-022-02933-7.

Liang, H., Xing, Y., Chen, J. (2012). Antimicrobial activities of endophytic fungi isolated from *Ophiopogon japonicus* (Liliaceae). *BMC Complement Altern Med* **12,** 238 https://doi.org/10.1186/1472-6882-12-238

Lu, H., Wei, T., Lou, H., Shu, X., Chen, Q. (2021). A Critical Review on Communication Mechanism within Plant-Endophytic Fungi Interactions to

Cope with Biotic and Abiotic Stresses. *J Fungi* (Basel), 7(9):719. doi: 10.3390/jof7090719. PMID: 34575757; PMCID: PMC8466524.

Madasi, A, Ajmeera, A., Renuka, G., Vemireddy, B., Vanteru, K.R. (2021). Production of Extracellular Enzymes, Antimicrobial and other Agriculturally Important Metabolites by Fungal Endophytes of *Litsea glutinosa* (Lour.) C.B. Rob. a Medicinal Plant. *J Pure Appl Microbiol,* (4):2317-2328. doi: 10.22207/JPAM.15.4.56

Maheshwari, R., Bhutani, N., Suneja, P. (2019). Screening and characterization of siderophore producing endophytic bacteria from *Cicer arietinum* and *Pisum sativum* plants. *J Appl Biol Biotech*.;7(05):7–14.

Malhadas, C., Malheiro, R., Pereira, J.A. *et al*. Antimicrobial activity of endophytic fungi from olive tree leaves. *World J Microbiol Biotechnol* **33**, 46 (2017). https://doi.org/10.1007/s11274-017-2216-7

Manganyi, M.C., Regnier, T., Tchatchouang, CD.K. (2019). Antibacterial activity of endophytic fungi isolated from *Sceletium tortuosum* L. (Kougoed). *Ann Microbiol* 69, 659–663. https://doi.org/10.1007/s13213-019-1444-5

Mengistu, A.A. (2020). Endophytes: Colonization, Behaviour, and Their Role in Defense Mechanism. Int J Microbiol. 2020 Jul 30;2020:6927219. doi: 10.1155/2020/6927219. PMID: 32802073; PMCID: PMC7414354.

Mohini, G.P., Jyoti, P., Sucheta, N.P., Amanpreet, K.S. (2015). International Journal of Current Microbiology and *Applied Sciences* ISSN: 2319-7706 Volume 4 Number 3 pp. 1035-1042.

Nair, D. N., and Padmavathy, S. (2014). Impact of endophytic microorganisms on plants, environment and humans. *Sci. World J*. 2014:250693. doi: 10.1155/2014/250693

Nurunnabi, T.R., Sabrin, F., Sharif, D.I. (2020).. Antimicrobial activity of endophytic fungi isolated from the mangrove plant *Sonneratia apetala* (Buch.-Ham) from the Sundarbans mangrove forest. *ADV TRADIT MED (ADTM)* **20**, 419–425 https://doi.org/10.1007/s13596-019-00422-9

Osama, S., El Sherei, M., Al-Mahdy, D.A. *et al*. Antimicrobial activity of spiculisporic acid isolated from endophytic fungus *Aspergillus cejpii* of *Hedera helix* against MRSA. *Braz J Microbiol* **55**, 515–527 (2024). https://doi.org/10.1007/s42770-023-01224-7

Prathyusha P., Rajitha Sri A.B., Ashokvardhan T., and Satya Prasad K (2015) Antimicrobial and Siderophore Activity of the Endophytic Fungus *Acremonium sclerotigenum* Inhabiting *Terminalia bellerica* Roxb. *Int. J. Pharm. Sci. Rev. Res.*, 30(1), 16, 84-87

Ray, S., Mishra, S., Bisen, K., Singh, S., Sarma, B.K., Singh, H.B., 2018. Modulation in phenolic root exudate profile of Abelmoschus esculentus expressing activation of defense pathway. *Microbiol. Res.* 207, 100–107. https://doi.org/10.1016/j.micres.2017.11.011.

Rodriguez, R.J., Jr White, J.F., Arnold, A.E., Redman, R.S., 2009. Fungal endophytes: diversity and functional roles. *New Phytol.* 182 (2), 314–330. https://doi.org/10.1111/j.1469–8137.2009. 02773.x.

Rustamova, N., Gao, Y., Zhang, Y., Yili, A. (2020). Biological Activity of Endophytic Fungi from the Roots of the Medicinal Plant *Vernonia anthelmintica, Microorganisms* 8, 586.

Salem, M.M., Mohamed, T.M., Shaban, A.M. (2024). Optimization, purification and characterization of laccase from a new endophytic *Trichoderma harzianum* AUMC14897 isolated from *Opuntia ficus-indica* and its applications in dye decolorization and wastewater treatment. *Microb Cell Fact* **23**, 266. https://doi.org/10.1186/s12934-024-02530-x

Sallam, N., Ali, E.F., Seleim, M.A.A. *et al.* Endophytic fungi associated with soybean plants and their antagonistic activity against *Rhizoctonia solani. Egypt J Biol Pest Control* **31**, 54 (2021). https://doi.org/10.1186/s41938-021-00402-9

Samira, Samadzadeh., Mohsen, Farzaneh., Zeinab, Shahsavari., Samad, Nejad, Ebrahimi., Mojtaba, Asadollahi., Mohammad, Hossein, Mirjalili. (2022). Characterization and antimicrobial activity of fungal endophytes from Crocus caspius (Iridaceae). *Biocatalysis and agricultural biotechnology*, 43:102429-102429. doi: 10.1016/j.bcab.2022.102429

Sandrawati, N., Ningsih, W., Layla, R. ., Eka Putra, A. ., Ismed, F. ., Ekawati Tallei, T. ., & Handayani, D. (2023). Endophytic Fungi from Mangrove Plant Acanthus ilicifolius L.: Antimicrobial, Anticancer, and Species Determination. *Trends in Sciences, 20*(7), 5089. https://doi.org/10.48048/tis.2023.5089

Santos, C., Ribeiro, A., Garcia, A., Polli, A., Polonio, J., Azevedo, J., Pamphile, J. (2019). Enzymatic and Antagonist Activity of Endophytic Fungi from Sapindus saponaria L. (Sapindaceae). *Acta Biológica Colombiana*. 24. 322-330. 10.15446/abc.v24n2.74717.

Sharaf, M.H., Abdelaziz, A.M., Kalaba, M.H., Radwan, A.A., Hashem, A.H. (2022). Antimicrobial, antioxidant, cytotoxic activities and phytochemical analysis of fungal endophytes isolated from *Ocimum basilicum. Appl Biochem Biotechnol.*, 194(3):1271–89.

Sharma, S., Dhar, K., Kaul, S. (2023). Antagonistic, plant growth promoting and extracellular hydrolytic enzyme activity of fungal endophytes of *Dioscorea bulbifera* L.. *Biocatalysis and agricultural biotechnology*, 50:102694-102694. doi: 10.1016/j.bcab.2023.102694

Shubha, J., and Srinivas, C. (2017). Diversity and extracellular enzymes of endophytic fungi associated with *Cymbidium aloifolium* L. *African Journal of Biotechnology*, 16(48), 2248-2258.

Sopalun, K., Iamtham, S. (2020). Isolation and screening of extracellular enzymatic activity of endophytic fungi isolated from Thai orchids, *South African Journal of Botany*, Volume 134, Pages 273-279, ISSN 0254-6299, https://doi.org/10.1016/j.sajb.2020.02.005.

Sumaiya, N., Hawar. (2022). Extracellular Enzyme of Endophytic Fungi Isolated from Ziziphus spina Leaves as Medicinal Plant. *International Journal of Biomaterials*, 2022:1-9. doi: 10.1155/2022/2135927.

Suman, A., Yadav, A.N., Verma, P. (2016b) Endophytic Microbes in Crops: Diversity and Beneficial Impact for Sustainable Agriculture. In: Singh DP, Singh HB, Prabha R (eds*) Microbial Inoculants in Sustainable Agricultural Productivity*: Vol. 1: Research Perspectives. Springer, New Delhi, pp 117-143. doi:https://doi.org/10.1007/978-81-322-2647-5_7

Sunitha, V.H., Nirmala, D., Srinivas, C. (2013). Extracellular Enzymatic Activity of Endophytic Fungal Strains Isolated from Medicinal Plants. *World Journal of Agricultural Sciences* 9 (1): 01-09.

Taghavi, S., van, D.L., Hoffman, A. (2010). Genome sequence of the plant growth promoting endophytic bacterium Enterobacter sp. 638," PLoS Genetics, vol. 6, 2010.

Taware, A.S., More, Y.W., Ghag, S.V., Rajurkar, S.K. (2017). Screening of endophytic fungi isolated from Azadirachta indica A. Juss. for production of enzymeBioscience Discovery, 8(4): 688-694, October 229-3469; 2231-024

Toghueo, R.M.K.I., Zabalgogeazcoa, B.R., Vázquez D.A, Boyom, F.F. (2017). Enzymatic activity of endophytic fungi from the medicinal plants *Terminalia catappa, Terminalia mantaly* and *Cananga odorata*, South African Journal of Botany, Volume 109, Pages 146-153, ISSN 0254-6299.

Suryanarayanan, T.S. (2023). Fungal Endophyte-Plant-Insect Interaction: A Tripartite Association Needing Attention. Indian journal of entomology. doi: 10.55446/ije.2023.1262.

Turbat, A., Rakk, D., Vigneshwari, A., Kocsubé, S., Thu, H., Szepesi, Á., Bakacsy, L., D. Škrbi , B., Species of the Pristine Sacred Forests of Meghalaya, India. *Biotechnology and Bioprocess Engineering*. 15. 33-40. 10.1007/s12257-011-0453-4.

Jigjiddorj, E. -A., Vágvölgyi, C., & Szekeres, A. (2020). Characterization of the Plant Growth-Promoting Activities of Endophytic Fungi Isolated from *Sophora flavescens. Microorganisms, 8*(5), 683. https://doi.org/10.3390/microorganisms8050683

Uzma, F., Chowdappa, S. (2017). Antimicrobial and Antioxidant Potential of EndophyticFungi Isolated from Ethnomedicinal Plants of Western Ghats, Karnataka. *J Pure Appl Microbiol.*, 11(2):1009-1025. doi: 10.22207/JPAM.11.2.43

Waqar, S., Bhat, A.A, Khan, A.A. (2024). Endophytic fungi: Unravelling plant-endophyte interaction and the multifaceted role of fungal endophytes in stress amelioration. *Plant Physiology and Biochemistry* 206;108174.

Wen, J., Okyere, S.K., Wang, J., Huang, R., Wang, Y., Liu, L., Nong, X., Hu, Y. (2023). Endophytic Fungi Isolated from *Ageratina adenophora* Exhibits Potential Antimicrobial Activity against Multidrug-Resistant *Staphylococcus aureus*. *Plants, 12,* 650. https://doi.org/10.3390/plants12030650

Wu, H., Yan, Z., Deng, Y., Wu, Z., Xu, X., Li, X., Zhou, X., Luo, H. (2020). *Acta Ecologica Sinica* 40 383–387

Yasser, M. M., Mousa, A.S.M., Marzouk, M.A., Tagyan, A.I. (2019). Molecular Identification, Extracellular Enzyme Production and Antimicrobial Activity of Endophytic Fungi Isolated from *Solanum Tuberosum* L. in *Egypt. Biosci Biotech Res Asia*;16(1).

Zhang, D., Sun, W., Xu, W., Ji, C., Zhou, Y., Sun, J., Tian, Y., Li, Y., Zhao, F., & Tian, Y. (2023). Antimicrobial and Cytotoxic Activity of Endophytic Fungi from *Lagopsis supina*. *Journal of microbiology and biotechnology, 33*(4), 543–551. https://doi.org/10.4014/jmb.2211.11055

Zhang, H., Sun, X., Xu, C. (2016) Antimicrobial activity of endophytic fungus *Fusarium* sp. isolated from medicinal honeysuckle. Plant. *Arch Biol Sci Belgrade* 68:25–30

Zhang, H., Sun, X., Xu, C. (2016). Antimicrobial activity of endophytic fungus Fusarium sp. isolated from medicinal honeysuckle plant. *Archives of Biological Sciences*. 68. 4-4. 10.2298/ABS140401004Z.

CHAPTER 16

A REVIEW ON AGERATINA ADENOPHORA AS A BEST HOST FOR AQUATIC HYPHOMYCETES

Anjali Koranga Gashyal[1a], S.C. Sati[2b], K. Khulbe[2c] and Himani Verma[2d*]

1 Department of Botany, L.S.M. Campus, S.S.J. University, Almora- 263601
2 Department of Botany, D.S. B. Campus, Kumaun University Nainital- 263001

Abstract

Ageratina adenophora (Spreng.) R.M. King & H. Rob. has emerged as a potential host for a variety of microorganisms, particularly aquatic hyphomycetes (aquatic fungi), acting as endophytes within its root tissues. The symbiotic interaction between *Ageratina adenophora* and aquatic hyphomycetes, exhibits its potential as a host for these microorganisms, and the implications of this symbiotic relationship in ecological and environmental contexts. The ecological importance of *Ageratina adenophora*, its ability to harbor aquatic hyphomycetes, and the impact on both plant health and aquatic fungal biodiversity is discussed in this paper.

Keywords: *Ageratina adenophora*, aquatic hyphomycetes, endophyte, Kumaun Himalaya

Introduction

Ageratina adenophora (Spreng.) R.M. King & H. Rob. (Synonym: *Eupatorium adenophora* Spreng.), a fast-growing perennial herb native to Central America, has become an invasive species in many parts of the world, particularly in tropical and subtropical regions including Indian Himalayan Region and peninsular India (Qiang 1998, Wan et al., 2010, Poudel et al. 2019, Kumar and Garkoti 2022). It is known for its robust growth and ability to thrive in disturbed habitats (Cronk and Fuller 1995, Tererai and Wood 2014, Yu et al. 2016). In traditional medicine, this plant is employed to address various ailments, including the common cold, wind-heat syndrome, insomnia, hepatitis, menstrual irregularities, and diabetes (Wang et al. 2017, Tripathi et al. 2018, Okyere et al. 2021). *A. adenophora* contains a diverse array of pharmacological compounds, encompassing sesquiterpenoids, terpenes, flavonoids, phenols, coumarins, steroids, and phenylpropanoids, along with their derivatives (Liu et al. 2015, Chen et al. 2018, Bhandari et al. 2021). Numerous studies have demonstrated that the production of these bioactive compounds is associated with the endophytes present in plant tissues (Sharma et al. 2021).

Plant-microbe interactions play critical roles in plant growth and development, prompting an increase in research on plant-associated microbiomes. Endophytes are microorganisms that inhabit the intercellular gaps of plant tissues for at least part of their life cycle without showing any visible illness (Wen et al. 2023). Endophytes are considered crucial for establishing and expanding the invasive host plant, nutritional allocation, and disease resistance (Rout et al. 2013, Tian et al. 2014, Vincent et al. 2016). Fungal endophytes require special consideration as they possess unique life cycles, morphological features, and host symbiosis characteristics from other endophytes producing powerful, distinctive physiological and ecological responses on their plant hosts and the surrounding communities (Schulz and Boyle, 2006, Rodriguez et al. 2009, Wani et al. 2015, Christian et al. 2015, Compant et al. 2021). The significance of endophytes in facilitating the transformation of nutrients and helping host plants overcome stressful environments has been the subject of recent research that has extended to other host plants, including economic crops, medicinal plants, and even invasive plants (Philippot et al. 2013, Zhalnina et al. 2018).

Aquatic hyphomycetes have also showed their occurrence in healthy plant parts as an endophyte (Waid 1954, Koranga et al. 2021). Aquatic hyphomycetes are peculiar group of fungi that are characterized by uncommon spore shape which mainly occur in submerged leaf litter in streams and rivers (Bärlocher 1982, Chauvet et al. 2016). These fungi are known for their bioactive potential including antimicrobial, antioxidant, nutrient solubilization properties and plant growth promoters (Singh and Sati 2020, Sati and Pant 2020, Pant and Sati 2021, Sati et al. 2022, Pant et al. 2023). Numerous studies have demonstrated the diversity of endophytic fungi in *A. adenophora* (Fang et al. 2019a, Wen et al. 2023). However, there is still a need for further studies on the relationship of invasive species with their endophytes.

Expansion of *A. adenophora* in Kumaun Himalayan region

Ageratina adenophora is a fast-growing invasive plant species. This species has spread rapidly and competes with native vegetation for resources. Its ability to grow in a wide range of habitats, combined with its prolific seed production, makes it a strong competitor in disturbed ecosystems. Its presence can reduce biodiversity, as it outcompetes local flora. *A. adenophora* have invaded the Kumaun Central Himalaya. The roots of *A. adenophora* release a variety of putative allelochemicals that are harmful to nearby plants (Kumar and Garkoti 2022, Khatri et al. 2024).

The invasive plants like *A. adenophora* may have high population expansion and extremely high invasion capacities due to the rhizospheric microorganisms (Zhao et al., 2019). Recently, Bard et al. (2024) also suggested that during the establishment of invasive alien species, the endophytes they acquire contribute to the plant's overall species richness, which highlights the role of endophytes in the adaptation and success of invasive alien species in new environments. But, the role of endophytes in invasiveness is poorly understood, and the fundamental processes by which endophytic communities aid in the invasion of alien plants remain unclear despite the apparent ecological significance of endophytes (Fang et al. 2019b, Jha et al. 2019, Zhao et al. 2019).

Occurrence of Aquatic hyphomycetes as endophyte in *Ageratina adenophora*

Aquatic hyphomycetes as an endophyte commonly occur in the plants growing around freshwater streams (Sridhar and Bärlocher 1992, Sati and Belwal 2005). Earlier studies have shown that these aquatic hyphomycetes mostly prefer *A. adenophora* growing in riparian areas of Kumaun Himalayan region as their host (Table 1.1). *Tetracladium breve, T. marchalianum* and *T. nainitalense* have shown antibacterial activity against pathogenic bacterial strains suggesting their potential roles in aiding host defense mechanisms (Arya and Sati 2011). The occurrence of root endophytic aquatic hyphomycetes in the healthy roots of *A. adenophora* growing in the riparian area of Kumaun Himalaya indicates their positive interaction in vigorous expansion despite of stress conditions.

Aquatic hyphomycetes in their preferred host plant (*A. adenophora*) may be more efficient at colonizing and surviving in the tissues of certain plant species due to their genetic traits, enzymatic capabilities, or ability to tolerate plant-specific defenses. Some plants produce chemical compounds that favor the growth of specific endophytes (Nair and Padmavathy 2014). The plant might produce certain secondary metabolites (such as alkaloids or phenolics) that attract or support specific microbial populations, and the endophyte may, in turn, help the plant metabolize these compounds or provide some other benefit. The genetic makeup of host plant may make it more receptive to specific endophytes, facilitating the establishment of the endophytic relationship.

Table 1.1: Aquatic hyphomycetes isolated as root endophytes from the roots of *Ageratina adenophora* in Kumaun Himalayan region

S. No.	Endophytes	Locality	Reference
1	*Acaulopage tetraceros*	J	Sati and Belwal 2005
2	*Alatospora acuminata* Ingold	BG, J, Jk, K RG, DG	Sati and Belwal 2005, Sati et al 2009b, Sati and Pathak 2017, Koranga et al 2021
3	*Campylospora chaetocladia* Ranzoni	BG, DG, J	Sati and Belwal 2005, Sati and Pathak 2017
4	*Cylindrocarpon aquaticum* (Nils) Marvanova and Descals	J, Jk, K, P, CG, GM, N, SV, VK, DG	Sati and Belwal 2005, Sati and Pathak 2017
5	*Flagellospora penicilloides* Ingold	K, DG, RG	Arya and Sati 2010
6	*Helicosporium lumbricoides* Linder	RG	Sati and Pathak 2017
7	*Lunulospora curvula* Ingold	J, Jk, K, GM	Sati and Belwal 2005, Sati et al 2009b
8	*Tetracladium apiense*	NT	Arya and Sati 2010
9	*Tetracladium breve* Roldan	GM, RG, SV	Arya and Sati 2010
10	*Tetracladium marchalianum* De Wild	CG, J, Jk, K, RT, SV, ST, P, RG	Sati and Belwal 2005, Sati et al 2009a, Sati and Pathak 2017
11	*Tetracladium nainitalense* Sati and Arya	GM, RG	Sati et al. 2009a

BG- Bageshwar, CG- Chorgalia, DG- Dogaon, GM- Gufa Mahadev, J- Jageshwar, Jk-Jeolikot, K-Kilbury, N- Niglat, NT-Nainital, RG- Ramgarh, RT- Ratighat, P- Pangot, ST- Saat Tal, SV-Snow View

Conclusion

The spread of *Ageratina adenophora* in the Kumaun Himalayan region is a growing concern. The vigorous growth of *A. adenophora* has reached to riparian areas of Kumaun Himalaya which could be due to the assistance of aquatic hyphomycetes as endophytes. Aquatic hyphomycetes as an endophyte play important role in regulating growth, development, stress tolerance, defence mechanism and allelopathic effects on their host plant. They can influence the growth, reproduction, and physiological processes of invasive plants, both directly and indirectly. Endophytes play a significant role in the growth and resource allocation of similar invasive plant and correlates with invasion effects on nitrogen, phosphate, and iron biogeochemical cycles.

Reviewing and highlighting past achievements, ongoing studies, and recent advancements in the involvement of root endophytic aquatic hyphomycetes in *Ageratina adenophora* is essential in attracting the attention of the scientific community to this new area. Further, their involvement also introduces new dimensions in understanding plant invasions, as alien plant interactions offer notable contributions to fungal life histories by facilitating host shifts to previously unassociated plants and exposing both hosts and endophytes to novel environments. Thus, providing insights into the potential exploitation of the novel sources for therapeutic applications in a variety of industries, including pharmaceutical, food, cosmetics, and medicine.

Acknowledgment

The authors are thankful to the Head, Department of Botany, Thakur Dev Singh Bisht Campus, Kumaun University, Nainital for providing necessary lab facilities.

References

Arya P. and Sati SC. (2011). Evaluation of endophytic hyphomycetes for their antagonistic activity against pathogenic bacteria. *Int. J. Microbiol.* 2(9): 343–347.

Arya, P., and Sati, S.C. (2010). Four species of aquatic hyphomycetes occurring as new root endophytes. *National Academy Science Letters-india, 33*(9-10), 299-301.

Bärlocher, F. (1982). On the ecology of Ingoldian fungi. *BioScience.* 32: 581-586.

Bard, N.W., Cronk, Q.C.B. and Davies, T.J. (2024), Fungal endophytes can modulate plant invasion. *Biol Rev,* 99: 1652-1671. https://doi.org/10.1111/brv.13085.

Bhandari, S., Khadayat, K., Poudel, S., Shrestha, S. and Marasini, B.P. (2021). Phytochemical analysis of medicinal plants of Nepal and their antibacterial and antibiofilm activities against uropathogenic *Escherichia coli*. BMC Complement. Med. Ther. 21:116.

Chauvet, E., Cornut, J., Sridhar, K. R., Selosse, M. A. and Bärlocher, F. (2016). Beyond the water column: aquatic hyphomycetes outside their preferred habitat. *Fungal Ecology,* 19: 112-127.

Chen, H., Zhou, B., Yang, J., Ma, X.H., Deng, S.H., Huang, Y., Wen, Y., Yuan, J. and Yang, X. (2018). Essential oil derived from *Eupatorium adenophorum* Spreng. mediates anticancer effects by inhibiting STAT3 and AKT activation to induce apoptosis in hepatocellular carcinoma. *Front. Pharmacol.* 9:483.

Christian, N., Whitaker, B. K. and Clay, K. (2015). Microbiomes: unifying animal and plant systems through the lens of community ecology theory. *Frontiers in Microbiology* 6, 153936.

Compant, S., Cambon, M. C., Vacher, C., Mitter, B., Samad, A. and Sessitsch, A. (2021). The plant endosphere world – bacterial life within plants. *Environmental Microbiology* 23, 1812–1829.

Cronk, Q.C.B. and Fuller, J.L. (1995) Plant Invaders. The Threat to Natural Ecosystems. Chapman and Hall, London.

Fang, K., Miao, Y.F., Chen, L., Zhou, J., Yang, Z.P., Dong, X.F. and Zhang, H.B. (2019a). Tissue-specific and geographical variation in endophytic fungi of

Ageratina adenophora and fungal associations with the environment. *Frontiers in microbiology*. 10: 2919.

Fang, K., Wang, Y. Z., and Zhang, H. B. (2019b). Differential effects of plant growth-promoting bacteria on invasive and native plants. *South African Journal of Botany*, 124: 94-101.

Jha, P.N., Gomaa, A.B., Yanni, Y.G., El-Saadany, A.E.Y., Stedtfeld, T.M., Stedtfeld, R.D., Gantner, S., Chai, B., Cole, J., Hashsham, S.A. and Dazzo, F.B. (2020). Alterations in the endophyte-enriched root-associated microbiome of rice receiving growth-promoting treatments of urea fertilizer and Rhizobium biofertilizer. *Microbial ecology*, 79: 367-382.

Khatri, K., Negi, B., Bargali, K. and Bargali, S.S. (2024). Toxicological assessment of invasive *Ageratina adenophora* on germination and growth efficiency of native tree and crop species of Kumaun Himalaya. *Ecotoxicology*. 33: 697-708.

Koranga, A., Pant, P. and Sati S.C. (2021). Biodiversity of asexually produced aquatic conidial fungi as root endophytes from Kumaun Himalaya. *The International Journal of Plant Reproductive Biology*. 13(1): 1-4.

Kumar, M. and Garkoti S.C. (2022). Allelopathy effects of invasive alien *Ageratina adenophora* on native shrub species of chir pine forest in the central Himalaya, India. *Journal of Forest Research*, 27(1): 53-62.

Liu, P.Y., Liu, D., Li, W.H., Zhao, T., Sauriol, F., Gu, Y.C., Shi, Q.W. and Zhang, M.L. (2015). Chemical Constituents of Plants from the Genus *Eupatorium* (1904-2014). *Chemistry and Biodiversity*. 12(10): 1481-515.

Nair, D.N. and Padmavathy, S. (2014). Impact of endophytic microorganisms on plants, environment and humans. Scientific World Journal. 2014:250693. doi: 10.1155/2014/250693.

Okyere, S.K., Wen, J., Cui, Y., Xie, L., Gao, P., Wang, J., Wang, S. and Hu, Y. (2021). Toxic mechanisms and pharmacological properties of euptox A, a toxic monomer from *A. adenophora*. *Fitoterapia*. 155: 105032.

Pant, P. and Sati, S.C. (2021). Antibacterial activity of root endophytic aquatic hyphomycetes against plant pathogenic bacteria. *Vegetos*. 34(4): 785-789.

Pant, P., Sati, S.C., Kumar, P. and Koranga, A. (2023). Mycochemical composition and antioxidant activity of root endophytic aquatic hyphomycetes isolated from the riparian area of Kumaun Himalaya, India. *Natural Product Research*, 1–5. https://doi.org/10.1080/14786419.2023.2297406

Philippot, L., Raaijmakers, J. M., Lemanceau, P. and Van Der Putten, W.H. (2013). Going back to the roots: the microbial ecology of the rhizosphere. *Nature reviews microbiology*. 11(11): 789-799.

Poudel, A.S., Jha, P.K., Shrestha, B.B. and Muniappan, R. (2019). Biology and management of the invasive weed *Ageratina adenophora* (Asteraceae): current state of knowledge and future research needs. *Weed research*. 59(2): 79-92.

Qiang, S. (1998). The history and status of the study on crofton weed (*Eupatorium adenophorum* Spreng.) a worst worldwide weed. J. Wuhan Bot. Res. 4(4): 366-372.

Rodriguez, R. J. Jr., White, J. F., Arnold, A. E. and Redman, R. S. (2009). Fungal endophytes: diversity and functional roles. *New Phytologist* 182, 314–330.

Rout, M.E., Chrzanowski, T.H., Westlie, T.K., DeLuca, T.H., Callaway, R.M. and Holben, W.E. (2013). Bacterial endophytes enhance competition by invasive plants. *American Journal of Botany*. 100(9): 1726-1737.

Sati, S.C., Arya, P., Belwal, M. (2009a) *Tetracladium nainitalense* sp nov, a root endophyte from Kumaun Himalaya, India. *Mycologia* 101: 692–695.

Sati, S.C. and Belwal, M. (2005). Aquatic hyphomycetes as endophytes of riparian plant roots. *Mycologia*. 97: 45-49.

Sati, S.C. and Pant, P. (2020). Two root endophytic aquatic hyphomycetes *Campylospora parvula* and *Tetracladium setigerum* as plant growth promoters. *Asian J. Agric. Res.* 14: 28-33.

Sati, S.C., Pant, P. and Vassilev, N. (2022) Significance of root endophytic aquatic hyphomycetes in phosphate solubilization. *Sydowia* 75: 103-109.

Sati, S.C., Pargaein, N. and Belwal, M. (2009b). Diversity of Aquatic Hyphomycetes as Root Endophytes on Pteridophytic Plants in Kumaun Himalaya. *Journal of American Science*, 5(4):179-182.

Sati, S.C. and Pathak, R. (2017). New root endophytic waterborne conidial fungi from Kumaun Himalayas. *Current Botany*, 8: 12-16.

Schulz, B. and Boyle, C. (2006). What are endophytes? In Microbial Root Endophytes, pp. 1–14. Springer, Berlin.

Sharma, H., Rai, A.K., Dahiya, D., Chettri, R. and Nigam P.S. (2021). Exploring endophytes for in vitro synthesis of bioactive compounds similar to metabolites produced in vivo by host plants. AIMS Microbiol. 7: 175–199.

Singh, L. and Sati, S.C. (2020). Antagonistic activity of isolated root endophytic freshwater fungus *Anguillospora longissima* against pathogenic fungi. *Natl. Acad. Sci. Lett.* 43(2): 201-205.

Sridhar, K.R. and Bärlocher, F. (1992). Endophytic aquatic hyphomycetes of roots of spruce, birch and maple. *Mycological Research*. 96(4): 305-308.

Tererai, F. and Wood, A.R. (2014). On the present and potential distribution of *Ageratina adenophora* (Asteraceae) in South Africa. *South African Journal of Botany*. 95: 152-158.

Tian, Y., Amand, S., Buisson, D., Kunz, C., Hachette, F., Dupont, J., Nay, B. and Prado, S. (2014). The fungal leaf endophyte *Paraconiothyrium variabile* specifically metabolizes the host-plant metabolome for its own benefit. *Phytochemistry*, 108: 95-101.

Tripathi, Y.C., Saini, N., and Anjum, N.A. (2018). Review of ethnomedicinal, phytochemical, pharmacological and toxicological aspects of *Eupatorium adenophorum* Spreng. *Asian J. Biomed. Pharmaceut. Sci.* 8: 25–35.

Vincent, J.B., Weiblen, G. D. and May, G. (2016). Host associations and beta diversity of fungal endophyte communities in New Guinea rainforest trees. Molecular ecology. 25(3): 825-841.

Waid, J.S. (1954). Occurrence of aquatic hyphomycetes upon the root surfaces of beech grown in woodland soils. *Trans. Brit. Mycol. Soc.* 37:420-421.

Wan, F., Liu, W., Guo, J., Qiang, S., Li, B., Wang, J., Yang, G., Niu, H., Gui, F., Huang, W., Jiang, Z. and Wang, W. (2010). Invasive mechanism and control strategy of *Ageratina adenophora* (Sprengel). Science China Life Sciences. 53(11): 1291-8.

Wang, C., Lin, H.L., Feng, Q.S., Jin, C.Y., Cao, A.C. and He, L. (2017). A New Strategy for the prevention and control of *Eupatorium adenophorum* under climate change in China. *Sustainability*. 9: 2037.

Wani, Z. A., Ashraf, N., Mohiuddin, T. and Riyaz-Ul-Hassan, S. (2015). Plant-endophyte symbiosis, an ecological perspective. Applied Microbiology and Biotechnology 99, 2955–2965.

Wen, J., Okyere, S.K., Wang, J., Huang, R., Wang, Y., Liu, L., Nong, X. and Hu, Y. (2023). Endophytic Fungi Isolated from *Ageratina adenophora* Exhibits Potential Antimicrobial Activity against Multidrug-Resistant *Staphylococcus aureus*. Plants (Basel). 12(3):650.

Yu, F., Akin-Fajiye, M., Tapa Magar, K., Ren, J. and Gurevitch, J. (2016). A global systematic review of ecological field studies on two major invasive plant species, *Ageratina adenophora* and *Chromolaena odorata*. Diversity and Distributions. 22: 1174-1185.

Zhalnina, K., Louie, K.B., Hao, Z., Mansoori, N., Da Rocha, U.N., Shi, S., Cho, H., Karaoz, U., Loqué, D., Bowen, B.P. and Firestone, M.K., (2018). Dynamic root exudate chemistry and microbial substrate preferences drive patterns in rhizosphere microbial community assembly. *Nature microbiology*. 3(4): 470-480.

Zhao, M., Lu, X., Zhao, H., Yang, Y., Hale, L., Gao, Q., Liu, W., Guo, J., Li, Q., Zhou, J. and Wan, F., (2019). *Ageratina adenophora* invasions are associated with microbially mediated differences in biogeochemical cycles. Science of the Total Environment, 677: 47-56.

CHAPTER 17

ETHNOBOTANY AND TRADITIONAL KNOWLEDGE OF INDIAN HIMALAYAN MEDICINAL PLANTS: CULTURAL HERITAGE, USES, AND CONSERVATION

Diksha Semwal[1], Ajay Hemdan[2]

[1] Wildlife Institute of India Dehradun, Uttarakhand

[2] College of Horticulture, VCSG Uttarakhand University of Horticulture and Forestry, Bharsar, Pauri Garhwal

---

Abstract

The study explores the ethnobotanical and traditional knowledge associated with medicinal plants in the Indian Himalayan Region (IHR). Highlighting its unique biodiversity, with over 1,748 documented medicinal plant species, the paper underscores the cultural and therapeutic significance of these species to indigenous communities. Through detailed accounts of plant diversity, traditional uses, and applications, it emphasizes the intricate relationship between biodiversity and traditional knowledge systems. It also examines threats such as habitat loss, overexploitation, and the erosion of traditional knowledge due to modernization and socioeconomic shifts. The chapter advocates for conservation strategies, sustainable harvesting practices, and the integration of traditional wisdom with modern science to preserve this invaluable cultural and biological heritage. Recommendations include community-driven conservation efforts, policy frameworks for equitable knowledge sharing, and fostering collaborations between traditional healers and researchers to achieve ecological sustainability and societal well-being.

Keywords: Traditional Knowledge, Therapeutic, Indigenous Medicinal Plants, Conservation

Introduction to Ethnobotany in Indian Himalayan Region

The Indian Himalayan Region (IHR) is a rich repository of ethnobotanical knowledge, particularly regarding the use of medicinal and edible plants by indigenous communities. This knowledge plays a crucial role in preserving cultural heritage, ensuring food sovereignty, and supporting traditional healthcare systems. The IHR holds a unique position among the world's mountain ecosystems. Spanning over 2500 km in length, with a width ranging from 80 to 300 km, and rising from low-lying plains to elevations exceeding 8000 m asl, it creates a distinctive climate that also impacts much of Asia. The region's diverse topographical features result in significant variability in climate and habitat conditions (Singh, 2006; Haq *et al.*, 2024). The IHR is home to 18,940 plant species, including Angiosperms, Gymnosperms, Pteridophytes, Bryophytes, Lichens, and Fungi. Among these, 6745 species of Angiosperms, spread across 225 families and 1768 genera, have been documented in the Trans, North West, and West Himalayas. This includes 604 tree species, 1049 shrubs, and 5092 herbs. The region also supports 675 wild edible plants and 1,748 medicinal plant species, many of which are endemic to the area (Samant, 2021).

Ethnobotany involves studying the traditional use of plants by indigenous communities. This field of knowledge dates back to ancient times and highlight the traditional applications of plant resources, which can aid in promoting tribal development. Dr. John William Harshberger, a botanist at the University of Pennsylvania, first introduced the term "ethnobotany" to describe the study of "plants used by primitive and aboriginal people" (Ford, 1994). Over time, ethnobotany has been defined as the traditional knowledge of indigenous communities regarding local plant diversity and how they utilize these plants in their daily lives. The term combines "ethno," meaning the study of people, with "botany," the study of plants and it focuses on the direct relationship between plants and humans (https://uou.ac.in/sites/default/files/slm/MSCBOT-608.pdf).

In the Indian Himalayan Region (IHR), the cultural groups include the Gaddis, Gujjars, Swangla, Lahulas, and Pangwalas (Trans, North West Himalaya); Bhotias, Rajees, Tharus, Buxas, and Jaunsarrees (West Himalaya); Bhutias and Lepchas (Central Himalaya); and Chakma and Nagas (East Himalaya). The region's rich plant diversity has been utilized by these communities for various purposes, including medicine, food, fuel, fodder, crafting agricultural tools, constructing houses, supporting small-scale enterprises (such as producing baskets, mats, hats, and kiltas), and performing religious rituals (Samant and Dhar, 1997; Samant, 2015).

Cultural Significance of Medicinal Plants and Traditional Knowledge Systems in the Himalayas

The cultural importance and traditional knowledge systems of ethnobotany in the Himalayas are closely connected to the health, nutrition, and ecological sustainability of local communities. These systems embody a rich legacy of plant use, with traditional knowledge playing a vital role in ensuring the survival and well-being of these communities. The local population, which includes ethnic groups such as Gujjar, Bakarwal, Pahari, and Kashmiri, possesses a unique and extensive repository of information (Haq *et al.*, 2023).

Ethnobotany is the scientific exploration of cultural practices and traditional knowledge related to the religious, medicinal, and other uses of plants. Knowledge of using plants as medicine has been accumulated through years of consistent practice. Tribal populations are the primary custodians of this traditional knowledge, which has historically been passed down orally by locals and traditional healers across generations. Ancient Indian texts like *Sushruta Samhita* (circa 500 BC), *Charaka Samhita* (ca 100 AD), and *Ashtanga Hridaya Samhita* (ca 200 AD) describe the use of approximately 1200 plant-based remedies, their properties, and therapeutic applications. Additionally, Indian Sanskrit literature, including the *Rigveda*, *Atharvaveda*, *Upanishads* (ca 1000–600 BC), *Mahabharata*, and *Puranas* (ca 700–400 BC), provides insights into the various uses of plants. These ancient sources are invaluable in understanding ethnomedicinal plants, with 148 medicinal plants identified in the *Rigveda* and *Atharvaveda*. Two of the most significant works in the Indian medical system are the writings of Charaka and Sushruta, with *Charaka Vedic literature* documenting 400–450 medicinal plants. The ancient text *Vriksha Ayurveda*, authored by Parasara, is a foundational work on the science of medicinal plants, detailing plants used for agriculture, food, fuel, and medicine. The specific roles of various Rudraksha (*Elaeocarpus ganitrus*) species, as described in *Lakshmipuran* and *Shivapuran*, are particularly noteworthy (https://uou.ac.in/sites/default/files/slm/MSCBOT-608.pdf).

Indian Systems of Medicine (ISM): Although Ayurveda, Siddha, and Unani are the traditional medical systems in India, the government of India established AYUSH (Department of Ayurveda, Yoga and Naturopathy, Unani, Siddha, and Homoeopathy) in March 1995 as the Department of Indian Systems of Medicines and Homoeopathy in the Ministry of Health and Family Welfare (https://uou.ac.in/sites/default/files/slm/MSCBOT-608.pdf).

WHO has defined traditional medicine as a comprehensive term used to refer both to traditional medical systems such as traditional Chinese medicine, the Indian subcontinent's Ayurveda and Siddha, and Arabic Unani medicine and also to various other forms of indigenous medicine (Bhattarai & Karki, 2004). Three traditional medical systems that are acknowledged in India are Ayurveda, Siddha, and Unani:

(A) Ayurveda: Ayurveda is a comprehensive and scientifically rooted medicinal system that originated in India. The term "Ayurveda" translates to "info of life", derived from two Indo-Aryan words: *Ayu* (life) and *Veda* (knowledge or science). The foundational knowledge of Ayurveda is found in the four Vedas, considered the oldest Indian scriptures (5000–1000 BC), which provide extensive insights into natural remedies. The classical texts *Sushruta Samhita* (focusing on surgery) and *Charaka Samhita* (focusing on internal medicine) were systematically compiled and are regarded as monumental works in Ayurveda. Their essential knowledge was later refined and consolidated in *Astanga Sangraha* and *Astanga Hrdaya*. Other notable ancient texts include *Madhava Nidana*, which emphasizes disease diagnosis, *Bhava Prakash*, detailing the properties of plants and dietary substances, and *Sarngadhara Samhita*, focusing on pharmacological formulations and dosage standards. Ayurveda is categorized into eight primary clinical disciplines:

1. **Kayachikitsa -** Internal medicine
2. **Shalya Tantra -** Surgery
3. **Shalakya Tantra -** Disorders above the clavicle (e.g., ENT and ophthalmology)
4. **Kaumarabhritya -** Pediatrics, obstetrics, and gynecology
5. **Bhuta Vidya -** Psychiatry
6. **Agada Tantra -** Toxicology
7. **Rasayana Tantra -** Rejuvenation and geriatrics
8. **Vajikarana -** Aphrodisiacs and eugenics

These subdivisions reflect the depth and scope of Ayurveda as a medical system, integrating diverse aspects of health and wellness (Mafuva *et al.*, 2014; Chavda *et al.*, 2022).

(B) Siddha: The Siddha system of medicine is considered a remarkable representation of Tamil culture, originating in India's southern and central regions. It has roots in Dravidian traditions and reached its zenith during the Indus Valley Civilization. Siddha medicine draws significant inspiration from Chinese alchemy, Taoism, and Taoist geophysics. Historically, it is believed to have been established by eighteen Siddhars, a revered group of Tamil sages. While the Siddha system shares similarities with Ayurveda, it maintains distinct philosophies, a holistic worldview, and unique lifestyle-focused practices (Shanmugam *et al.*, 2024).

(C) Unani: The Unani system of medicine is a synthesis of ancient medical traditions from Egypt, Syria, Iran, China, India, and other Eastern countries. It originated in Greece and later spread to the Arabian Peninsula. Unani medicine was introduced to India by Arab and Persian settlers in the 11th century and gained widespread acceptance and prominence under Mughal rule (Ansari, 2021).

The relationship between humans and nature is deeply interconnected. Plant-based knowledge has been an essential aspect of societies throughout human history, extending beyond their role as a food source to their use in medicine. Plants have consistently served as a reliable foundation for traditional healthcare systems across diverse cultures. It is estimated that approximately 50,000 plant species, representing 18.9% of the world's flora, have been documented for their use in traditional medicinal systems globally (Bhat *et al.*, 2013; Baydoun *et al.*, 2015; Parmesan, 2006; Ibrar *et al.*, 2007).

Herbal remedies account for about 85% of traditional medicines worldwide, as reflected in folklore medicine practices (Cohen & Ernst, 2010). Factors such as low per capita income, limited access to modern healthcare, trust in herbal treatments, and the easy availability of medicinal plants contribute to the widespread reliance on herbal medicine (Loya *et al.*, 2009; Ganesan *et al.*, 2008; Sullivan & Shealy, 1997).

Growing scientific research supports the integration of traditional knowledge with modern scientific perspectives to develop culturally appropriate and environmentally sustainable approaches for improving local communities' well-being (Ayeni & Basiri, 2018). Despite being considered less advanced compared to modern healthcare, traditional ethnomedicinal practices remain indispensable. According to the WHO, about 80% of the population in developing countries relies on herbal medicines for treating ailments (Tangjitman *et al.*, 2015; Hussain *et al.*, 2024).

Traditional Uses and Applications of Key Medicinal Plants

Traditional medicinal plants have been integral to healthcare systems worldwide, offering diverse applications based on centuries of indigenous knowledge. These plants are utilized for their therapeutic properties, including anti-inflammatory, antimicrobial, and analgesic effects. A total of 1,748 species of medicinal plants, classified under 915 genera and 223 families, have been documented in the Indian Himalayan Region (IHR). The diversity and distribution of these species are summarized in Table 1. The families with the highest representation of medicinal plants include Asteraceae (129 species), Fabaceae (107 species), Lamiaceae (63 species), Rubiaceae (55 species), Euphorbiaceae (51 species), Ranunculaceae (48 species), Rosaceae (41 species), Poaceae (40 species), Orchidaceae (37 species), Polygonaceae (32 species), and Gentianaceae (27 species). Among genera, *Polygonum* (19 species), *Euphorbia* (16 species), *Piper* (16 species), *Ficus* (13 species), *Aconitum* and *Swertia* (12 species each), *Artemisia* (11 species), and *Solanum*, *Berberis*, *Desmodium*, and *Allium* (10 species each), as well as *Saussurea* (9 species), are notably rich in medicinal plant diversity (Samant *et al.*, 1998).

Of the total medicinal plant species recorded, 25.8% are native to the Himalayan region, 5.66% are native to the Himalayas and adjoining biogeographic domains, 3.55% are endemic to the region, and 11.9% are classified as near-endemic to the IHR. The altitudinal distribution of these species indicates that the majority are found in areas up to 1,800 m elevation, while the fewest occur at altitudes above 3,800 m. Biogeographically, the Central Himalaya (including Sikkim and Darjeeling Hills) hosts the highest number of medicinal plant species, followed by the Western Himalaya (Uttarakhand) and the Trans and North Western Himalaya (encompassing Himachal Pradesh, Jammu & Kashmir, and the Union Territories of Ladakh) (Samant, 2021).

Table 1: Medicinal Plant Diversity and Distribution in the Indian Himalayan Region (IHR)

Category	Details	Number/Percentage
Total Species in IHR		1,748
Total Genera in IHR		915
Total Families in IHR		223
Families with Maximum Medicinal Plants	Asteraceae	129 spp.
	Fabaceae	107 spp.
	Lamiaceae	63 spp.
	Rubiaceae	55 spp.
	Euphorbiaceae	51 spp.
	Ranunculaceae	48 spp.
	Rosaceae	41 spp.
	Poaceae	40 spp.
	Orchidaceae	37 spp.
	Polygonaceae	32 spp.
	Gentianaceae	27 spp.
Genera with Maximum Medicinal Plants	Polygonum	19 spp.
	Euphorbia	16 spp.
	Piper	16 spp.
	Ficus	13 spp.
	Aconitum	12 spp.
	Swertia	12 spp.
	Artemisia	11 spp.
	Solanum	10 spp.
	Berberis	10 spp.
	Desmodium	10 spp.
	Allium	10 spp.
	Saussurea	9 spp.

Category	Details	Number/Percentage
Distribution of Medicinal Plant Species	Native to the Himalayas region	25.8%
	Native to the Himalayas and neighbouring domains	5.66%
	Endemic to IHR	3.55%
	Near endemic to IHR	11.9%

(Source: Samant *et al.*,1998; Samant, 2021)

The notable medicinal plants of the IHR are *Aegle marmelos, Azadirachta indica, Berberis asiatica, B. aristata, Boerhaavia diffusa, Justicia adhatoda, Gloriosa superba, Syzygium cumini, Withania somnifera, Costud speciosa, Coleus forskohlii, Didymocarpus pedicellata, Rauwolfia serpentina, Asparagus racemosus, Oroxylum indicum, Tinospora cordifolia, Terminalia arjuna, chebula, T. bellirica, Glycyrrhiza glabra, Mucuna pruriens, M. bracteata, Pholidota articulate, Swertia angustifolia, S. chirayita, S. cordata, Hypericum perforatum, Heracleum lanatum, Cinnamomum tamala, Viola canescens, Bergenia ligulata, Hedychium spicatum, Phyllanthus emblica, Plantago major, Acorus calamus, Paris polyphylla, Taxus wallichiana, Thalictrum foliolossum, valeriana wallichii, Adiantum venustum, Angelica glauca, A. nubigena, Lilium polyphyllum, Trillium govanianum, Corydalis govaniana, Inula racemosa, Bunium persicum, Saussurea costus, S. obovata, Arnebia benthamii, A. euchroma, Picrorhiza kurroa, Dactylorhiza hatagirea, Aconitum heterophyllum, Sinopodophyllum hexandrum, Rheum australe, Nardostachys jatamansi, Panax pseudogeinseng, Fritillaria roylei, Rhododendron anthopogon, Ephedra gerardiana, Hippophae rhamnoides ssp. turkestanica, Coptis teeta, etc. (Samant et al., 1998). Seventeen medicinal plants i.e., Saussurea costus (Endangered), Aconitum deinorrhizum, A. falconeri var. latilobum, A. ferox, Allium stracheyi, Berberis affinis, Coptis teeta, Dioscorea deltoidea, Inula racemosa, Nardostachys jatamansi, Panax pseudoginseng and Picrorhiza kurroa (All Vulnerable), Berberis kashmiriana, Codonopsis affinis,* and *Saussurea bracteata* (All Rare) and *Angelica nubigena and Pittosporum eriocarpum* (Both Indeterminate) are recorded in the Red Data Book of Indian Plants (Nayar and Sashtry, 1987, 1988, 1990; Samant *et al.*, 1998a). A total of 120 species have been evaluated under various threat categories, including Critically Endangered, Endangered, Vulnerable, Near Threatened, Least Concern, and Data Deficient. This assessment was conducted through Conservation Assessment and Management Plan (CAMP) workshops, involving subject matter experts from the Indian Himalayan Region (IHR) (Ved *et al.*, 2003, 2005).

The majority of medicinal plant species, which serve as the primary mode of treatment in most developing nations and are widely utilized in traditional and ethnomedicinal practices, are predominantly harvested from wild populations. The World Health Organization (WHO) has documented approximately 20,000 medicinal plant species used globally, while other estimates suggest this number could range between 35,000 and 70,000 species worldwide. It is estimated that

70–80% of the global population primarily relies on herbal medicine for their primary healthcare needs. In South Asia alone, over 800 million individuals depend on plant-based remedies. In China, plant-based medicine is relied upon by 40% of urban patients and over 90% of rural patients. Approximately 10,000–11,250 species are used in ethnomedicine in China, 6,000–7,500 in India, 2,237 in Mexico, 2,572 by North American Indigenous populations, and 1,600 species in Nepal. With ongoing fieldwork across various developing regions, including research among tribal, ethnic, and indigenous communities, the documented number of medicinal plants used for therapeutic purposes is expected to rise significantly in the future (Bhattarai & Karki, 2004).

The World Health Organization (WHO) defines traditional medicine as the cumulative body of knowledge, practices, and techniques whether scientifically explicable or not applied to the diagnosis, prevention, and treatment of physical, mental, or social imbalances. These practices rely heavily on empirical experience and observations passed down through generations, either orally or in written form. Traditional medicine represents a comprehensive integration of dynamic medical knowledge and ancestral wisdom (Bhattarai & Karki, 2004). Some of the ethnobotanically used medicinal plants in IHR are as follows:

Table 2: Description of Ethnobotanically used medicinal plants in IHR

S. No.	Botanical Name	Local Name	Uses
1	*Acorus calamus* L.	Vacha	Leaves and flower decoction juice are given for cough, and fever. The root powder is used to cure Worm infestation among children. The root powder is given with milk in general debility
2	*Allium fedtschenkoanum* Regel	Bhook	Skin infection
3	*Allium humile* Kunth	Pari Pyaz	Gastrointestinal disorder
4	*Amaranthus spinosus* L.	Ghanhar	Backache
5	*Angelica cyclocarpa*	Murchar	Respiratory tract infection, gastrointestinal disorder
6	*Emblica officinalis* Gaertn.	Amla	Fruit juice is given to increase the flow of urine, act as a diuretic, and also given in diarrhoea, dysentery and to cure Jaundice
7	*Rhododendron arboreum* Sm.	Burans	The decoction of corolla mixed with 1 table of spun sugar is used to cure cardiovascular diseases. Dried flower powder is given in dysentery. Leaf paste is applied on the forehead in headache

S. No.	Botanical Name	Local Name	Uses
8	*Heracleum canescens* Lindl.	Palhar	Wound healing
9	*Inula orientalis* Lam.	Sabir Jari	Asthma, urinary tract infection
10	*Jurinea himalaica*	Gugal Dhoop	Backache, bleeding wounds, rheumatism, aphrodisiac
11	*Saussurea costus*	Kuth	Stomach worms, backache, rheumatism, obesity
12	*Berberis lycium* Royle	Sumbel	Backache, bone fractures
13	*Chenopodium album* L.	Bathwa	Constipation, flatworms
14	*Cannabis sativa* L.	Bhang	Diarrhea, stomach disorder
15	*Equisetum arvense* L.	Banda Kaya	Acidity, kidney infection
16	*Fragaria nubicola*	Khanmarch	Jaundice, fever, cholera
17	*Aconitum chasmanthum*	Mohri	Asthma, cough, skin issues
18	*Rheum webbianum* Royle	Chotyeal	Wounds, urinary tract infection, constipation
19	*Rumex nepalensis*	Hoola	Allergy, constipation, menstrual disorders
20	*Thymus linearis*	Ban-Ajwain	Pneumonia, throat infection, menstrual disorders
21	*Valeriana pyrolifolia*	Mushk Bala	Constipation
22	*Plantago major* L.	Chamchi Patar	Toothache, constipation
23	*Picrorhiza kurroa*	Kaur	Abdominal pain, blood pressure issues
24	*Taxus baccata* L.	Thooni	Stomach ulcer, asthma
25	*Juglans regia* L.	Akhroot	Brain tonic, toothache
26	*Oxyria digyna*	Khatkara	Dysentery
27	*Skimmia laureola*	Nera	Abdominal pain, kidney stone
28	*Trillium govanianum*	Tray Patra	Stomach ulcer
29	*Verbascum Thapsus*	Gadi Kan	Respiratory tract infection, jaundice
30	*Bergenia ciliate*	Batbewa	Wound healing, stomach ulcer, urinary tract infection
31	*Heracleum cachemiricum*	Chora	Asthma, gastrointestinal disorder
32	*Angelica archangelica*	Chora	Asthma, gastrointestinal disorder
33	*Achillea millefolium*	Sultani Booti	Stomach disorder, antiseptic, toothache
34	*Taraxacum officinale*	Hund	Kidney and liver disorders, constipation
35	*Betula utilis*	Bhurj	Rheumatism, gallbladder stone

S. No.	Botanical Name	Local Name	Uses
36	*Mentha longifolia*	Poodna	Asthma, gastric problem
37	*Fritillaria pallidiflora*	Rech-Eend	Broncho-asthma, tuberculosis
38	*Impatiens edgeworthii*	Buntil	Pneumonia
39	*Silene vulgaris*	Murkan	Constipation
40	*Amaranthus hybridus*	Ghanhar	Backache
41	*Rosa moschata*	Bramboos	Sore throat, skin infections
42	*Berberis aristate*	Kashmal	Diabetes, wounds, infections
43	*Cuscuta reflexa*	Neela Dahari	Wound infections
44	*Equisetum ramosissimum*	Banda Kaya	Acidity, kidney infection
45	*Aesculus indica*	Ban-Khoor	Indigestion
46	*Dactylorhiza hatagirea*	Nar Mada	Aphrodisiac, depression
47	*Rhamnus virgate*	Khargu	Skin problems, laxative
48	*Polygonum aviculare*	Dhrubra	Constipation
49	*Anaphalis nepalensis*	Jarri	Wound healing
50	*Oxytropis cachemiriana*	Kathra	Toothache
51	*Primula denticulate*	Mamera	Eye disorders
52	*Pinus wallichiana*	Kail	Wound healing, skin infection
53	*Viola fedtschenkoana*	Bnafsha	Stomach acidity, throat infection

(Source: Das *et al.*, 2012; Hussain *et al.*, 2024)

Threats to Ethnobotanical Knowledge and Plant Resources

The ethnobotanical knowledge and plant resources in the Himalayan region are under substantial threat due to various anthropogenic pressures and shifting socioeconomic conditions. These challenges not only endanger the survival of endemic plant species but also threaten the traditional knowledge systems associated with their utilization. The use of medicinal plants in healthcare is predominantly preserved by the elder members of communities, while younger generations are often unaware of the value and applications of this natural heritage (Rahman *et al.*, 2020).

As younger individuals migrate for education and employment opportunities, indigenous knowledge systems are progressively eroding. Consequently, the traditional ethnomedicinal practices of native communities face a serious risk of decline or complete loss, underscoring the urgent need for their documentation and conservation (Tamang *et al.*, 2023). The loss of ethnomedicinal knowledge is closely linked to the rapid depletion of botanical resources, a phenomenon widely recognized. Indigenous knowledge regarding medicinal plants is particularly vulnerable, as it diminishes in tandem with species extinction. The systematic documentation of ethnomedicinal knowledge is essential not only to safeguard this cultural heritage but also to draw global attention to the need for conservation measures aimed at preserving both the plants and the associated traditional wisdom (Singh *et al.*, 2014; Hussain *et al.*, 2024).

These are some key challenges impacting ethnobotanical knowledge and plant resources:

1. **Loss of Traditional Knowledge**: Ethnobotanical knowledge, typically transmitted orally through generations, is at risk due to modernization and urbanization. The younger generation is moving away from indigenous practices, leading to diminished knowledge transfer.
2. **Habitat Destruction:** Natural habitats are being destroyed by deforestation, industrial growth, agricultural expansion, and urbanization, causing a significant loss of plant biodiversity. Many ethnobotanically significant species are endangered or extinct.
3. **Overexploitation of Resources:** Unsustainable harvesting of plants for medicinal, nutritional, or commercial purposes is depleting many vital plant resources relied upon by local communities.
4. **Climate Change:** Alterations in temperature and precipitation patterns are disrupting ecosystems, making certain plants less accessible or entirely unavailable in their native regions.
5. **Cultural Assimilation:** Globalization and integration into mainstream cultures lead to the dilution of indigenous identities and traditions, including unique botanical knowledge and practices.
6. **Lack of Documentation and Research:** A significant amount of ethnobotanical knowledge remains undocumented, increasing the likelihood of permanent loss. The absence of systematic efforts to catalog such knowledge exacerbates the problem.
7. **Commercial Exploitation:** The commodification of ethnobotanical knowledge, often through unfair practices like biopiracy, exploits indigenous communities and discourages them from sharing their knowledge.

Conclusion and Future Perspectives

Ethnobotanical knowledge has been a vital part of human civilizations, offering insights into traditional plant usage for food, medicine, and rituals. However, with the increasing threats posed by modernization, habitat destruction, and the loss of traditional knowledge systems, there is an urgent need to prioritize conservation and sustainable practices. The documentation and integration of indigenous knowledge with modern scientific research could open new avenues in fields like pharmacology, agriculture, and biodiversity conservation. Governments and organizations must work towards creating frameworks that encourage the equitable sharing of benefits derived from this knowledge while ensuring that the cultural heritage of Indigenous communities remains protected.

Looking ahead, establishing community-led conservation programs and investing in educational initiatives to promote ethnobotanical studies are crucial. Efforts like ex-situ conservation through seed banks and botanical gardens, alongside the development of sustainable harvesting practices, can safeguard medicinal and aromatic plants (MAPs). Furthermore, fostering collaborations between traditional healers and modern researchers can bridge gaps in understanding and lead to innovations in healthcare and sustainable resource management. Ethnobotany's role in linking cultural traditions with contemporary science makes it a cornerstone for achieving both ecological and socio-economic sustainability.

References

Ansari, A. P. (2021). 'Il j bi'l-Tadb r (regimenal therapy): a core mode of Unani treatment. *Journal of Complementary and Integrative Medicine, 18*(3), 449-458.

Ayeni, E. A. and Basiri, B. (2018). Ethnoveterinary survey of plants used in treating livestock among the Fulani people of Girei, Adamawa State, Nigeria. *World News of Natural Sciences.* 16: 53–66

Baydoun, S., Chalak, L., Dalleh, H. and Arnold, N. (2015). Ethnopharmacological survey of medicinal plants used in traditional medicine by the communities of Mount Hermon, Lebanon. *Journal of Ethnopharmacology*. 173:139-156

Bhat, J. A., Kumar, M. and Bussmann, R. W. (2013). Ecological status and traditional knowledge of medicinal plants in Kedarnath Wildlife Sanctuary of Garhwal Himalaya, India. *Journal of Ethnobiology and Ethnomedicine*. 2(9): 1-18.

Bhattarai, N. & Karki, M. (2004). MEDICINAL, FOOD AND AROMATIC PLANTS | Medicinal and Aromatic Plants: Ethnobotany and Conservation Status.

Chavda, N., Riddhi, R., & Hitesh, S. (2022). An Overview of Historical Background of Ethnobotany And Indigenous Culture of India. *İnternational Journal of Creative Research Thoughts, 10*(4), 738-751.

Cohen, P. A. and Ernst, E. (2010). Safety of herbal supplements: a guide for cardiologists. *Cardiovascular therapeutics*. 28(4):246-53

Das, T., Mishra, S. B., Saha, D., & Agarwal, S. (2012). Ethnobotanical survey of medicinal plants used by ethnic and rural people in Eastern Sikkim Himalayan region. *African Journal of Basic & Applied Sciences*, 4(1), 16-20.

Ford, R. I. (Ed.). (1994). *The Nature and Status of Ethnobotany, 2nd ed*. University of Michigan Press. https://doi.org/10.3998/mpub.11396367

Ganesan, S., Chandhirasekaran, M. and Selvaraj, A. (2008). Ethnoveterinary healthcare practices in southern districts of Tamil Nadu. *Indian Journal of Traditional Knowledge,* vol. 7, no. 2:347–354.

Haq, S. M., Khoja, A. A., Waheed, M., Siddiqui, M. H., Alamri, S., Alfagham, A. T., AL-Humaid, L. A. & Bussmann, R. W. (2024). Food ethnobotany of forest resource in the high altitude Himalaya Mountains: Enhancing the food sovereignty of ethnic groups. *Forest Policy and Economics, 164,* 103247.

Haq, S. M., Waheed, M., Khoja, A. A., Amjad, M. S., Bussmann, R. W., & Ali, K. (2023). A cross-cultural study of high-altitude botanical resources among diverse ethnic groups in Kashmir Himalaya, India. *Journal of Ethnobiology and Ethnomedicine, 19*(1), 12.

https://uou.ac.in/sites/default/files/slm/MSCBOT-608.pdf

Hussain, M., Alam, J., Shah, G. M., Gul, A., Majid, A., Shafqat, N., & Khan, S. M. R. (2024). Assessment of traditional knowledge of medicinal plants practiced by rural communities residing around Musk Deer National Park, Kashmir Himalaya, Pakistan. *Ethnobotany Research and Applications,* 28, 123.

Ibrar, M., Hussain, F. and Sultan, A. (2007). Ethnobotanical studies on plant resources of Ranyal hills, District Shangla, Pakistan. *Pakistan Journal of Botany.* 39(2):329

Loya, A. M., Gonzalez-Stuart, A. and Rivera, J. O. (2009). Prevalence of polypharmacy, polyherbacy, nutritional supplement use and potential product interactions among older adults living on the United States-Mexico border. *Drugs & aging.* 26: 423-436.

Mafuva, C., & Marima-Matarira, H. T. (2014). Towards professionalization of traditional medicine in Zimbabwe: A comparative analysis to the South African policy on traditional medicine and the Indian Ayurvedic system.

Parmesan, C. (2006). Ecological and evolutionary response to recent climate change. *Annual Review of Ecology, Evolution, and Systematics.* 37:637-69

Rahman, I. U., Ijaz, F. Z., Afzal, A., Ali, N., Afzal, M. and Asif, M. A. (2020). Novel survey of the ethno-medicinal knowledge of dental problems in Manoor Valley (Northern Himalaya). Pakistan. *Journal of Ethnopharmacology.* 194:877-894.

Samant, S. S. (2021). Floristic diversity of the Himalaya in relation to climate change: status, values and conservation. *Journal of Graphic Era University,* 3154.

Samant, S.S. & Dhar, U. 1997. Diversity, endemism and economic potential of wild edible plants of Indian Himalaya. *International Journal of Sustainable Development & World Ecology.* 4, 179–191.

Samant, S. S. (2015). Assessment, Valuation and Conservation Prioritization of Floristic Diversity in Trans, North Western and Western Himalaya. D. Sc. Thesis, Kumaun University, Nainital, Uttarakhand.

Samant, S. S., Dhar, U. & Palni, L. M. S. (1998). Medicinal plants of Indian Himalaya: Diversity, Distribution, Potential Values. Gyanodaya Prakashan, Nainital.

Shanmugam, C., Mani, R., & Selvakumar, S. (2024). Siddhar Science: Nurturing Intellectual Properties for A Meaningful Life. *Educational Administration: Theory and Practice, 30*(4), 1928-1932.

Singh, H., Husain, T., Agnihotri, P., Pande, P. C. and Khatoon, S. (2014). An ethnobotanical study of medicinal plants used in sacred groves of Kumaon Himalaya, Uttarakhand, India. *Journal of Ethnopharmacology*. 154(1):98-108

Singh, J. S. (2006). Sustainable development of the Indian Himalayan region: Linking ecological and economic concerns. *Current science*, 784788.

Sullivan, K. and Shealy, C. N. (1997). Complete Natural Home Remedies. *African Journal of Pharmacy and Pharmacology*. 3:621-5

Tamang, S., Singh, A., Bussmann, R. W., Shukla, V. and Nautiyal, M. C. (2023). Ethno-medicinal plants of tribal people: A case study in Pakyong subdivision of East Sikkim, India. *Acta Ecologica Sinica*. 43(1):34-46

Tangjitman, K., Wongsawad, C., Kamwong, K., Sukkho, T. and Trisonthi, C. (2015). Ethnomedicinal plants used for digestive system disorders by the Karen of northern Thailand. *Journal of Ethnobiology and Ethnomedicine*. 1(1):1-3

Ved, D. K., Kinhal, G. A., Ravikumar, K., Prabhakaran, V., Ghate, U., Vijaya Shankar, R. & Indresha, J. H. (2003). Conservation assessment and management prioritization for the medicinal plants of Jammu and Kashmir, Himachal Pradesh, and Uttaranchal. Foundation for Revitalisation of Local Health Traditions, Bangalore, India.

Ved, D.K., Kinhal, G.A., Ravikumar, K., Vijaya Shankar, R. & Haridasan, K. (2005). Conservation assessment and management prioritization (CAMP) for the wild medicinal plants of North-East India. *Medicinal Plants Conservation*, 11, 40–44.

CHAPTER 18

INTRODUCTION ABOUT MEDICINAL AND AROMATIC PLANTS

Ajay Hemdan[1], Diksha Semwal[2], Jaidev Chauhan[3], B.P. Nautiyal[1]

1 College of Horticulture Veer Chandra Singh Garhwali Uttarakhand University of Horticulture and Forestry Bharsar Pauri Garhwal Uttarakhand
2 Wildlife Institute of India Dehradun, Uttarakhand
3 High Altitude Plant Physiology Research Centre Hemwati Nandan Bahuguna Garhwal University (A Central University) Srinagar Garhwal Uttarakhand

---

Abstract

Medicinal plants have been used for centuries and may possibly be considered the genesis of modern medicine. Medicinal plants are a large plant category that has a lot of attention because of their pharmacological, cosmetic, and nutritional applications. Furthermore, they provide an alternative to traditional crops with species that are in great demand on the present worldwide market. The chapter includes an introduction to medicinal and aromatic plants, as well as their classification and significance as map plants. A medicinal plant is any plant that includes chemicals that can be utilized for therapeutic reasons or that are precursors for chemo-pharmaceutical semi-synthesis in one or more of its organs. When a plant is labeled as medicinal, it implies that the plant is valuable as a medication or therapeutic agent or an active element in a medical preparation. Herbal medications are in significant demand for basic health care in both developed and developing nations because to their broad biological and pharmacological activity, better safety margins, and lower prices. The rising demand for healthful foods rich in antioxidants, vitamins, and minerals, as well as those with therapeutic value, has resulted in the over-exploitation of important agricultural and medicinal plants in the modern age. This load can be lessened by making better use of underused plants with nutritional and therapeutic value.

Keywords: Medicinal plants, Pharmaceutical, Cosmetic, utilization, antioxidant.

Introduction

Plants are the most conspicuous members of the biosphere. They are nature's genetically regulated bioreactors. Throughout the earth's long history, sophisticated chemical processes have evolved that provide a variety of molecular structures utilizing very basic molecules as starting materials and under relatively mild reaction conditions. Green plants (autotrophs) use water and ample light energy to convert carbon dioxide (inorganic carbon) to the reduced form of carbon (i.e. carbs) through the extraordinary process of photosynthesis. In the process, oxygen is liberated, which is critical for the survival of life in our eco-system. Almost all biological functions obtain their energy directly or indirectly from the reduced carbon molecules that are generated. The extent of photosynthetic activity is astounding. Tens of billions of tons of carbon are extracted from the atmosphere each year and converted into incredibly complex primary and secondary metabolites such as nucleic acids, carbohydrates, proteins, amino acids, heterocycles, isoprenoids, phenolics, alkaloids, lipids, and so on. An insight technology which Nature uses and may be called as "Natural technology", will offer entirely new concepts for the development of eco technologies; it may be termed as phytotechnology when based on plants **(Banerji, 1992).**

Medicinal plants are those that are high in secondary metabolites and are potential medication sources. Secondary metabolites include alkaloids, glycosides, coumarins, flavonoids, steroids, and other compounds. These plants are the primary source of drugs used in the Indian system of medicine (ISM) (Ayurveda, Unani, and Siddha) and Homeopathy. These plant s are found in various parts of the country in different environmental and climatic conditions. WHO has listed over 21,000 plant species used around the world for medicinal purpose. In India 8000 species listed is available. ISM used nearly 2500 plant species belonging to more than 1000 genera. Of the nearly 800 species used in the industry 25% are cultivated. Man relied on the healing abilities of medicinal plants prior to the arrival of artificial medications. Some people appreciate these plants because of an old belief that plants were meant to provide man with food, medicinal treatment, and other benefits. It is estimated that around 80% of the world's 5.2 billion people reside in less developed nations, and the World Health Organization believes that approximately 80% of these people rely nearly entirely on traditional medicine for their basic healthcare requirements. Medicinal plants form the "backbone" of traditional medicine, which implies that more than 3.3 billion people in developing countries regularly use medicinal plants **(Hunt, 2000).** There are nearly 2000 ethnic groups in the world, and almost every group has its own traditional medical knowledge and experiences **(Liu** et al.**, 2009; Kebriaee, 2003).**

Many medicinal plants used by the trade are mostly harvested from wild growth (almost 80%), depriving the flora of its important medicinal plant wealth (e.g. Rauwolfia, Dioscorea). Many therapeutic plant species in our nation have gone extinct or endangered as a result of this practice. This should be avoided, and valuable therapeutic plants should be conserved by establishing herbal gardens and gene banks. Other major human activities threatening biodiversity and medicinal plant riches in this nation include deforestation, shifting farming, overgrazing, and overexploitation. The Commerce Ministry is in charge of the economy.

GOI has banned export of 52 medicinal plants in order to conserve them in nature and semi-wild conditions. World Conservation Union (IUCN) has identified following categories:

Category	Brief description
Extinct	The last remaining member of the species has died or is presumed beyond reasonable doubt to have been dead.
Extinct in the wild	Captive individuals survive, but there is no natural population
Critically endangered	Faces an extremely high risk of extinction in the immediate future
Endangered	Faces high risk of extinction in the immediate future
Vulnerable	Faces high risk of extinction in the medium term
Rare	Not available plenty in nature

Aromatic plants are characterized as those that contain essential oils. These essential oils are the aromatic plants' odoriferous stream volatile components. They are mostly composed of a complicated blend of acyclic and/or cyclic monoterpenoids. These terpenoids are essentially secondary metabolites with no known role in the plant's core metabolism. These essential oils derived from spices and condiments that add flavour and improve food taste are utilized in a variety of flavour sectors. Some of these types' unique characteristics. They are often found in plant aerial parts such as flowers, fruits, and leaves. They can also accumulate in roots and wood (for example, sandal wood). The composition and amount of the essential oil's different monoterpenoids. The type and amount of the numerous monoterpenoids in the essential oil is a feature of both the genus and the species. The majority of commercially available essential oil-bearing plants are members of the families Labiatae, Myrtaceae, Rutaceae, Compositae, Rosaceae, Umbelliferae, Graminae, and Pinaceae. The accumulation of essential oils in a plant is determined by the developmental stage of the concerned organ/plant component. It is connected with the early growth stage in the majority of aromatic plants. Aromatherapy is a type of complementary medicine that use volatile plant materials and other aromatic molecules to affect a person's mind, mood, cognitive function, or health. It is now gaining popularity among tourists and the wealthy segments of society due to its general health benefits. Aromatherapy use in the world is currently tiny (200-300 tons), but it is predicted to expand if paired with medicines **(Kumar, 2020)**.

Importance of Medicinal and Aromatic Plants

Medicinal and aromatic plants have a significant role in human life. Here are some of their key benefits:

1. **Medicinal Uses:** These plants have been used for thousands of years for their therapeutic properties. They are a vital source of organic compounds, used in pharmaceuticals, herbal remedies, and other healthcare products. Many modern medicines are derived from the compounds found in these plants.
2. **Aromatic Uses:** Aromatic plants are valued for their fragrance and are often used in perfumes, aromatic oils, and cosmetics. They are also used in aromatherapy for their therapeutic benefits.
3. **Culinary Uses:** Many aromatic plants are used as spices and flavoring agents in cooking. They not only enhance the taste and aroma of food but also have digestive and other health benefits.
4. **Industrial Uses:** Some aromatic plants are used in industrial applications such as making natural dyes, resins, and insecticides.
5. **Cultural and Spiritual Uses:** Many medicinal and aromatic plants have cultural and spiritual significance. They are used in rituals, ceremonies, and traditional medicine systems worldwide.
6. **Environmental Benefits:** These plants play a crucial role in maintaining and improving biodiversity. They can be used in soil conservation and reforestation efforts.

Classification of Medicinal and Aromatic Plants

Of the 2,50,000 higher plant species on earth, more than 80,000 species are reported to have at least some medicinal value and around 5000 species have specific therapeutic value. They are classified according to the part used, habit, habitat, therapeutic value etc, besides the usual botanical classification.

Medicinal Plants

1. Based on part used

1. **Whole plant:** Some medicinal plants which whole parts viz., Panchang (Flowers, Stems, Roots, leaves, seeds and fruits etc.) is used eg. *Abutilon hirsutum, Boerhaavia diffusa, Phyllanthus neruri, Bacopa monnieri, Andrographis paniculata, Achillea millefolium, Arabidopsia thaliana, Artemisis annua, Borago*

officinalis, Hydrocotyle asiatica, Pholidota articulata, Reseda odorata, Urtica dioica, Veronica officinalis.

2. **Roots:** A root is a plant part that grows underground and anchors the plant in place. Roots absorb water and minerals from the soil, store food, and synthesize plant growth regulators eg. Aconitum sp., *Butea monosperma, Cassia fistula, Catharanthus roseus, Rauvolfia serpentina, Withania somnifera, Amaranthus spinosus, Angelica glauca, Cimicifuga racemosa, Codonopsis ovata, Nardostachys grandiflora, Onosma emodi, Panax ginseng, Plumbago zeylanica, Polygonatum biflorum, Potentilla fulgens, Rheum emodi, Rumex obtusifolia, Saussurea lappa, Trillium erectum, Tylophora indica.*
3. **Stem:** A stem is the main trunk of a plant. It is a primary plant axis that develops buds and shoots. It supports other plant parts such as leaves or fruits eg. *Tinospora cordifolia, Ephedra gerardiana, Lactuca sativa, Tdzgdaraxacum officinale.*
4. **Bark:** Bark is the tough exterior covering of a woody root or stem. Specifically, it includes the tissues outside the cambium that consist of an inner layer, especially of secondary phloem, and an outer layer of periderm eg. *Saraca asoca, Alstonia scholaris, Cinchona officinalis.*
5. **Leaves**: A leaf is a green above-ground plant organ. Its main functions are photosynthesis and gas exchange. A leaf is flat so it absorbs the most light, and thin, so that the sunlight can get to the chloroplasts in the cells. Most leaves have stomata, which open and close eg. *Indigofera tinctoria, Lawsonia inermis, Aloe vera, Adhatoda vasica, Eclipta alba, Adiantum capillus veneris, Ammania baccilera, Andropogon citratus, Anisomeles indica, Calendula officinalis, Coleus blumei, Enhydra fluctuans,Euphorbia hirta, Melilotus officinalis, Mentha piperita, Rhyncostylis retusa, Trianthema monogyna, Xanthium strumarium.*
6. **Flowers:** A flower, also known as a bloom or blossom, is the reproductive structure found in flowering plants (plants of the division Angiospermae). Flowers consist of a combination of vegetative organs – sepals that enclose and protect the developing flower, petals that attract pollinators, and reproductive organs that produce gametophytes, which in flowering plants produce gametes eg. *Biophytum sensityvum, Mimusops elengi, Anthemis nobilis, Jasminum officinale, Jasminum sambac, Rosa alba, Rosa centifolia, Rosa damascena. Tagetes erecta.*
7. **Fruits:** A fruit is the soft, pulpy part of a flowering plant that contains seeds. It is formed from the ovaries of angiosperms and is exclusive only to this group of plants1. Fruits are seed-bearing structures formed from the ovaries of flowering plants after flowering1. Fruits are classified into four types namely: Simple fruits, Aggregate fruits, Multiple fruits, and Accessory fruits eg. *Solanum dulcamara, Solanum indicum, Embelia tsjeriam-cottam, Cordia dichotoma, Terminalia bellirica, Capsicum frutescens, Diospyros*

melanoxylon. Hurmulus lupulus, Papaver somniferum, Physalis minima, Trapa bispinosa.

8. **Seeds:** A seed is a mature ovule that comprises an embryo or a miniature undeveloped plant and food reserves, all enclosed within a protective seed coat eg. *Datura stramonium, Vernonia anthelmintica, Ammi majus, Anethum sowa, Brassica nigra, Brassica juncea, Foenculum vulgare, Plantago ovata.*

9. **Rhizome:** A rhizome, also known as rootstocks, is a type of plant stem that can be situated either at the soil surface or underground. It contains nodes from which roots and shoots originate eg. *Alpinia galanga, Cephaelis ipecacuanha, Glycyrrhiza glabra, Podophyllum hexandrum, Zingiber officinale, Acorus calamus, Coptis teeta, Curcuma aromatica, Nymphaea alba.*

10. **Pulp:** The term "pulp" often refers to the soft, fleshy part of a plant, particularly the fruit. The pulp of certain medicinal plants can have various uses and properties eg.Agave sisalana.

11. **Bulb:** Bulbs are specialized organs used by plants for the storage of nutrients and reserve substances that the plant uses for its growth and development, especially when conditions are not favorable. eg. *Allium cepa, Allium sativum, Urginea indica.*

12. **Tubers:** A tuber is a type of enlarged structure used as storage organs for nutrients in some plants. Tubers are typically short, thickened, mostly underground stems that constitute the resting stage of various plants. They are largely composed of starch-storing parenchyma tissue. They are used for the plant's perennation (survival of the winter or dry months), to provide energy and nutrients for regrowth during the next growing season, and as a means of asexual reproduction eg. *Dioscorea bulbilera, Gloriosa superba, Habenaria acuminata, Habenaria intermedia, Helianthus tuberosus, Kaempferia galanga, Orchis latifolia.*

2. Based on habit

1. **Grasses:** Grasses are a type of plant with narrow leaves growing from the base. They are monocotyledon herbaceous plants and include the "grass" of the family Poaceae eg. *Cynodon dactylon, Citronella sp., Cymbopogon sp.*

2. **Sedges:** Sedges are a type of plant that are usually found in marshy areas. They are part of the family Cyperaceae. Sedges are distinguished from grasses by their three-sided, solid stems and by leaves with three ranks instead of two. They are usually tufted monocotyledonous marsh plants eg. *Cyperus rotundus.*

3. **Herbs:** Herb is a seed-producing plant that does not develop persistent woody tissue but dies down at the end of a growing season. This term is in contrast to shrubs and trees which possess a woody stem eg. *Vernonia cineria, Aconitum heterophyllum, Aconitum nepellus, Acorus calamus, Alpinia galanga, Andrographis paniculata, Atropa acuminata, Bacopa monnieri, Boerhaavia diffusa,*

Catharanthus roseus, colchicum luteum, Coptis teeta, Costus speciosus, Datura strumonium, Digitalis purpurea, Euphorbia hirta, Ferula narthax, Glycyrrhiza glabra, Hyoscyamus niger. Lobelia nicotianaefolia, Mentha arvensis, Ocimum sanctum, Plantago ovata, Podophyllum hexandra, Rheum emodi, Saussurea lappa, Solanum xanthocarpum, Swertia chirayita, Trianthema monogyna, Tribulus terrestris, Urginea indica, Valerian officinalis, Abutilon hirsutum, Acampe papillosa, Achillea millaefolium, Achyranthes aspera, Allium cepa, Allium sativum, Aloe barbadensis, Aloe vera, Alpinia galanga, Amaranthus tricolor, Ammania baccifera, Ammi majus, Angelica glauca, Anethum sowa, Anthriscum cerefolium, Argemone mexicana, Artemisia vulgaris, Calendula officinalis, Carthamus tinctorius, Chenopodium album, Chlorophytum arundinarea, Crocus sativus, Coriandrum sativum, Dendrobium nobile, Eclipta alba, Enhydra fluctuans, Erigeron canadensis, Erodium cicutarium, Eulophia dabia, Equisetum arvense, Geranium wallichianum, Habenaria accuminata, Malaxis muscifera, Malva parviflora, Marsilea minuta, Melilotus officinalis, Orchis latifolia, Panax ginseng, Papaver somniferum, Pholidota articulata, Polycarpa corymbosa, Polygonum aviculare, Portulaca oleracea, Primula reticulata, Rhyncostylis reticulata, Ruta graveolens, Thymus serphyllum, Tagetes erecta, Trillium erectum, Typha latifolia, Urtica dioica, Vanda tessellata.

4. **Shrubs:** A shrub, often also called a bush, is a small-to-medium-sized perennial woody plant . Unlike herbaceous plants, shrubs have persistent woody stems above the ground. They can be either deciduous or evergreen. Shrubs are distinguished from trees by their multiple stems and shorter height, typically less than 6–10 m (20–33 ft) tall eg. *Adhatoda vasica, Aristolochia indica, Berberis aristata, Ephedra gerardiana, Gaultheria fragrantissima, Holarrhenna antidysenterica, Lawsonia inermis, Rauvolfia serpentina, Sida cordifolia, Withania somnifera, Nerium oleander, Hibiscus rosa sinensis, Hibéscias mutabilis, Jasminem homile, Jassminton officinale, Jasminum multiflorum, Lavandula officinalis, Plumbago zeylanica, Calotropis gigantea, Woodfordia fruiticosa, Vitex negundo.*

5. **Climbers:** Climbers, in the context of botany, are plants that grow upwards by attaching themselves to other plants or objects. They are known for their ability to climb up walls, trees, or fences. Climbers use various means to secure themselves, including twining their stems around a support, sending out tendrils, or using aerial roots eg. *Asparagus racemosus, Hydrocotyle asiatica, Hemidesmus indicus, Ipomoea hederacea, Piper nigrum, Tinospora cordifolia, Tylophora indica, Coccinia indica, Gloriosa superba, Humulus lupulus, Melothria heterophylla, Dioscorea alata, Basella rubra, Paederia foetida, Cissus quadrangularis, Mucuna pruriens, Ipamora aquatica, Passiflora edulis, Clitorea ternatea, Merremia hederacea, Artigonon lepropus, Citrullus colocynthis*

6. **Trees:** tree is a perennial plant with an elongated stem, or trunk, usually supporting branches and leaves. In some usages, the definition of a tree may be narrower, including only woody plants with secondary growth

eg. *Azadirachta indica, Aegle marmelos, Alstonia scholaris, Butea monosperma, Cassia fistula, Cinchona officinalis, Cinnamomum zeylanicum, Emblica officinalis, Hydnocarpus kurzu, Madhuca indica, Mallotus philippensis, Pinus roxburghii, Saraca indica, Santalum album, Sterculea urens, Eugenia jambolana, Tamarindus indica, Pterocarpus marsupium, Nyctanthes arbortristis, Bauhinia variegata, Diospyros melanoxylon, Citrus medica, Messua ferrea, Michelia champaka, Moringa oleifera, Acacia nilotica, Olea europea, Terminalia arjuna, Phyllanthus acidus, Averrhoea carambola.*

3. Based on climatic zones

1. **Tropical:** The tropical area of India is the region that lies south of the Tropic of Cancer. This region experiences high temperatures throughout the year with a small daily and annual range. The tropical zone is nearer to the equator and therefore experiences high temperatures throughout the year eg. *Andrographis paniculata, Acorus calamus, Azadirachta indica Adhatoda vesica, Centella asiatica, Ocimum basilicum, Paederia foetida, Withania somnifera.*
2. **Sub-tropical:** Subtropical areas in India are primarily found in the middle and northern regions of the country. These regions experience a subtropical climate due to their latitude, which results in moderately high temperatures throughout the year, with distinct seasonal changes eg. *Mentha arvensis, Costus speciosus, Artemisia maritima, Hydnocarpus kurzii, Lavendula vera, Litsea cubeba, Mucuna pruriens, Pogostemon calelin, Zanthoxylum armatum, Curcuma caesia, Leonotis nepetaefolia.*
3. **Temperate:** The temperate areas of India are primarily found in the northern regions, particularly in the Himalayan range. This includes states like Jammu & Kashmir, Himachal Pradesh, Uttarakhand, and the northeastern states. These regions experience a temperate climate due to their high altitude, which results in cooler temperatures compared to the plains eg. *Atropa belladonna, Coptis tita, Ephedra gerardiana, Geranium nepalensis, Panax pseudoginseng, Swertia chirata, Satyrium nepalensis, Rubia cordifolia, Taxus baccata, Orchis latifolia.*
4. **Alpine:** Alpine areas in India are primarily found in the higher altitudes of the Himalayan range. These regions experience an alpine climate due to their high altitude, which results in cooler temperatures and a harsher climate. The alpine forests occur all along the Himalayas at altitudes ranging between 2,900 to 3,500 m or even up-to 3800 m above sea level, depending upon the location and the variety of species. These forests can be divided into:

Sub-alpine forests: These occur at the upper limit of tree forest adjoining alpine scrub and grasslands and comprise of dense growth of small crooked trees and large shrubs with coniferous overwood.

Moist alpine scrub: This is a low evergreen dense growth of rhododendron, birch, berberis and honeysuckle which occurs from 3,000 metres and extends up to the snowline.

Dry alpine scrub: This is the uppermost limit of scrub xerophytic, dwarf shrubs, over 3,500 metres above sea level and found in the dry zone.

Species found in Alpine zone are eg. *Aconitum ferox, A. heterophyllum, Illicium griffithii, Berberis spp., Podophyllum hexandrum, Rheum emodi, Delphinium subulatum Angellica glauca, Nardostachys jatamansi, Picrorrhiza kurooa.*

5. Based on life cycle

1. **Annual herbs:** Anise, Basil, Borage, Calendula (Pot Marigold), Chamomile, Chervil, Corlander.
2. **Blennial herbs:** Caraway seeds, Prime rose, Bai Zhi, Mullein, Teasel, Viper's Bugloss.
3. **Aromatic follages:** *Allium spp., Artemisia spp, Foeniculum vulgare, Tagetes minuta*
4. **Perennial herbs:** Aloe vera, Fennel, mints.

6. Botanical classification

Most of the medicinal and aromatic plants belong to the following familles:

1. **Acanthaceae:** *Adhatoda vasica, Andrographis paniculata, Hygrophila auriculata.*
2. **Agavaceae**: *Agave sisalana.*
3. **Amaranthaceae:** *Achyranthes aspera, Amaranthus spinosus.*
4. **Apocynaceae:** *Alstonia scholaris, Catharanthus roseus, Holarrhena antidysenterica.*
5. **Araceae:** *Acorus calamus.*
6. **Boraginaceae:** *Borago officinalis, Onosma echioides.*
7. **Caryophyllaceae:** *Arenaria serpyllifolia*
8. **Compositae:** *Artemisia, Calendula, Achillea millefolium, Anaphalis neelgerriana, Anthemis nobilis, Arnica sororia, Enhydra fluctuans, Erigeron canadensis. Helianthus tuberosus, Lagera alata, Saussurea lappa, Tagetes erecta.*
9. **Cruciferae:** *Armoracia rusticana, Brassica nigra, Arabidopsis thaliana.*
10. **Cucurbitaceae:** *Bryonia alba, Coccinia indica, Cucumis trigonus.*
11. **Euphorbiaceae:** *Emblica officinalis, Euphorbia hirta, Phyllanthus niruri.*
12. **Labiatae:** *Mentha arvensis, Ocimum sanctum, Lavandula, Pogostemon, Ajuga bracteosa, Anisomeles indica, Coleus blumei.*
13. **Leguminosae:** *Cassia fistula, Glycyrrhiza glabra, Saraca asoca, Melilotus officinalis, Trifolium pratense.*

14. **Liliceae:** *Colchicum autumnale, Asparagus aphyllus, Aloe vera, Allium cepa, Allium sativum, Chlorophytum arundinaceium, Gloriosa superba, Polygonatum biflorum, Urginea indica.*
15. **Malvaceae:** *Sida cordifolia, Malva sylvestris, Abelmoschus moschatus.*
16. **Meliaceae:** *Azadirachta indica*
17. **Nyctaginaceae:** *Boerhavia diffusa, Mirabilis jalapa*
18. **Oleaceae:** *Nyctanthes arbortristis, Jasminum grandiflorum, Jasminum officinale, Jasminum sambac.*
19. **Orchidaceae:** *Acampe papillosa, Dendrobium nobile, Vanda tessellata, Coelogyne cristata, Cypripedium calceolus, Eria muscicola, Eulophia dabia, Habenaria spp., Malaxis cylindrostachya, Orchis latifolia, Pholidota articulata, Rhyncostylis retusa Satyrium nepalense.*
20. **Papaveraceae:** *Papaver somniferum, Argemone Mexicana.*
21. **Phytolaccaceae:** *Phytolacca Americana.*
22. **Poaceae:** Cymbopogon, Citronella
23. **Ranunculaceae:** *Aconitum, Coptis teeta, Anemone pulsatilla, Cimicifuga americana, Hepatica acutiloba.*
24. **Rosaceae:** *Rosa gollica, Potentilla fulgens, Rosa alba, Rosa centifolia, Rosa damascene*
25. **Rubiaceae:** Ipecacuanha, Cinchona, *Rubia cordifolia.*
26. **Rutaceae:** *Aegle marmelos, wood apple.*
27. **Scrophulariceae:** Bacopa monnieri, Digitalis purpurea, Picrorhiza kurrooa, Scoparia dulcis, Veronica officinalis.
28. **Solanaceae:** *Atropa belladonna, Datura, Withania somnifera, Hyoscyamus niger, Physalis minima, Solanum indicum, Solanum dulcamara.*
29. **Umbelliferae:** *Hydrocotyle asiatica, Ferula narthex, Ammi majus, Anethum sowa, Anthriscus cerefolium, Archangelica officinalis.*
30. **Verbenaceae:** *Vitex agrus-castus.*
31. **Zingiberaceae:** *Alpinia galanga, Elettaria cardamomum, Curcuma amada, Curcuma aromatica, Curcuma caesia, Kaempferia galanga, Zingiber officinale.*

Aromatic Plants

Out of a total of about 1500 species of aromatic plants known, only a little over 300 species have been studied in some detail. Of the 50 species which find use as commercial source of essential olls and aroma-chemicals, the number of those having regular and large scale utilization hardly exceeds two dozen.

Based upon growth habit aromatic plants are classified into four categories:

1. **Aromatic grasses:** Lemongrass, palmarosa, citronella, vetiver.

2. **Aromatic herbs and shrubs:** Mints, ocimums, patchoull, rosemary, clarysage, thyme, celery, coriander, cumin, fennel, ajowan, davana, chamomile, geranium, cardamom, ginger, kacholam.
3. **Aromatic trees:** Sandalwood, eucalyptus, clove, camphor, cinnamon, nutmeg. Linaloe
4. **Aromatic flowers:** Rose, jasmine, tuberose, marigold, champaca.

Conclusion

Despite their recognized significance, medicinal plants are underutilized because of lack of awareness about their planting material and cultivation techniques. Therefore, there is an immediate need to highlight the implications and utilization of these underutilized plants and conserve their genetic resources to ensure nutritional and therapeutic supply for future generation. These tools for in vitro regeneration of medicinal plants possess incredible potential for the production of secondary metabolites in cell suspension cultures. The key for successful commercial production of secondary metabolites by plant biotechnology approach is the establishment of bioreactors. The connection between human and their search for drugs in nature dates from the far past, of which there are enormous evidence from different sources (written documents, preserved monuments, and even original plant medicines). Awareness of medicinal plants' usage is a result of the many years of struggles against diseases and man learned to pursue drugs in barks, seeds, fruits, and other parts of the plants. Contemporary science has acknowledged and considered in modern pharmacotherapy the active actions of plant origin drugs, known by ancient civilizations and used throughout the millennia. The knowledge of the development of ideas related to the usage of medicinal plants with awareness has increased the ability of pharmacists and physicians to respond to the challenges that have emerged with the spreading of professional services in facilitation of man's life.

Refrences

Davidson-Hunt, I. Ecological Ethnobotany: Stumbling Toward New Practices and Paradigms.

Liu, Y., Dao, Z., Yang, C., Liu, Y., & Long, C. (2009). Medicinal plants used by Tibetans in Shangri-la, Yunnan, China. *Journal of Ethnobiology and Ethnomedicine*, *5*(1), 1-10.

Kebriaee-zadeh, A. (2003). Overview of national drug policy of Iran. *IJPR*, *1*(2).

Kumar, N. (1997). *Introduction to spices, plantation crops, medicinal and aromatic plants*. Oxford and IBH Publishing.

De, L. C. (2017). *Cultivation and Breeding of Medicinal and Aromatic Plants*. Pointer Publishers.

Banerji, A. (1992). Biotechnical potential of natural products. *Bioelectrochemistry and bioenergetics*, *27*(2), 105-113.

CHAPTER 19

CRITICAL ANALYSIS OF A SOCIAL MOVEMENT CHIPKO UTTARAKHAND

Mohit Upreti

Assistant Professor, Department of Sociology Government Degree College Talwari, Chamoli, Uttarakhand

In this critical study, we examine specific issues of a given region from the viewpoint of achieving dedicated aim, related objectives and much needed sustainable development. To be precise towards the topic, work for water conservation, to save seeds & traditional cultivation and against plastic & polythene pollution, migration and forest fire are the thrust areas for change and sustainable development in the region. Ethnographically, the living legends of Chipko movement, intelligent and conscious youth of the region see these issues as connected with the Chipko movement. Water conservation & save seeds campaigns are taken as conformity and plastic pollution, migration & forest fires as deviance from the epic movement. In other words, opinion makers are attached, emotionally too, with the nucleus and the main forces of the Chipko movement.

Against this background, it is felt that if basic infrastructure in terms of do's and don'ts, feasible and visible, is laid, the emotional cord of the movement can play a decisive role of a pull factor, which may finally play a positive role towards progressive sustainable development, besides fulfilling other aim and objectives. In this context, it may be added that Uttarakhand is not an exception. All societies have such epic movements. As a matter of fact in the West also, of all the „new" social movements which emerged from the student movements of the late 1960"s, it is environmental movements which have had most enduring influence on politics and which have undergone the most wide ranging institutionalization in terms of both of professionalization of their activities and of the regularization of their access to policy makers (Christopher Rootes, 2007). „Eco-warriors" have become folk heroes, not only with young people, but even with the sections of a generally conservative press (Paterson, 1997). Today, Greens are established and dominating players in most advanced societies. In a nutshell, action aspect of research is too strong to be ignored. You don"t need lobbies when the full house is with you. This is true of all societies.

The work, thus, attempts to raise an issue that may generate a debate to evolve a development strategy that may help to contribute in building an equitable and just society by accepting, countering and overcoming the challenges and by working for sustainable development.

Hypothesis:

- It is strongly felt that activists, opinion makers and conscious masses in garhwal region of Uttarakhand see current movements like water conservation & save seeds as continuation.

Review of relevant literature

There is no dearth of literature on Chipko movement. We have hundreds of studies, projects, dissertations, thesis and books and thousands of articles on various aspect of the movement. To mention a few, Ramachandra Guha (The Unquite Woods, 1980) presents a historically rooted subaltern analysis of the Chipko movement. Vandana Shiva (Staying Alive, 1988) gives an eco-feminist view of the movement. Paul Routledge (Terrains of Resistance, 1993) goes to the extent of calling hill region as an internal colony of Uttar Pradesh and India. Somen Chakraborty (A Critique of Social Movement in India, 1999) sees it as a movement for a social change. Amrita Basu (Women, Political Parties and Social Movements in South Asia, 2005) and Naveen Joshi (Dawanal Devbhoomi Developers) attaches Chipko movement with Gandhian thought of non-violent mass movement.

The works quoted above are the masterpiece of related jurisdiction. However, the sentiments of the local people, wherein chipko is internalized as an epic and the present campaigns (mini movements) are taken as a continuation or conformity of this epic did not get a mention in these and other important works. Maybe, the context was different or emphasis was on chipko as a case study with the belief that the movement cooled off by the end of the century. The new generation and the old also actively involved in the new campaigns see new campaigns as further offshoots of chipko. These offshoots are vital and success in them is more vital for the life and sustainability.

Normally it is believed that the great movement cooled down in 1990s. However, when the area was visited, living legends of the movement in their late 60s, 70s and 80s emphatically said, "No." The same was the passion of new generation. According to them, it has only changed its form. Present movements; be it save seeds, conservation of water or efforts against forest fire, migration, plastic and polythene etc. are its new manifestations. Migration is the most important issue, all talk with utmost worry. They see these as, connects and conformity of chipko, on the one hand and disconnects and deviance of the movement, on the other. Chipko, it seems, is emerging as a new „Epic" and why not.... especially when adherence can generate a debate to evolve a new developmental strategy.

There is no research work on above stated viewpoint. The proposed research thus is a modest attempt to fill in this gap. More importantly, it intends to add new literature. Who knows that similar efforts at different places, especially in a developing world may contribute by playing its part in making environmental movements from local to global? And thence contribute towards sustainable development goals (SDGs).

Conclusion

In the study, we seek to confirm that an epic movement is not an event. Its continuity embraces both conformity and deviance. People are attached, emotionally too, with the nucleus of a movement and its main forces. Deviance, though followed, is not and cannot be attributed to the nucleus of the movement and its main forces. Against this background, this work seeks to utilise the emotional cord people have with the movement, towards current desirable objectives and goal of the society. This exercise starts after the proper infrastructure and physical conditions are created.

The pattern may suggest some sort of universal validity. As suggested earlier, the proposed study, thus attempts to generate an issue that may generate a debate to evolve a development strategy that may help to contribute in building an equitable and just society by accepting, countering and overcoming the challenges and by working for genuine sustainable development.

References

Basu, A. (2005). Women, Political Parties and Social Movements in South Asia. United Nations Research Institute for Social Development.

Bishnoi, R. (1992). A Blueprint for Environment: Conservation as Creed. Dehradun: Surya Publications.

Chakraborty, S. (1989). A Critique of Social Movements in India: Experiences of Chipko, Uttarakhand, and Fishworkers' Movement. New Delhi: Indian Social Institute.

Doherty, B. (2000). Direct Action in British Environmentalism. Routledge. Feyerabend, P. (1975). Against Method. Humanities Press.

Gottlieb, R. S. (1996). This Sacred Earth: Religion, Nature, Environment. Psychology Press.

Guha, R. (1990). The Unquiet Woods: Ecological Change and peasant resistance in the himalyas. University of CALIFORNIA PRESS.

Guha, R., & Gadgil, M. (1995). Ecology and Equity: The Use and Abuse of Nature in Contemporary India. London: Routledge.

Kamat, J. (2006). The Bishnoi Community. Geographia India online.

Kedzior, S. (2006). A Political Ecology of The Chipko Movement . University of Kentucky.

Khator, R. (1989). Forests: The People and the Government. New Delhi: National Book Organization. Männer, K. (2010). From livelihood cooperatives to right based work . Swallows IB.

Ranjan, A., & Others. (2015). Pani aur Beej . Khadi: UJJS.

Rootes, C. (1999). Environmental Movements: Local, National and Global.

Routledge, P. (1993). Terrains of Resistance: Nonviolent Social Movements and the Contestation of Place in India. Michigan: Praeger.

Sharma, K., Nautiyal, K., & Pandey, B. (1987). Women in Struggle: The Role and Participation of Women in the Chipko Movement. New Delhi: Centre for Women's Development Studies.

Shiva, V. (1988). Staying Alive : Women ecology and development. New Delhi: Zed books.

Shiva, V., & Bandyopadhyay, J. (1986). The Evolution, Structure, and Impact of the Chipko Movement. International Mountain Society .

Vasan, S. (2006). Living with Diversity: Forestry Institutions in The Western Himalaya. Indian Institute of Advanced Studies.

अध्याय 20

शीर्षक: "हिंदी साहित्य में पर्यावरण चेतना"

डॉ पुष्पा रानी

सहायक प्राध्यापक-हिन्दी

राजकीय महाविद्यालय तलवाड़ी चमोली उत्तराखंड

शोध सारांश

"द्यौ; शांतिरन्तरिक्षं शांतिः पृथ्वी।
शांति रापः शांतिरौषधयः शांति।
वनस्पतयः शांति विश्वेंदेवा शांति ब्रह्म शांतिः
सर्व शांति देव शांतिः सामा शांतिरेधि॥(1)

हमारे चारों ओर का आवरण ही 'पर्यावरण' है। पर्यावरण के बिना मानव जीवन की कल्पना भी संभव नहीं है। मनुष्य का जीवन पूर्णतया पर्यावरण पर ही निर्भर है। परंतु हम मनुष्यों ने अपने फायदे के लिए पर्यावरण के साथ छेड़छाड़ की है। जिसका परिणाम ग्लोबल वार्मिंग, प्रदूषण, भूस्खलन, बाढ़, भूकंप आदि प्राकृतिक प्रकोप के रूप में देखा जा सकता है। हिंदी साहित्य में पर्यावरण चिंतन पर छायावादी स्तंभ जयशंकर प्रसाद की 'कामायनी' हिंदी साहित्य की पहली रचना है, जिसमें प्रकृति और पर्यावरण असंतुलन की समस्या पर विचार किया गया है। इसके अतिरिक्त वेद की ऋचाएं हो या आदिकवि वाल्मीकि की क्रौंच पक्षी को देखकर प्रस्फुटित प्रथम श्लोक, रामायण ,महाभारत ,कालिदास की 'मेघदूत' संस्कृत साहित्य ,हिंदी साहित्य सभी में पर्यावरण चिंतन व्यक्त किया गया है।

समकालीन हिंदी साहित्य के सभी कवियों, लेखकों, उपन्यासकारों ने पर्यावरण के प्रति अपनी चिंता व्यक्त की है। हिंदी साहित्य में आदिकाल से लेकर रीतिकाल तक के कवियों ने अपने काव्य में पर्यावरण के प्रति चेतना जागृत की है। वहीं छायावाद से लेकर आधुनिक काल तक के कवियों ने भी अपनी काव्य रचनाओं में पर्यावरण के प्रति प्रौढ़ दृष्टिकोण रखा है। कबीर, सूर, तुलसी, जायसी, प्रसाद, निराला, महादेवी वर्मा, हजारी प्रसाद द्विवेदी, प्रेमचंद आदि कवियों व लेखकों के साहित्य में पर्यावरण के प्रति चिंतन व्यक्त हुआ है।

भारतीय संस्कृति में कई पर्व पर्यावरण संरक्षण व प्रकृति प्रेम को प्रकट करतें हैं। ये पर्व है, जुलाई माह में मनाए जाने वाला 'हरेला' इस दिन सभी व्यक्ति एक वृक्ष अवश्य लगतें हैं। इसके अतिरिक्त 'हरियाली तीज', 'वट सावित्री' ये पर्व भारतीय चिंतन में वृक्षों को लगाने, पूजन करने से मानव जीवन को आध्यात्मिक शांति की ओर ले जाते हैं। साथ ही वट वृक्ष की पूजा से पति परमेश्वर को दीर्घायु मिलेगी, ऐसा हमारे शास्त्रों में लिखा है। इसलिए भारतीय स्त्रियां वटवृक्ष की पूजा- अर्चना करतीं हैं। प्रकृति ने हमें जीवनदायिनी ऑक्सीजन से लेकर फल-फूल, सब्जी, अनाज, लकड़ी आदि सभी कुछ प्रदान किया है। अब हमारी नैतिक जिम्मेदारी है कि हम भी पर्यावरण को स्वस्थ व स्वच्छ बनाए रखने हेतु कारगर प्रयास अवश्य करें।

आज पर्यावरण विनाश और पूंजीवाद ने भारतीय संस्कृति, समाज को प्रभावित किया है, तो साहित्य इससे अछूता कैसे रह सकता है। पूंजीवाद के विस्तार, औद्योगिकरण, उदारीकरण,उपभोक्तावाद, बाजारीकरण, भूमण्डलीकरण, सड़कों का चौड़ीकरण,से भूस्खलन बढ़ा, बर्फबारी में कमी आई है इन सभी का प्रभाव हमारे पर्यावरण पर पड़ा है। इसी प्रतिरोध को साहित्यकारों ने अपने साहित्य में खोलकर रखा है। मानव के अंतर्द्वंद्व, बैचेनी, चिंता, सामाजिक सरोकारों को समकालीन हिंदी साहित्य बखूबी प्रस्तुत करता है।

प्रस्तावना

पर्यावरण का अर्थ है- "हमारे आस-पास, हमारी धरती हमारा आकाश। पर्यावरण उन सभी स्थितियों और प्रभाव का मेल है, जो जीव- समूह, जीवन और विकास पर ही नहीं वरन् वनस्पति जगत और प्रकृति पर भी प्रभाव डालते हैं। मनुष्य प्रकृति से अधिक से अधिक ऊर्जा, ज्यादा उत्पादन प्राप्त करने के लिए प्रकृति प्रदत्त संसाधनों को निचोड़ने में लगा है। सड़कों पर दौड़ती अनगिनत गाड़ियां, धुआं उगलते कारखाने, 5G इंटरनेट इन सबने मिलकर हमारे पर्यावरण को भयभीत स्थिति तक पहुंचा दिया है। हिंदी काव्य धारा में सन 1960 के बाद देशकल परिवेश में आए परिवर्तनों के कारण काव्य का स्वर भी बदला- बदला दिखाई पड़ा है।

प्रस्तुत शोध के अध्ययन का उद्देश्य -:

1. प्रस्तुत शोध के अध्ययन का उद्देश्य भूमण्डलीकरण के कारण बढ़ते पर्यावरण संकट को चित्रित करना है।
2. प्रस्तुत शोध पत्र के माध्यम से हिंदी साहित्य में आदिकाल से लेकर वर्तमान काल तक की काव्य रचनाओं में पर्यावरण चेतना पर दृष्टिपात करना है।
3. प्रस्तुत शोध पत्र के माध्यम से हिंदी साहित्य में पर्यावरण चेतना को केंद्र में रखकर भविष्य के लिए सुझाव प्रस्तुत करना है।

शोध प्रविधि -: प्रस्तुत शोध पत्र में हिंदी साहित्य में पर्यावरण चेतना का अध्ययन करने के लिए प्रथम स्रोत के रूप में ऐतिहासिक व विश्लेषणात्मक पद्धतियों का प्रयोग किया गया है । द्वितीयक स्रोतों के रूप में लेखकों की पुस्तकों, शोध पत्र- पत्रिकाओं एवं इंटरनेट द्वारा अध्ययन किया गया है।

बीज शब्द -: काव्य, गद्य, व अन्य विधाओं में, पर्यावरण संरक्षण,प्रदूषण एवं पवित्रता, निष्कर्ष।

काव्य हिंदी साहित्य में आदिकालीन काव्य से लेकर समकालीन काव्य में पर्यावरण चेतना दृष्टिगत हुई है। आचार्य रामचंद्र शुक्ल के अनुसार "हिंदी साहित्य के प्रथम महाकाव्य माने जाने वाले चंदबरदाई कृत "पृथ्वीराज रासो" में प्रकृति की सुंदरता का खूब बखान किया गया है-:" **पत्त पुरातन झरिग पत्त अंकुरित उट्ठि तुछ।**

ज्यों सैसव उत्तरिच चढिय बैसव किशोर कुछ।।

सीतल मंद सुगंध आइ रितुराज अचानं।

रोमराइ अंग कुछ नितम्ब तुच्छ सरसानं॥"(2)

आदिकालीन कवि विद्यापति की रचित "पदावली" में पर्यावरण चेतना का सुंदर चित्रण किया गया है-: **"- मौली रसाल मुकुल भेल ताब, समुखहिं कोकिल पंचम गाय।"**(3)

संत तुलसीदास जी ने समाज के समग्र पर्यावरण की विशुद्धि पर बल दिया है। **रामचरितमानस में** राम और लक्ष्मण जनकपुरी देखने के लिए प्रस्थान करते हैं। जनकपुरी के उद्यानों में फल और फूलों से संपन्न सुंदर वृक्षों की फलती हुई लताओं को दोनों राजकुमार देखते हैं, ओर अवलोकन करते हैं, लेकिन वे उस वाटिका के पुष्प और फलों को कुलषित नहीं करते और न ही अपनी इच्छा से तोड़ते हैं, अपितु गुरु के लिए वांछित पुष्पों को मालीगणों से पूछकर ही ग्रहण करते हैं। यथा-**"नवपल्लव फल सुमन सुहाए। निज संपति सुर रूख लजाए।**

चातक कोकिल कोर चकोरा कूजत विहग नटत कल मोरा।"(४)

कृष्ण भक्त कवियों ने राधा - कृष्ण के सौंदर्य चित्रण के साथ-साथ प्रकृति व पर्यावरण चेतना की रूपरेखा को जोड़ा है। कृष्ण भक्त कवि सूरदास की पंक्तियां दृष्टव्य हैं -**" तब ये लगती अति सीतल।**

अब भी विषम ज्वाल की पुंजें॥" (5)

बिहारी ने "बिहारी सतसई" में प्रकृति का सुंदर चित्रण किया है - **"चुवत स्वेद मकरंद कर , तरु दोतरह बिरमाय।**

आवत लच्छिन देव ते, थक्यौ बटोही बार।"(6)

आधुनिक काल में खड़ी बोली को काव्य भाषा के रूप में प्रतिष्ठित करने वाले आचार्य महावीर प्रसाद द्वि वेदी के कृपा पात्र राष्ट्रभक्ति भाव से साहित्यिक ओजस्विता पाने वाले कवि मैथिलीशरण गुप्त रचित काव्य में पर्यावरण संरक्षण की जो गूंज है वो किसी से छुपी नहीं है। "पंचवटी" में मैथिलीशरण गुप्त कहते हैं- **"पंचवटी की छाया में, सुंदरपर्ण कुटीर बना।**

उसके सम्मुख स्वच्छ शिला पर, धीर वीर निर्भीकमना॥"(7)

महाकवि जयशंकर प्रसाद ने अपने काव्य को मूलत: प्रकृति को केंद्र मानकर लिखा है। प्रसाद कृत " कामायनी" महाकाव्य की कुछ पंक्तियां जो उनके प्रकृति संरक्षण प्रेम को दर्शाती है- **"हिमगिरि के उत्तुंग शिखर पर, बैठ शिला की शीतल छांह।**

एक पुरुष भीगे नयनों से, देख रहा था प्रलय प्रवाह॥"(8)

गद्य में पर्यावरण चेतना हमें गोविंद मिश्र के उपन्यास " पांच आंगनों वाला घर" में सवेरे का समय प्राकृ. तिक परिवेश का मनोरम दृश्य होता है - " **सुबह की आवाजें शुरू हो गई थी। तरह-तरह की छोटी - छोटी घंटियों की तरह चिलकती आवाजें.... चिक- चिक , चिख-चिख, टिक टूं.... चि- चि- चि.... प्रतीक्षा कब सुबह हो और वे पेड़ की कैद से निकलें- आसमान के प्रांगण में चिलकते दौड़ते फिरें। "**(9)

काशीनाथ सिंह की कहानी " जंगल जातकर्म"पर्यावरण को बचाने के लिए यह कहानी चिपको आंदोलन के आसपास की है। भारतीय साहित्य और दर्शन स्वयं में संपूर्ण रूप से पर्यावरण केन्द्रित रहे हैं। वैदिक काल के अध्ययन से पता चलता है कि तत्कालीन सभ्यता के व्यक्ति प्रकृति की पूजा करते थे। उनके लिए पृथ्वी उनकी माता थी जिसकी रक्षा के लिए तथा जिस पर जीवन बनाए रखने के लिए वे अपने सभी कर्तव्यों का पालन करते थे। हजारी प्रसाद द्विवेदी का निबंध **"कुटज" में कहा गया है - "यह धरती**

मेरी माता है और मैं इसका पुत्र हूं इसलिए मैं सैदव इसका सम्मान करता हूं और मेरी धरती माता के प्रति नतमस्तक हूं।"(10) **मेहता नगेंद्र सिंह** समकालीन हिंदी कथा के प्रमुख हस्ताक्षरों में से एक है। पर्यावरण चेतना और जागरूकता से संबंधित उनकी एक लघु कथा है -**"वृक्ष ने कहा"** जिसमें एक बरगद के वृक्ष ने किस प्रकार सीख दी है कि **"हरे- भरे कोमल वृक्षों को काटा जाना किसी प्रकार भी उचित नहीं है। चाहे उसकी लकड़ियां पूजा- अर्चना के कार्य में ही क्यों ना लाई गई हो।"**(11)

जय श्री राय की कहानी " खारा पानी" गोवा के समुद्री जीवन पर आधारित उन मछुआरों की कहानी है जिसके समुद्र पर कंपनी वालों का अधिकार हो गया है। समुद्री मछली पर जीवन यापन करने वाले मछुआरों को विस्थापित होना पड़ रहा है। सैलानियों के द्वारा फैलाए गए प्रदूषण से समुद्री जीव मार रहे हैं। कहानी में रामा कहता है **"कुदरत ने हमें जो दिया था, हम उसी से संतुष्ट थे। सर पर आकाश था, नीचे धरती का बिछौना.... अपने जल, जंगल, जमीन से हमें सब कुछ मिल जाता था.... मगर अब तो सब छीन गया है।"**(12)

रश्मि भार्गव की कहानी **"कजरी और एक जंगल"** पर्यावरण चेतना को केंद्र में रखकर लिखी गई है जिसमें जंगल को देवता माना गया है परंतु कुछ लोग पेड़ों को काटकर तस्करी करते हैं कजरी अपने बाबा से कहती है- **" बाबा देखों ये लोग जंगल के पीछे ही पड़े हैं , जब देखों तब ये आते हैं और जानवरों को मार देते हैं और फिर उनकी खाल बेच देतें है। कभी चोरी से पेड़ काटकर लकड़ियां ले जाते हैं।"**(13) कविता, कहानी, उपन्यास के अतिरिक्त संस्मरण और यात्रा वृत्तांत जैसी विधाओं में पर्यावरण चेतना की संवेदना घनीभूत करती है। प्रेमचंद जैसे युग प्रवर्तक साहित्यकार के साहित्य में पर्यावरण चेतना के दर्शन होते हैं **"पूस की रात" कहानी में** प्राकृतिक परिवेश का पर्दा खोलते हैं - **"रात को शीत ने धधकना शुरू किया।"**(17) **प्रेमचंद ने "दो बैलों की कथा** "के माध्यम से प्रकृति के संवेदनात्मक पक्ष को उभारा है। यात्रा वृत्तांत के अंतर्गत निर्मल वर्मा के " चीड़ो पर चांदनी", अज्ञेय के "यायावर रहेगा याद" महत्वपूर्ण रचनाएं हैं। प्रकृति के छिपे हुए इतने सौंदर्य स्तरों की खोज की है जो उनकी बौद्धिकता का नहीं उनकी रचनात्मकता का भी परिचायक है।

निष्कर्ष

लेखक समय की मांग को संवेदनात्मक धरातल पर प्रस्तुत करता है। पर्यावरण प्रदूषण की समस्या गंभीर विषय है। आज की ज्वलंत समस्या है। प्रकृति हमारी मां है और हम प्रत्येक भारतवासी को अपनी मां रूपी प्रकृति की रक्षा करनी चाहिए। तभी हमें अपने चारों ओर शुद्ध पर्यावरण प्राप्त हो सकेगा। हमें पर्यावरण के प्रति संवेदना रखेंगे तो विनाश, प्रलय से बचें रहेंगे। वर्तमान में मनुष्य को प्रकृति एवं पृथ्वी को बचाने का संकल्प लेना होगा। जन- जागृति अभियान तथा विश्व नागरिक कर्तव्य भावना से प्रेरित होकर पर्यावरण को संरक्षित करना होगा।

संदर्भ ग्रंथ सूची

1. यजुर्वेद शक्ति सूक्त (36/17)
2. चंदबरदाई, पृथ्वीराज रासो
3. विद्यापति, पदावली
4. रामचरितमानस, हनुमान प्रसाद पोद्दार, प्रकाशक, मोती लाल जालान, गीता प्रेस गोरखपुर दोहा- 227.
5. सूरदास, सूरसागर।
6. बिहारी, बिहारी सतसई।
7. मैथिलीशरण गुप्त, पंचवटी
8. जयशंकर प्रसाद, कामायनी, चिंता सर्ग पृष्ठ संख्या -05.

9. गोविंद मिश्र, पांच आंगनों वाला घर, राधाकृष्ण प्रकाशन, दिल्ली सं- 1995 पृष्ठ संख्या- 07.
10. हजारी प्रसाद द्विवेदी, निबंध, कुटज, पृष्ठ संख्या - 32.
11. नगेन्द्र सिंह मेहता, वृक्ष ने कहा, पृष्ठ संख्या- 07.
12. जय श्री राम , खारा पानी
13. डॉ रश्मि भार्गव, कजरी और एक जंगल, मधुमती मार्च - अप्रैल 2012, पृष्ठ संख्या - 45।
14. पर्यावरण और समकालीन हिंदी साहित्य, डॉ प्रभाकरन हेब्बार इल्लत।

अध्याय: 21

पर्यावरण संरक्षण और स्थिरता में महिलाओं का योगदान

डॉ० अर्चना

असिस्टेंट प्रोफेसर, भूगोल विभाग,

स्वतन्तत्रा संग्राम सेनानी थान सिंह रावत राजकीय महाविद्यालय नैनीडांडा पौड़ी गढ़वाल

सारांश

पर्यावरण मानव जीवन शैली के लिए एक अभिन अंग है। तो इसका संरक्षण और बचाव भी मनुष्य का प्रमुख कर्तव्य बन जाता है। पर्यावरण सरंक्षण मनुष्य जीवन के लिए एक अनिवार्य विषय बन गया है। पर्यावरण के सरंक्षण के लिए अनेक प्रयास व आन्दोलन जारी है। पर्यावरण को बचाने के लिए महिलाओं का भी योगदान रहा है। महिलाएँ वैदिक काल से ही पर्यावरण सरंक्षण के पक्ष में रही है। इसका उदाहरण- तुलसी पूजा व वटवृक्ष पूजा है। हमारे देश की महिलाऐं हमेशा से ही इस दिशा में कार्यशील रही है। हमारी भारतीय संस्कृति को देखने पर रीति-रिवाजो परम्पराओं में पर्यावरण संरक्षण के प्रति जागरुता दिखाई देती रही है। हमारे देश की स्त्रियां हमेशा से ही पर्यावरण सरंक्षण में पुरुषो से आगे रहती है। इसका पता हमें कही पर्यावरण सरंक्षण के आंदोलनों का अध्ययन करने पर चला है महिलाओं ने पर्यावरण व वनो की रक्षा के लिए विभिन्न आंदोलनों में भाग लिया इनमें अग्रणी भूमिका निभायी है और अपने प्राणों का बलिदान तक दे दिया, कही विद्वानो का मानना है कि समाज में परिवर्तन महिलाओं के बिना नहीं हो सकता। अत: महिलाओं ने सदैव ही पर्यावरण संरक्षण के लिए अपने कदम सबसे पहले आगे रखा है। वर्तमान समय में महिलाएँ पर्यावरण संरक्षण में न केवल प्रमुख भूमिका निभाती है बल्कि पर्यावरण की एजेन्ट भी है। वे संरक्षण प्रयासों में अद्वितीय दृष्टिकोण ज्ञान और कौशल विकास भी लाती है। और स्थायी भविष्य प्राप्त करने के लिए उनका योगदान आवश्यक है। वे स्थायी संसाधन प्रबंधन मे महत्वपूर्ण भूमिका निभाती है। सुनीता नारायण जन्म 1961 भारत की प्रसिद्ध पर्यावरणविद है। वे हरित राजनीति और अक्षय विकास की महान समर्थक हैं। कु० सुनीता नारायणन सन 1982 से भारत स्थित विज्ञान एवं पर्यावरण केन्द्र से जुडी रहीं हैं। और आज हर क्षेत्र में अपनी सहयोग दे रही है। उनकी भूमिका पयार्वरण संरक्षण के लिए और अधिक बढ जाती है, हमारे देश में अमृता देवी, गौरा देवी, मेघा पाटेकर वन्दना शिवा तथा सुनीता नारायण जैसी महिलाओं ने खेज लड़ी आंदोलन, चिपको आंदोलन, नर्मदा आंदोलन, नवदन्या, मघ मदर्न अंदोलन तथा उत्तराखण्ड नदी बचाओ अभियान की शुरुआत की।

प्रस्तावना

पर्यावरण संरक्षण के महत्व को कम करके नहीं गिना जा सकता जैव विविधता को सुरक्षित रखने से लेकर जलवायु परिवर्तन तक पर्यावरण संरक्षण वर्तमान और भविष्य में आने वाली पीढी के लिए एक रहने योग्य स्थान बनाए रखना जरुरी है। लेकिन इस बात को अनदेखा कर दिया जाता है। कि पर्यावरण संरक्षण के प्रयासो में महिलाओ की महत्वपूर्ण भूमिका है। महिलाऐ न केवल पर्यावरण संरक्षण के बारे मे सोचने वाली है, वह इस की एजेन्ट भी है। पर्यावरण सरंक्षण महिलाओं के लिए हरित अर्थव्यवस्था में भाग लेने का मौका भी दे सकती है। जिससे नवीनीकरण ऊर्जा पारिस्थितिकी पयर्टन और टिकाऊ कृषि जैसे क्षेत्र समलित है। महिलाओं को आर्थिकीकरण में वढावा देकर उनके जीवन को सफल बनाने के लिए कही प्रयास किये जा रहे है। जिससे वह अपने जीवन को सुरक्षित व सुन्दर बनाने में मदद कर सकते है। जबकि पयार्वरण स्थिरता को भी बढ़ावा दिया जा रहा है। महिलाऐ व्यवसाय प्राकृतिक संसाधनो के संरक्षण को आगे बढ़ाने में सफल रही है। जबकि स्थानीय लोगों के समुदाय की आर्थिक लाभ भी देते हैं।

शोध पद्धति

- अमृता देवी भारत की पहली प्रलेखित महिला पर्यावरण कार्यकर्ता हैं। लगभग 300 साल पहले, अमृता देवी ने जोधपुर के महाराजा के लिए महल बनाने के लिए पेड़ों की कटाई के खिलाफ राजस्थान, भारत में एक विरोध प्रदर्शन का नेतृत्व किया था। उनके प्रयास की विफलता ने समुदाय में काफी आक्रोश पैदा किया। लोककथाओं के अनुसार, राजा ने स्थानीय समुदायों से फिर कभी लकड़ी न मांगने की कसम खाई थी। अमृता देवी बिश्नोई समुदाय से थीं, जो अपनी श्रद्धा और प्रकृति के प्रति प्रेम के लिए प्रसिद्ध है।

- चिपको (जिसका अर्थ है "गले लगाना") आंदोलन की अग्रणी गौरा देवी ने महिलाओं से पेड़ों की रक्षा करने और उन्हें कटने से रोकने के लिए उन्हें गले लगाने की वकालत की। रेनी गांव में, जहां उन्होंने महिला मंगल दल की नेता के रूप में काम किया, उन्होंने 27 महिलाओं का नेतृत्व किया और लकड़हारों का विरोध किया, जिस दिन उन्हें पेड़ काटने थे। यह आंदोलन शांतिपूर्ण प्रतिरोध के गांध ीवादी दर्शन पर आधारित था, ताकि पारिस्थितिक संतुलन के विनाश का विरोध किया जा सके। गांधी के अहिंसक प्रतिरोध की वकालत से प्रेरित होकर, आंदोलन का मूल ग्रह के नाजुक पारिस्थितिक सामंजस्य के विनाश को शांतिपूर्ण तरीके से रोकने का प्रयास था। जनवरी 1974 में सरकार का फैसला, अलकनंदा नदी के किनारे 2,500 पेड़ों की नीलामी करने का, क्योंकि वन संपत्ति खेल के सामान के निर्माता को दी जा रही थी, चिपको आंदोलन के लिए प्रेरणा का काम किया, जब उत्तराखंड के चमोली गांव की महिलाओं ने धमकी के बावजूद पेड़ों को काटने से रोकने के लिए उन्हें गले लगाना शुरू कर दिया। वे पूरी रात पेड़ों को देखती रहीं जब तक कि लकड़हारे हार मानकर चले नहीं गए। जैसे ही आंदोलन की खबर आस-पास के गांवों तक पहुंची, और अधिक लोग इसमें शामिल हो गए। उत्तराखंड के अन्य भागों में भी विरोध का ऐसा ही तरीका अपनाया गया, जो दर्शाता है कि महिलाएं पर्यावरण के लिए अभिय चला रही थीं। 1980 के दशक में चिपको आंदोलन से प्रेरित होकर कर्नाटक के उत्तर कन्नड़ में जंगल की सुरक्षा के लिए अप्पिको आंदोलन की शुरुआत हुई थी। इस आंदोलन का नेतृत्व पांडुरंगा हेगड़े ने किया था। कागज की मिलों और प्लाईवुड कारखानों जैसे कई उद्योगों ने वन संसाधनों का अत्यधिक दोहन किया और जलविद्युत बांधों की एक श्रृंखला विकसित हुई जिसने विशाल वन और कृषि क्षेत्रों को जलमग्न कर दिया। 1980 तक, इन सभी गतिविधियों के परिणामस्वरूप जंगल अपने मूल आकार के लगभग एक चौथाई तक सिकुड़ गए।

इस आपदा के मद्देनजर, पश्चिमी घाटों की रक्षा और संरक्षण के इरादे से अप्पिको आंदोलन उभरा 26 । 1983 और 1990 के दौरान, भारत के कर्नाटक निवासियों ने वनों की कटाई को रोकने के लिए को ष्पेडों गले लगानाष (अप्पिको) नामक एक तकनीक इस्तेमाल किया ।

- नर्मदा बचाओ आंदोलन एक सामाजिक आंदोलन है जिसका आयोजन और नेतृत्व स्थानीय लोगों, पर्यावरणविदों और मानवाधिकार कार्यकर्ताओं द्वारा किया जाता है। यह नर्मदा नदी पर कई विशाल बांध परियोजनाओं का विरोध करने के लिए आयोजित किया गया था, क्योंकि उनके महत्वपूर्ण पारिस्थितिक और सामाजिक-आर्थिक निहितार्थ हैं। नर्मदा नदी पर सबसे बड़े बांधों में से एक सरदार सरोवर बांध है, जो गुजरात, महाराष्ट्र और मध्य प्रदेश को सिंचाई के बिजली और पानी प्रदान करता है।
- नवदान्या आंदोलन की स्थापना 1984 में पर्यावरणविद वंदना शिवा ने भारत में पारंपरिक कृषि पद्धतियों को बढ़ावा देने के लिए की थी। ष्नवदान्याष सबसे बड़ा जैविक खेती आंदोलन है जिसका मतलब है नौ फसलें। यह एक गैर-लाभकारी संगठन है जो जैविक खेती के तरीकों और जैव विविधता के संरक्षण का समर्थन करता है। इस संगठन ने किसानों को बाजार स्थापित करने में मदद करने के अलावा उपभोक्ताओं को प्रीमियम जैविक खाद्य पदार्थों को बढ़ावा दिया है। नवदान्या का प्राथमिक उद्देश्य बायोपाइरेसी के सामने बीज विविधता की रक्षा करना है और इसके लिए उन्होंने 17 भारतीय राज्यों में 111 सामुदायिक बीज बैंक स्थापित किए हैं। वे जीएमओ (जेनेटिकली मॉडिफाइड ऑर्गेनिज्म) बीजों के खिलाफ लड़ रहे हैं और जैव विविधता संरक्षण प्रयासों में शामिल हैं। नवदान्या आंदोलन के अधि कांश सदस्य पूरे भारत से महिला किसान हैं।
- गंधमर्दन आंदोलन: गंधमर्दन पहाड़ियों को व्यापक रूप से ष्आयुर्वेद स्वर्गष के रूप में जाना जाता है, क्योंकि वहां औषधीय पौधों, ऑर्किड और अन्य असामान्य प्रजातियों की असाधारण विविधता पाई जाती है। इसके अलावा, पहाड़ियों में बॉक्साइट प्रचुर मात्रा में है। पहाड़ियाँ, जो आदिवासियों की आजीविका का समर्थन करती हैं और क्षेत्र के पारिस्थितिकी तंत्र को बनाए रखने के लिए महत्वपूर्ण हैं, भारत के सबसे मुखर व्यक्तियों द्वारा जंगलों और उनके जीवन के तरीके की रक्षा के लिए लड़ी गई लड़ाइयों में से एक का स्थल रही हैं। गंधमर्दन पहाड़ी की रक्षा के लिए संघर्ष 1980 के दशक में बाल्को (भारत एल्युमिनियम कंपनी) के खनन कार्यों से शुरू हुआ, जो आदिवासी लोगों द्वारा किया गया था, जिनका अस्तित्व बॉक्साइट भंडार के निष्कर्षण से सीधे प्रभावित था। स्थानीय लोगों द्वारा पाँच साल तक चलाए गए लगातार अभियान के परिणामस्वरूप 213 मिलियन टन बॉक्साइट निकालने के लिए बाल्को का संचालन बंद कर दिया गया था। यह स्थानीय पर्यावरण और पश्चिमी उड़ीसा के निवासियों, जो जंगलों पर निर्भर थे, दोनों के लिए एक बड़ी जीत थी। बाल्को परियोजना ने पहाड़ी ढलानों को रेगिस्तानी बनाने और वनों की कटाई का खतरा पैदा कर दिया, जहाँ लोग अपनी धार्मिक प्रथाओं का पालन करते थे। इसने हजारों स्वतंत्र किसानों को, जिनके परिवार पीढ़ियों से एक ही भूमि पर खेती करते आ रहे थे, अपनी जीवन शैली खोने के खतरे में डाल दिया गंधमर्दन पहाड़ियों के स्थानीय समुदायों ने क्षेत्र मे प्रचुर जैव विविधता की रक्षा के लिए एक अभियान शुरू किया।
- राधा भट्ट ने 2008 में उत्तराखंड नदी बचाओ अभियान की शुरुआत की थी, ताकि कई जलविद्युत परियोजनाओं के निर्माण को रोका जा सके, जिससे हिमालय राज्य की नाजुक, भारी वनों की कटाई वाली पारिस्थितिकी में गंगा और उसकी अधिकांश सहायक नदियों का प्रवाह बाधित हो सकता था। उन्होंने महिला सशक्तिकरण और चिपको आंदोलन को बढ़ावा दिया। उन्होंने हिमालय में खुले खनन के खिलाफ विरोध प्रदर्शनों में अग्रणी भूमिका निभाई है। उनके प्रयासों से क्षेत्र के 50 से अधिक गांवों में भूस्खलन को रोका गया। राधा भट्ट को उनके असाधारण कार्य के लिए इंदिरा प्रियदर्शिनी पर्यावरण पुरस्कार मिला

पर्यावरण अध्ययन का क्षेत्र

पर्यावरण अध्ययन का क्षेत्र अत्यधिक विस्तृत है, जिसके अन्तर्गत सूक्ष्म पारिस्थितिक तन्त्र से लेकर वृहद् जीव मण्डलीय पारिस्थितिक तन्त्र के अध्ययन को सम्मिलित किया जाता है। पर्यावरण के अन्तर्गत चार प्रमुख

घटकों को सम्मिलित किया जाता है- स्थल मंडल, जल मंडल, वायु मंडल एवं जैव मंडल।

स्थल मण्डल में समस्त बाहरी भूपटल, सागर व महासागर की सतह भी सम्मिलित है। पर्यावरण का सर्वाधिक महत्वपूर्ण अवयव है। इसके अंतर्गत धरातल की रचना, मृदा, चट्टानों, भू-आकृतियों, भूमिगत जलस्रोतों और प्राकृतिक संसाधन सम्मिलित हैं। इन सभी से मिलकर पर्यावरण का निर्माण होता है और ये सभी जीवन के हर पहलू को प्रभावित करते हैं। इसकी मोटाई लगभग 35 किलोमीटर मानी जाती है।

जल मण्डल के अन्तर्गत पृथ्वी पर पाए जाने वाले सभी जल स्रोतों, सागरों, महासागरों, नदियों, झीलों, तालाबों, कुओं, बावड़ियों और पोखरों को सम्मिलित करते हैं। जल जीवन के लिए सर्वाधिक महत्वपूर्ण घटक है। जल मण्डल व स्थलमण्डल के आयतन में 7:3 का अनुपात है।

वायु-मण्डल के अंतर्गत भू-भाग और जलमण्डल के चारों ओर पाया जाने वाला वायु का आवरण शामिल है। इसमें स्थल एवं जल मण्डल दोनों ही समाहित हैं। वायु मण्डल पर्यावरण का गतिशील घटक है, जिसमें निरन्तर मौसमी परिवर्तन होते रहते हैं। भू-भाग की सतह से इसकी अधिकतम सीमा लगभग 500 किलोमीटर तक स्वीकार की गई है। इसमें ऑक्सीजन, नाइट्रोजन, ऑर्गन, कार्बन डाईऑक्साइड, हाइड्रोजन, हीलियम और ओजोन गैसें विभिन्न अनुपात में पाई जाती हैं। वायुमण्डल में क्षोभमण्डल, समताप मण्डल, मध्यमण डल और आयनमण्डल चार परतें पाई जाती हैं।

जैव मण्डल पृथ्वी के स्थल मंडल, जल मंडल एवं वायुमंडल के उस भाग को कहते हैं जिसमें जीवन पाया जाता है। यह परत वायु मण्डल, स्थल मण्डल और जल मण्डल के मिलने से बनी एक पतली पट्टी के रूप में पाई जाती है। जिसमें सभी पौधों और जीवों का जीवन पाया जाता है। इसकी मोटाई भू-पटल की सतह से लगभग 7 किलोमीटर गहराई तक है।

पर्यावरण अध्ययन का महत्व

पृथ्वी का निवासी होने के कारण, हमारा कार्य का तरीका इस उपग्रह तथा उसके निवासियों पर प्रभाव डालता है। चक्रवात, भूकम्प, ज्वालामुखी विस्फोट जैसी कुछ प्रमुख आपदाऐं हमारे पर्यावरण को अत्यधिक प्रभावित करती हैं। मानवीय गतिविधियाँ जैसेकि पर्यावरण में प्रदूषण का प्रभाव, वनों का कटना व नदियों पर बांधों को निर्माण ने हमारे पर्यावरण को नुकसान पहुँचाया है। पर्यावरण अध्ययन वह विज्ञान है जिसके माध्यम से मनुष्य तथा उसके पर्यावरण में सम्बन्धों पर अध्ययन किया जाता है। वर्तमान शताब्दी में जीवन को बनाए रखने के लिए पर्यावरण की गुणवत्ता को बनाए रखना आवश्यक है। यह तभी सम्भव है, जब हम सभी पर्यावरण के महत्व को समझें।

- सुनीता नारायण ने 1990 के दशक में जलवायु परिवर्तन पर काम करना शुरू किया। 1991 में उनकी पुस्तक, ग्लोबल वार्मिंग इन एन अनइक्वल वर्ल्ड: ए केस ऑफ एनवायरनमेंटल कोलोनियलिज्म ने जलवायु परिवर्तन सम्मेलन में समानता के सिद्धांत की पुष्टि करने में महत्वपूर्ण भूमिका निभाई। वर्ष 2005 में उन्हें पद्म श्री पुरस्कार मिला । उन्हें वर्षा जल संचयन पर उनके काम के लिए जाना जाता है, जिसके लिए उन्हें स्टॉकहोम जल पुरस्कार मिला। उन्होंने जल प्रबंधन के लिए नई नीतियां तैयार करने में भारत सरकार के साथ काम किया। वह सेंटर फॉ

मुख्य बिन्दु -

1. राष्ट्रीय पर्यावरण नीति हमारे समाज की सामाजिक-आर्थिक, राजनीतिक, सांस्कृतिक और पर्यावरणीय चुनौतियों को पहचानती है (2006)। व्यावसायिक सुरक्षा, स्वास्थ्य सेवा प्रणाली, शिक्षा, हाशिए पर पड़े और कमजोर व्यक्ति के सशक्तिकरण और लैंगिक समानता के विभिन्न रूपों के माध्यम से, सामूहिक गरीबी को कम करने के मूल उद्देश्य को ध्यान में लाया जाता है। आर्थिक सुरक्षा, स्वास्थ्य सेवा प्रणाली, शिक्षा, वंचित व्यक्ति के सशक्तिकरण और लैंगिक समानता की विविध अभिव्यक्तियों के माध्यम से, सामूहिक गरीबी को कम करने के मूल उद्देश्य को संबोधित किया जाता है।

2. राष्ट्रीय वन नीति (1988) और संयुक्त वन प्रबंधन (1990 के दशक) पहल के तहत वन सम रक्षा समितियों की सदस्यता में महिलाओं की संख्या 33% होनी चाहिए थी। इसकी स्थापना के पीछे तर्क यह था कि राज्य वन विभाग वनों की कटाई को प्रभावी ढंग से नियंत्रित कर सकते हैं यदि वे क्षतिग्रस्त जंगलों को फिर से लगाने के लिए स्थानीय आबादी के साथ सहयोगात्मक प्रबंधन समझौतों पर बातचीत करते हैं।
3. 2002 का जैव विविधता अधिनियम पारंपरिक ज्ञान के सहयोगी और संरक्षक के रूप में महिलाओं की भूमिका के महत्व पर जोर देता है। ऐतिहासिक रूप से, कृषि समुदायों में बीज भंडार के संरक्षण के लिए हमेशा महिलाएँ जिम्मेदार रही हैं।
4. भारत में सरकार और गैर सरकारी संगठन जल संचयन परियोजनाओं में अधिकाधिक महिलाओं को शामिल करने के लिए मिलकर काम कर रहे हैं।
5. ग्रामीण क्षेत्रों में अक्षय ऊर्जा को बढ़ावा देने के प्रयास अक्सर महिलाओं को बायोगैस संयंत्र और खाना पकाने के लिए सौर पैनल प्रदान करके उन्हें सशक्त बनाने पर केंद्रित होते हैं। आर्थिक सीढ़ी के निचले पायदान पर रहने वालों को रसोई के लिए ऊर्जा के सुरक्षित और टिकाऊ स्रोतों तक पहुँच से वंचित नहीं किया जाना चाहिए। ऊर्जा क्षेत्र में महिलाओं के योगदान को नजरअंदाज नहीं किया जाना चाहिए, बल्कि नीति निर्माताओं को नीतियों को तैयार करने और परियोजनाओं को डिजाइन करने की प्रक्रिया में उन्हें सक्रिय रूप से शामिल करना चाहिए।

निष्कर्ष

पर्यावरण संरक्षण महत्वपूर्ण है क्योंकि पर्यावरण का क्षरण मनुष्यों, जानवरों और पौधों के लिए अत्यधिक हानिकारक है। क्षतिग्रस्त पारिस्थितिकी तंत्र को बहाल होने में सैकड़ों साल लग सकते हैं। घरों और समुदायों में प्राकृतिक संसाधन प्रबंधन रणनीतियाँ पर्यावरणीय क्षरण से प्रभावित होती हैं। पूरी दुनिया में, महिलाएँ वनों और कृषि भूमि के प्रबंधन और पानी, भोजन और ईंधन के प्रावधान की देखरेख करती हैं। महिलाएँ पर्यावरण के मुद्दों को लेकर अधिक चिंतित हैं। जैव विविधता संरक्षण में उनका योगदान महत्वपूर्ण है।

महिलाओं का प्रकृति और पर्यावरण समर्थक दृष्टिकोण और विश्वासों के साथ एक मजबूत रिश्ता है। उन्होंने पर्यावरण संबंधी चिंताओं पर मूल्यों का एक अनूठा सेट विकसित किया है। प्रकृति के प्रति महिलाओं का दृष्टिकोण इस बात को प्रभावित करता है कि वे पर्यावरणीय चुनौतियों का सामना कैसे करती हैं।

सन्दर्भ ग्रन्थ-

1. वानिकी एवं पर्यावरण विज्ञान
2. पर्यावरण अध्ययन
3. भौमिकी एवं पर्यावरण विज्ञान
4. वीकिपीडिया

अध्याय: 22

जयशंकर प्रसाद के महाकाव्य "कामायनी में पर्यावरणीय चेतना"

डा० राधा रावत

सहायक प्राध्यापक- हिन्दी

डा0 शिवानन्द नौटियाल राजकीय स्नातकोत्तर महाविद्यालय कर्णप्रयाग (चमोली)

शोध-सार

प्रकृति मानव जीवन की सहचरी है। प्रकृति (पर्यावरण) के साथ मानव का घनिष्ठ संबंध हैं। प्रकृति के बिना मानव जीवन की कल्पना नहीं की जा सकती है। आधुनिक युग में मानव अपने समाज और व्यक्तिगत विकास की चाह में प्रकृति के साथ खिलवाड़ करता जा रहा हैं, जिसके परिणाम स्वरूप वह प्राकृतिक आपदाओं का शिकार होता रहता है। जबकि मानव प्रकृति प्रदत्त पर्यावरणीय चेतना को जीवन से जोड़कर उसे संतुलित करते हुए उच्च विकसित जीवन व्यतीत कर सकता है। हिन्दी साहित्य के छायावादी युग में कवि जयशंकर प्रसाद, सूर्यकांत त्रिपाठी निराला, सुमित्रानंदन पंत व महादेवी वर्मा का प्रकृति के साथ गहरा संबंध रहा है। यहाँ कवियों ने प्रकृति को मानव जीवन के ऐसे उपादान के रूप में स्वीकार किया है जिसके अभाव में जीवन की कल्पना संभव नही है। प्रकृति कवियों को अपनी ओर आकर्षित ही नहीं करती अपितु मानवीय भावों और संवेदनाओं के साथ तारतम्य स्थापित करते हुए नये आयामों को प्राप्त करती हुई प्रतीत होती है। वह मानव को जीवन जीने की प्रेरणा प्रदान करती है। प्रस्तुत शोध पत्र में छायावाद के प्रर्वतक जयशंकर प्रसाद के महाकाव्य 'कामायनी' में उनकी पर्यावरणीय चेतना को उल्लेखित किया गया है। प्रसाद ने प्रकृति को सौंदर्य और सौंदर्य को प्रकृति स्वीकार किया है। उन्होंने प्रकृति का मानवीकरण कर उसे जीवन के सुख-दुःख का परिचायक माना है। प्रकृति के साथ अपना संबंध जोड़ते हुए वे प्रकृति के चरित्रों को मानव का सच्चा सहयोगी स्वीकारते हैं।

जयशंकर प्रसाद के महाकाव्य 'कामायनी' में पर्यावरणीय चेतना

'पर्यावरण' शब्द से अभिप्राय है-चारों ओर से घिरा हुआ। पर्यावरण शब्द 'परि' और 'आवरण' दो शब्दो से मिलकर बना है। 'परि' का अर्थ- चारों ओर तथा 'आवरण' का अर्थ- ढ़का हुआ या घिरा हुआ है। अर्थात् जो चारो ओर से ढका हो वह पर्यावरण कहलाता है। मानव जीवन को चारों तरफ से घेरे हुए और उसके परिवेश (जैविक, अजैविक तत्व, प्रक्रियाओं और घटनाओं) को निर्मित करने वाले कारक को पर्यावरण कहा

जाता है। इस आवरण और परिवेश में रहकर ही मानव अपना जीवन व्यतीत करता है। उसके विकास व विनाश की व्यापकता इस पर्यावरण में रहकर ही गतिमान होती है। जयशंकर प्रसाद की रचना 'कामायनी' मनु, श्रद्धा, इड़ा के माध्यम से एक तरफ मानव सभ्यता के विकास की गति को दर्शाती है तो दूसरी ओर इन पात्रों के जीवन व मनोभावों को प्रकृति के साथ जोड़कर उसे गतिशील बनाती है। प्रकृति की संवेदना और आत्मिक चेतना के सामंजस्य-पूर्ण शिखर से ही रस का स्रोत फूटता है और कामायनी इसी रस की गंगा है। प्राचीन कथावस्तु के क्षीण सूत्रों से प्रसाद ने मानव-विकास की गाथा को जिस प्रकार गुम्फित किया है वह मानव मन की प्राकृतिक तथा सांस्कृतिक वृत्तियों के विकास-क्रम की तात्विक और मनोवैज्ञानिक व्याख्यना करने में सहज ही सफल है।[1]

वर्तमान समय में मानव विकास की जितनी सीढ़ियां चढ़ रहा है, पर्यावरणीय स्तर पर वह उतना ही विनाश की ओर अग्रसर हो रहा है। जो मानव अस्तित्व के लिए सबसे बड़ी चुनौती है। प्रकृति ही मानव जीवन की संजीवनी है। इसके अस्तित्व से मानव का जीवन और उसकी चेतना जुड़ी हुई है। मानव जीवन एवं पर्यावरण एक दूसरे के पर्याय हैं। इस पर्यायता के पश्चात् जब प्रकृति का दोहन किया जाता है तो वह विनाशकारी रूप में मानव के सम्मुख प्रकट होती है जो आने वाले समाज और उसके भविष्य के लिए चिंतनीय है। मानव द्वारा निरंतर किए जा रहे पर्यावरण के विनाश से भविष्य की चिंता स्वतः मानव को दुःखी करने लगती है। छायावाद के स्तम्भकार जयशंकर प्रसाद की 'कामायनी' का चिन्ता सर्ग मनु के प्रलयकालीन प्रकृति संबंधी चिंतन से प्रारम्भ होता है जो वर्तमान में प्रासांगिक एवं चिन्तनीय है-

हिमगिरि के उत्तंग शिंखर पर, बैठ शिला की शीतल छांह,

एक पुरूष भीगे नयनों से देख रहा था, प्रलय प्रवाह।

नीचे जल था ऊपर हिम था, एक तरल था एक सघन,

एक ही तत्व की प्रधानता कहो उसे जड़ या चेतन।

मनु की यह चिंता प्राकृतिक असंतुलन और अतिदोहन के संदर्भ में प्रासांगिक है। प्रकृति सदैव न्याय करती है। उसके साथ हम जैसा करेंगे वैसा ही परिणाम हमें प्राप्त होगा। सृष्टि के बाद कालान्तर में प्रलय परिर्वतन का सहज क्रम है किन्तु इस प्रलय अथवा प्राकृतिक आपदा के लिए मनुष्य की स्वार्थी और अर्थ लोलुप प्रवृत्ति जिम्मेदार है। जिसके कारण जल-प्रलय के रूप में प्राकृतिक विनाश उपस्थित हुआ। कामायनी में प्रसाद जी ने स्पष्ट किया है कि यह प्राकृतिक विनाश प्रकृति के प्रति मानव के असंवेदनशील व्यवहार का ही परिणाम है। अपनी विलासिता के मद में मानव प्रकृति के वास्तविक स्वभाव को भूल जाता है-

प्रकृति रही दुर्जेय, पराजित हम सब थे भूले मद में,

भोले थे, हाँ तिरेत केवल सब विलासिता के नद में।

इसके साथ ही प्रसाद जी ने स्पष्ट किया है कि प्रकृति केवल मानव के सुख और प्रेम में ही उसकी सहयोगी बनकर नही चलती अपितु निराशा के समय में भी परस्पर उसका सहयोग करती है। वह अपने दुःख के क्षणों को प्रकृति के साथ आरोपित करने में सुख की अनुभूति करता है। प्रकृति (पर्यावरण) का यह रूप ही मानव को जीवन देता है और उसमें आशा का संचार करता है। कामायनी के 'इड़ा' सर्ग में मानव जीवन की निराशा एवं पर्यावरण के महत्व को स्पष्ट करते हुए जयशंकर प्रसाद ने लिखा है-

नभ नील लता की डालों में उलझा अपने सुख से हताश,

कलियाँ जिनको मैं समझ रहा वे काँटे बिखरे आसपास।

कितना बीहड़ पथ चला और पड़ रहा कहीं थक कर नितांत,

उन्मुक्त शिखर हँसते मुझ पर रोता मैं निर्वाचित अशांत।

इस नियति नटी के अति भीषण अभिनय की छाया नाच रही,

पावस रजनी में जुगनू गण को दौड़ पकड़ता मैं निराश ।

उन ज्योति कणों कर विनाश।4

छायावादी कवि प्रसाद के काव्य में तो प्रकृति ही मानव को जीवन देती है उसकी सहचरी बनकर चलती है, और उसे जीवन की ओर उन्मुख करती है इसी कारण वे अपने जड़ जीवन को सूर्य व वायु के समान गतिशील बनाकर चराचर जगत् में निरन्तर गतिशील बने रहना चाहते है-

देखे मैंने वे शैल श्रृंग।

जो अचल हिमानी से रंजित, उन्मुक्त, उपेक्षा भरे तुंग,

अपने जड़-गौरव के प्रतीक वसुधा का कर अभिमान भंग।

अपनी समाधि में रहे सुखी, बह जाती हैं नदियाँ अबोध,

कुछ श्वेत बिंदु उसके लेकर, वह स्तिमित-नयन गत शोक-क्रोध।

स्थिर-मुक्ति, प्रतिष्ठा मैं वैसी चाहता नहीं इस जीवन की,

मैं तो अबाध गति मरूत-सदृश, हूँ चाह रहा अपने मन की।

जो चूम चला जाता अग-जग प्रति-पग में कम्पन की तरंग,

वह ज्वलनशील गतिमय पतंग।

आशा सर्ग में प्रकृति का जो चित्र उदित हुआ है उसका प्रकाश मनु के मन का अंधकार मिटाकर उसमें नई आशा का संचार करता है-

उषा सुनहल तीर बरसती जय-लक्ष्मी सी उदित हुई।

उधर पराजित काल रात्रि सी जल में अंतर्निहित हुई।

जयशंकर प्रसाद की चेतना प्रकृति के प्रति इतनी स्वच्छ व निर्मल है कि वे मानव अंगों को निर्मित करने में प्रकृति के योगदान को स्वीकारते हैं। इस स्वीकार्य में कवि को नील परिधान में सुशोभित श्रद्धा के सुकुमार अंग ऐसे लग रहे है मानो बादलों के जंगल में गुलाबी रंग का बिजली का फूल खिला दिखाई हुआ हो-

नील परिधान बीच सुकुमार, खुल रहा मृदुल अधखुला अंग,

खिला हो ज्यों बिजली का फूल, मेघवन बीच गुलाबी रंग।

सांसारिक जगत में मानव के सुख दुःख में जितने सहयोगी उसके आस-पास रहने वाले मनुष्य है उससे कहीं अधिक प्रकृति (पर्यावरण) उसे प्रभावित करती है-

मानस का स्मृति शतदल खिलता, झरते बिंदु मरंद घने,

मोती कठिन पारदर्शी ये इनमें कितने चित्र बने?

आँसू सरल तरल विद्युतकण नयनालोक विरह तम में,

प्राण पथिक यह संबल लेकर लगा कल्पना जग रचने।

प्रसाद के काव्य में प्राकृतिक सौन्दर्य की भी अद्भुत छटा विद्यमान है। वे प्रकृति का मानवीकरण करते हुए उसकी अपूर्व रमणीयता का चित्र अंकित करते हैं। संध्याकालीन बादलों की रंग-बिरंगी चूनर धारण किए हुए हिमालय की श्रेणियां तुषार का किरीट धारण किए प्रतीत हो रही है-

संध्या घनमाला की सुन्दर, ओढ़े रंग बिरंगी छींट।

गगन चुम्बिनी शैल श्रेणियां, पहने हुए तुषार किरीट।

कवि प्रसाद जी ने प्रकृति के कोमल एवं भयानक दोनों रूपों का चित्रण किया है। कामायनी के 'चिन्ता' सर्ग में जहां प्रलय की विभीषिका का चित्रण है वहां समुद्र की लहरों का विकराल रूप देखेते ही बनता है-

उधर गरजती सिंधु लहरियां, कुटिल काल के जालों सी।

चली आ रही फेन उगलती, फन फैलाये ब्यालों सी।

'कामायनी' जयशंकर प्रसाद द्वारा रचित छायावादी युग की सर्वश्रेष्ठ कृति है। जिसमें आधुनिक जीवन की प्रमुख समस्याओ का समाधान प्रस्तुत करने के साथ-साथ प्रकृति के सौंदर्य व महत्व का भी अंकन किया गया है। प्रकृति दर्शन से छायावादी कवि को यह जो नया आलोक मिला, उसने सामान्य रूप से उसकी सम्पूर्ण जीवन दृष्टि बदल दी और विशेष रूप से प्रकृति संबंधी सौंदर्य दृष्टि। उसे पहली बार प्रकृति की स्वतंत्र सत्ता का बोध हुआ। प्रसाद की कामायनी में जो कुछ भी सौंदर्य से परिपूर्ण है वह मानव और उसका पर्यावरण ही है। यह सर्वथा सत्य है कि 'प्रसाद जी वर्ड्सवर्थ की भाँति प्रकृति के अन्तःकरण में प्रविष्ट हो उसका कोना-कोना झाँक आये थे। अतः इनके प्रकृति चित्रण में केवल रूपानुभव ही नहीं होता, इनकी अद्‌भुत कल्पनाशक्ति के स्पर्श मात्र से प्रकृति हँसती-बोलती, इठलाती मान करती, दौड़ती तथा विश्राम करती हुई भी दृष्टिगत होती है। 'प्रसाद' ने मानव की सम्पूर्ण चेतना प्रकृति में उड़ेलकर उसमें अद्‌भुत प्राण-प्रतिष्ठा कर दी है। प्रकृति को मानवता के समग्र उत्थान और चेतना का प्राण तत्व माना है।

संदर्भ संकेत

1. प्रसाद और कामायनी, गंगाप्रसाद पाण्डेय, प्रकाशन केन्द्र लखनऊ, पृ0-29
2. कामायनी (चिंता सर्ग), जयशंकर प्रसाद, पृ0-01
3. उपरिवत, पृ0-02
4. कामायनी (इड़ा सर्ग), जयशंकर प्रसाद, पृ0-39
5. उपरिवत, पृ0- 39-40
6. कामायनी (आशा सर्ग), पृ0-5
7. कामायनी (श्रद्धा सर्ग), पृ0-11
8. कामायनी (स्वप्न सर्ग), पृ0-49
9. कामायनी (आशा सर्ग), पृ0-7
10. कामायनी (चिंता सर्ग), पृ0-4
11. नामवर सिंह, छायावाद, नई दिल्ली, राजकमल प्रकाशन, 2015 (16वां संस्करण), पृ0- 39
12. जयशंकर प्रसाद के काव्य में पकृति-चित्रण, डॉ0 मुकुल खण्डेलवाल, भूमिका, कला प्रकाशन, वाराणसी

INDEX